The Application of Computer Techniques to ECG Interpretation

The Application of Computer Techniques to ECG Interpretation

Editor

Peter W. Macfarlane

MDPI • Basel • Beijing • Wuhan • Barcelona • Belgrade • Manchester • Tokyo • Cluj • Tianjin

Editor
Peter W. Macfarlane
Institute of Health and Wellbeing
University of Glasgow
Glasgow
United Kingdom

Editorial Office
MDPI
St. Alban-Anlage 66
4052 Basel, Switzerland

This is a reprint of articles from the Special Issue published online in the open access journal *Hearts* (ISSN 2673-3846) (available at: www.mdpi.com/journal/hearts/special_issues/TACTEI).

For citation purposes, cite each article independently as indicated on the article page online and as indicated below:

LastName, A.A.; LastName, B.B.; LastName, C.C. Article Title. *Journal Name* **Year**, *Volume Number*, Page Range.

ISBN 978-3-0365-3141-0 (Hbk)
ISBN 978-3-0365-3140-3 (PDF)

Contents

About the Editor

Peter W. Macfarlane is Emeritus Professor and Hon Senior Research Fellow at the University of Glasgow. He was Professor in Medical Cardiology from 1991–1995 and Professor of Electrocardiology from 1995–2010. He obtained a DSc in 2000 for a compilation of publications on Computer Assisted Reporting of Electrocardiograms (CARE).

The work of his team has been adopted commercially and the University of Glasgow ECG interpretation program developed in his laboratory is currently used worldwide. He is particularly interested in differences in ECG appearances due to age, gender and ethnicity, and as a result, he has influenced international guidelines for the ECG definition of acute myocardial infarction.

In addition, he has established a Central ECG Laboratory for handling ECGs recorded in national and international clinical trials and epidemiological studies including the landmark WOSCOPS. He has published over 400 scientific papers and 14 books, some of which are conference proceedings.

He was also jointly awarded the 1998 Rijlant International Prize in Electrocardiology by the Belgian Royal Academy of Medicine. In January 2014, he was awarded a CBE for Services to Healthcare.

Preface to "The Application of Computer Techniques to ECG Interpretation"

The aim of this book is to present a group of articles which outline the different areas where automated analysis of ECGs can be extremely beneficial in diagnosis and treatment of patients. The historical introduction and basic papers on the latest international standards for signal processing and communications set the scene for the various ways in which the understanding of the electrical activity of the heart as recorded on the body surface can be enhanced through modelling, both forward and inverse, to produce diagnostic data bases and assist clinical decision making. Artificial intelligence is also featured in several papers encompassing standard 12 lead electrocardiography as well as ambulatory monitoring. The use of data bases for prediction of outcomes in large population studies and for improving the efficacy of automated patient monitoring by ECG are also covered. The value of comprehensive body surface mapping allied to advanced electrophysiological modelling and statistical analysis is presented. The role of electrocardiographic imaging (ECGI) linked with speckle tracking echocardiography in the selection of patients for cardiac resynchronisation therapy is outlined. An update on the latest concepts in atrial abnormalities as determined only from the ECG is also included.

The editor would like to thank all of the authors for their contributions to this publication. He also wishes to thank Julie Kennedy for assistance in producing the final version of this book.

Peter W. Macfarlane
Editor

Editorial

The Application of Computer Techniques to ECG Interpretation

Peter W. Macfarlane

Institute of Health and Wellbeing, University of Glasgow, Glasgow G12 8QQ, UK;
Peter.Macfarlane@glasgow.ac.uk

Citation: Macfarlane, P.W.
The Application of Computer
Techniques to ECG Interpretation.
Hearts **2022**, *3*, 1–5. https://
doi.org/10.3390/hearts3010001

Received: 4 January 2022
Accepted: 5 January 2022
Published: 11 January 2022

Publisher's Note: MDPI stays neutral
with regard to jurisdictional claims in
published maps and institutional affiliations.

1. Introduction

It is over 120 years since Einthoven introduced the electrocardiogram. The technology has changed dramatically from recording three signals (or leads) through to recording over 300 leads, but paradoxically at the other extreme, the use of one lead is becoming popular in the area of wearables, such as a wristwatch. The ECG can be used to monitor heart rate and rhythm in its simplest form or to confirm a myocardial infarction or even to suggest left ventricular diastolic dysfunction at the other. In all of these areas nowadays, the digital computer or microprocessor plays a very significant role.

The concept behind this publication of many articles related to different aspects of ECG analysis was essentially to highlight the various areas where computer techniques support electrocardiology. The topics vary widely and should be of interest to clinicians, biomedical engineers and computer scientists with an interest in electrocardiography.

2. Historical Links

The compendium opens with a review of the history of automated ECG interpreta-tion [1] which had its beginnings around 1960 and is still in the process of development at the present time. Whereas large digital computers were involved in ECG processing at the outset, today's technology allows automated ECG interpretation in extremely small devices. The future may well see more centralisation of interpretation which would have its benefits in terms of updating a single copy of a specific analytical program undertaking the analysis compared to updating thousands of individual ECG machines. On the other hand, given the rate at which technology advances, it is not difficult to foresee that updated software could be downloaded to individual ECG machines, though that would mean more costly equipment.

One of the most significant projects in the field of automated ECG analysis was the work entitled Common Standards for Quantitative Electrocardiology, often known as the CSE project. Some of the output from that project is still very relevant today, particularly in relation to a database of ECGs where the clinical details for each patient are documented. It is now a little outdated, but it is still used as a yardstick for comparing the performance of different algorithms for ECG interpretation. Various definitions for ECG component amplitudes and durations were set out as a result of this work [1].

3. Updated Standards

An important spin off from the CSE project was the creation of a Standard Communica-tion Protocol (SCP). The concept was that there should be a common format in which ECG data was stored so that the same data could be analysed by different software packages which would accept the data in a well-defined format. This was not widely used, possibly for commercial reasons, but Rubel has persisted with the concept and has contributed in a major way to a recently updated protocol [2].

There are significant regulations related to software being regarded as suitable for inclusion in an electrocardiograph for general use. Through the years, different standards have been proposed by the international electrotechnical commission (IEC) to which software for automated ECG interpretation and hardware should conform. For many years the IEC 60601-2-51 standard was in place, but this was replaced by IEC 60601-2-25. The more recent standard omitted any reference to the diagnostic performance of the software much to the amazement of those in the field who could not understand how this came about!

In order to resolve the shortcoming of the current standard, Young, Schmid and colleagues have been extremely active over the past few years in preparing a new IEC standard for automated ECG interpretation and the broad principles are outlined [3]. It is hoped that this new standard will be accepted by the various bodies and then become operational in the near future. There are still shortcomings in the sense that there is no currently available database of cardiac arrhythmias which can be used to test software, but work is in progress to remedy this situation.

4. Modelling

Given that deficiencies exist in well-defined data bases for evaluating performance of ECG software, the work of Doessel et al. in modelling the electrical activity of the heart is of particular relevance [4]. This detailed paper illustrates how modelling has reached the point of being able to generate a wide variety of ECG abnormalities including arrhythmias, conduction defects as well as other abnormalities including myocardial infarction and ischaemia. This is part of a wider European collaboration involving a number of centres interested in modelling. It could ultimately lead to the establishment of a large database that could be used as a yard stick against which interpretative software could be evaluated.

5. Big Data

Ribeiro and colleagues in Brazil have been prominent in recording large numbers of ECGs in Health Centres within one state of that vast country. Their work is now expanding into other states in Brazil and in all probability they have the biggest networking facility worldwide for the routine collection of ECGs.

The authors have made use of almost 2.5 million ECGs from individuals followed up for 3.7 years. The group had a mortality rate of 3.3% and the authors examine the predictive value of a variety of ECG abnormalities with respect to overall mortality [5]. This is an excellent example of big data and computer techniques enhancing the value of ECG analysis.

6. Artificial Intelligence

One of the more recent developments in ECG interpretation has been the use of artificial intelligence to facilitate ECG classification. This approach, while extremely promising, is still in its infancy with many aspects remaining to be carefully evaluated.

One of the groups which has been most active in this area is from the Mayo Clinic, Minnesota, USA. Rafie et al. describe an approach to 12 lead ECG interpretation which uses one form of artificial intelligence [6]. This is one of the earliest examples of 12 lead interpretation using this newer technique, and the authors frequently refer to the "potential" use of the approach in routine ECG interpretation.

AI has also been used in ambulatory ECG analysis. Xue and Yu [7] outline the areas in ambulatory monitoring where the AI methodology is likely to be of greatest help. Ambulatory ECG has problems in the sense that patients may be active at any time during a recording leading to the increased possibility of noise contaminating the ECG, while the amount of data that can be generated is clearly enormous, particularly if, as is sometimes the case nowadays, the recording is made for up to 14 days. It is clear, therefore, that automated methods for detecting abnormalities within such a long period are required.

AI in routine ECG interpretation, in the editor's opinion, still has some hurdles to climb before being fully accepted for basic 12 lead ECG interpretation. Principal amongst these is the positive predictive value of a diagnostic statement. This is not always considered in publications relating to the technique, but it is gradually being acknowledged that it is one aspect of AI that has to be reviewed very carefully [1].

7. Monitoring

Monitoring of the ECG in the coronary care unit, an intensive care unit or high dependency unit is commonplace. Many monitoring devices will have in-built computer algorithms to facilitate detection of life-threatening arrhythmias. However, the biggest problem which these devices have suffered in recent years has been dealing with artefact due to patient movement etc., which is inevitable during long term monitoring.

The article by Pelter et al. [8] highlights the problem and outlines an attempt to produce a database of ECGs from patients being monitored that will ultimately lead to an enhancement of algorithms for accurate detection of significant arrhythmias. Current algorithms have a very high percentage of false positive alarms often to the extent that nursing staff simply turn off the alarms in order to avoid continuous interruption for checking what frequently turns out to be a false alarm. Of course, this raises the obvious problem that a genuine alarm can be missed by manual monitoring. This scenario is known as alarm fatigue. The authors outline the creation of a data base of ECGs where alarms are genuine and illustrate how three experienced individuals may be required to agree an interpretation before it can be included in the database. This work suggests that there is much that can still be carried out to enhance the accuracy of alarm detection by patient monitors. There are many interesting examples of life-threatening arrhythmias in this article.

8. Body Surface Mapping

Body surface mapping (BSM) has been available since the initial work of Taccardi in the 1960s. Nowadays, as many as 300 electrodes can be placed on the anterior and posterior surfaces of the thorax in order to obtain as detailed a pattern as possible of the cardiac excitation as it appears on the body surface. Clearly automated methods for processing such a large amount of data are invaluable and essentially the technique would not have been followed to any significant extent had it not been for the revolution in technology over the past 50 years.

BSM can be used in different ways. The most basic is simply to visualise the spread of excitation on the body surface and from that infer the nature of any abnormality. Another application is to utilise the activity on the body surface to determine the electrical activity on the surface of the heart and the spread of excitation with the myocardium. This so-called inverse modelling is gaining importance and has led to clinically available devices for investigation of cardiac arrhythmias by localising the source of the abnormal rhythm. Bergquist et al. [9] outline the leading edge methodology which they use to process body surface maps.

The use of the inverse modelling approach has given rise to the term of electrocardiographic imaging (ECGI). Currently, there is a need for a patient to undergo computed tomography (CT) or magnetic resonance imaging (MRI) in order to obtain accurate cardiac "geometry" prior to ECGI. This is then linked with the mathematical model which allows the cardiac excitation to be determined from the body surface potentials. The authors refer to their recent work on "imageless" ECGI, which represents a significant advance in inverse modelling with CT and MRI being unnecessary prior to ECGI.

The authors conclude that BSM is predominantly a research tool, but nevertheless its use has led to ECGI among other things and it will continue to be of very significant benefit to electrocardiological developments in general.

9. Application of ECGI

Andrews et al. describe the use of ECGI in facilitating the understanding of a cardiac resynchronisation therapy. [10]. When ECGI is combined with speckle tracking echocardiography, it is possible to look at the motion of the left ventricle and at the same time link this with the spread of electrical activation. ECGI was in large part developed by one of the authors of this article (Rudy) some years ago.

The study describes remodelling of the heart after bi-ventricular pacing has been initiated. Using ECGI, epicardial electrograms can be determined and from this, various other measures of electrical activity are computed. This allows the investigators to assess the effects of pacing over time. The link between myocardial activation and muscular contraction can be assessed, using this technique, against the effectiveness of biventricular pacing. The authors conclude that the ECGI measured delay in activation of the left ventricle is an excellent index for selecting patients for cardiac resynchronisation therapy.

10. Interatrial Block

Interest in interatrial blocks has increased recently due to the work of Bayes-de-Luna and his colleagues. They have described different types of interatrial block which in general terms have not been frequently reported as part of an ECG interpretation. The work presented [11] is therefore of significance in highlighting these electrocardiographic abnormalities and hence underlying myocardial problems.

It would indeed be difficult to have a detailed automated interpretation of P wave abnormalities, together with PR interval changes, given that a P wave can often be one of the most difficult components of the ECG to measure accurately. Its projection onto multiple leads will result on occasions in the P wave essentially not being seen in some leads making analysis even more difficult. The door is therefore open for those who wish to undertake further work in automated ECG interpretation to use their talents in this particular area.

11. Conclusions

It is hoped that the articles in this book will show the reader that there is still life in electrocardiographic research! Although it is a relatively old investigational technique, the ECG still remains of paramount importance in clinical investigation of patients. Advances in technology are gradually leading to advances in understanding of various aspects of the ECG although there are still many areas where knowledge is incomplete. Perhaps in due course there will be a complete understanding of the genesis of the ECG, and automated techniques will be able to give a fully detailed interpretation of the individual spread of cardiac activation in every patient.

Funding: This editorial received no external funding.

Conflicts of Interest: The author declares no conflict of interest.

References

1. Macfarlane, P.W.; Kennedy, J. Automated ECG Interpretation—A Brief History from High Expectations to Deepest Networks. *Hearts* **2021**, *2*, 433–448. [CrossRef]
2. Rubel, P.; Fayn, J.; Macfarlane, P.W.; Pani, D.; Schlögl, A.; Värri, A. The History and Challenges of SCP-ECG: The Standard Communication Protocol for Computer-Assisted Electrocardiography. *Hearts* **2021**, *2*, 384–409. [CrossRef]
3. Young, B.; Schmid, J.J. The New ISO/IEC Standard for Automated ECG Interpretation. *Hearts* **2021**, *2*, 410–418. [CrossRef]
4. Doessel, O.; Luongo, G.; Nagel, C.; Loewe, A. Computer Modeling of the Heart for ECG Interpretation—A Review. *Hearts* **2021**, *2*, 350–368. [CrossRef]
5. Paixao, G.M.M.; Lima, E.M.; Gomes, P.R.; Oliveira, D.M.; Ribeiro, M.H.; Nascimento, J.S.; Ribeiro, A.H.; Macfarlane, P.W.; Ribeiro, A.L.P. Electrocardiographic Predictors of Mortality: Data from a Primary Care Tele-Electrocardiography Cohort of Brazilian Patients. *Hearts* **2021**, *2*, 449–458. [CrossRef]
6. Rafie, N.; Kashou, A.H.; Noseworthy, P.A. ECG Interpretation: Clinical Relevance, Challenges, and Advances. *Hearts* **2021**, *2*, 505–513. [CrossRef]
7. Xue, J.; Yu, L. Applications of Machine Learning in Ambulatory ECG. *Hearts* **2021**, *2*, 472–494. [CrossRef]
8. Pelter, M.M.; Mortara, D.; Badilini, F. Computer Assisted Patient Monitoring: Associated Patient, Clinical and ECG Characteristics and Strategy to Minimize False Alarms. *Hearts* **2021**, *2*, 459–471. [CrossRef]
9. Bergquist, J.; Rupp, L.; Zenger, G.; Brundage, G.; Busatto, A.; MacLeod, R.S. Body Surface Potential Mapping: Contemporary Applications and Future Perspectives. *Hearts* **2021**, *2*, 514–542. [CrossRef]

10. Andrews, C.M.; Singh, G.K.; Rudy, Y. Excitation and Contraction of the Failing Human Heart in Situ and Effects of Cardiac Resynchronization Therapy: Application of Electrocardiographic Imaging and Speckle Tracking Echo-Cardiography. *Hearts* **2021**, *2*, 331–349. [CrossRef]
11. Bayes-de-Luna, A.; Fiol-Sala, M.; Martinez-Selles, M.; Baranchuk, A. Current Aspects of Interatrial Block. *Hearts* **2021**, *2*, 419–432. [CrossRef]

Review

Automated ECG Interpretation—A Brief History from High Expectations to Deepest Networks

Peter W. Macfarlane * and Julie Kennedy

Institute of Health and Wellbeing, University of Glasgow, Glasgow G12 8QQ, UK; Julie.Kennedy@glasgow.ac.uk
* Correspondence: Peter.Macfarlane@glasgow.ac.uk

Abstract: This article traces the development of automated electrocardiography from its beginnings in Washington, DC around 1960 through to its current widespread application worldwide. Changes in the methodology of recording ECGs in analogue form using sizeable equipment through to digital recording, even in wearables, are included. Methods of analysis are considered from single lead to three leads to twelve leads. Some of the influential figures are mentioned while work undertaken locally is used to outline the progress of the technique mirrored in other centres. Applications of artificial intelligence are also considered so that the reader can find out how the field has been constantly evolving over the past 50 years.

Keywords: electrocardiogram (ECG); automated ECG analysis; CSE study; age; sex; race; historical aspects

Citation: Macfarlane, P.W.; Kennedy, J. Automated ECG Interpretation—A Brief History from High Expectations to Deepest Networks. *Hearts* **2021**, 2, 433–448. https://doi.org/10.3390/hearts2040034

Academic Editor: Matthias Thielmann

Received: 24 August 2021
Accepted: 14 September 2021
Published: 23 September 2021

Publisher's Note: MDPI stays neutral with regard to jurisdictional claims in published maps and institutional affiliations.

1. Introduction

Augustus Waller (1856–1922) was the first person to record a single lead electrocardiogram (ECG), which is a recording of the electrical activity of the human heart, in St Mary's Hospital, London in May 1887 [1]. Ventricular depolarisation and repolarisation were demonstrated using the Lippmann capillary electrometer. Waller had been a medical student in Aberdeen and Edinburgh and was made professor in the University of Aberdeen, Scotland, in 1881. He made further observations on ECGs on his dog Jimmie, who was often used in his lectures.

Around the same time, in Aberdeen, Scotland, John A. MacWilliam, Professor of the Institute of Medicine, introduced the term atrial flutter and the concept of ventricular fibrillation [2]. James Mackenzie, a Scottish Physician, published his own work in 1902 on the study of the pulse [3]. He used a home-made polygraph to record the action of all four chambers of the heart. His contribution to ECG development was acknowledged by Karel Frederik Wenckebach, the much-honoured Dutch physician who researched irregularities of cardiac rhythm.

Shortly thereafter, Willem Einthoven, based in Leiden in the Netherlands, introduced the three standard bipolar limb leads with the use of his own galvanometer, as a result of which he was awarded the Nobel Prize for Medicine in 1924 [4]. The first commercial version of the Einthoven electrocardiograph was produced in 1908 by the Cambridge Instrument Company in England. It recorded Einthoven's three leads, I, II and III and used pails of conducting solution as electrodes.

Thomas Lewis, one of Mackenzie's former junior staff members, published 'The Mechanism and Graphic Registration of the Human Heart'. In its third edition, in 1925, it summarized early work on cardiac arrhythmias based on the use of Einthoven's three limb leads.

It is unthinkable that these pioneers could have projected forwards over 100 years to predict what electrocardiography would be like at the present time. Electronic computers had not been invented and electrical circuitry had certainly not been miniaturised to the extent which it is nowadays.

Electrocardiography itself advanced from the use of 3 leads to what became known as the standard 12 lead configuration, although only 10 electrodes were required to derive the 12 signals [5]. Further advances were made through the use of multiple electrodes on the body surface which allowed the spread of electrical activation to be mapped [6]. Thereafter, modelling became extremely sophisticated and the cardiac electrical activity could be calculated from inverse modelling with body surface potentials as the input to the model [7].

In 1920, Hubert Mann of the Cardiographic Laboratory, Mount Sinai Hospital, New York, described the derivation of a 'monocardiogram' later to be called 'vectorcardiogram' or VCG [8]. Waller had also been interested in the concept of the vector force. The VCG displays P, QRS and T waves, in the form of 'loops' which are determined from vectors representing successive instantaneous mean electrical forces from the heart throughout an entire cardiac cycle. These forces are oriented three-dimensionally during each heartbeat and can be represented by a time sequence of vectors which display their magnitude and direction. A number of electrode arrays have been developed, aimed at recording the three components of the resultant cardiac electrical force, the so-called heart vector, in three mutually perpendicular directions. This approach became known as three orthogonal lead electrocardiography. The most popular lead system, a 'corrected orthogonal lead system', which recorded leads X, Y and Z, was developed by Ernest Frank in 1956 [9]. The vector loops were initially viewed on oscilloscopes.

Advances in data collection and in miniaturisation of ECG amplifiers and recording equipment greatly facilitated such studies. It goes without saying that the introduction of the digital computer itself also led to a transformation in the way that analogue ECG signals could be processed, interpreted and stored.

Other electrical circuits were used to record additional (unipolar and augmented unipolar) ECG leads, and the basic 12 lead ECG, i.e., a recording of 12 different electrical signals of which 8 were independent, was established by the early 1940s. A fuller account can be found elsewhere [10]. Since that time, many other developments have emerged in the interim, but the 12 lead ECG still remains the major approach to recording the electrical activity of the heart in hospitals and healthcare facilities worldwide. Its use is complemented by other techniques which have emerged more recently but, nevertheless, the 12 lead ECG has remained the rock on which other advances have been built.

From its inception until the present day, the 12 lead ECG has been interpreted by cardiologists and other specialists worldwide by visually reviewing the patterns recorded. Attempts to automate the process commenced in the late 1950s and gathered momentum until the bulk of ECG recording machines being sold today have the ability to offer an interpretation of the ECG. Such a report may not always agree with the interpretation of a cardiologist, though it increasingly does, but the approach offers clinicians and others support in handling one of the most commonly used tests in medicine. Interpretative software has advanced through the years and the emergence of techniques for ECG interpretation using machine learning, which in itself is a branch of artificial intelligence, have grown considerably. The purpose of this article is to review the development of automated ECG processing from its beginnings to the present day, with the hope that between the time the article was written and published, there will not have been some world-shattering development to make the content of this paper outdated.

2. The Early Days

The use of computers for ECG interpretation was first evaluated using the orthogonal lead ECG and the 12 lead ECG. The first approach to automating analysis of ECGs commenced in 1957 in the laboratory of Dr. Hubert V. Pipberger (Figure 1) using three simultaneously recorded orthogonal leads [11]. The Veteran's Administration (VA) Hospital in Washington DC established a special research programme for medical electronic data processing as medical electronics began a period of growth, and Pipberger was appointed director [12].

Figure 1. Dr. C. Caceres (**left**) and Dr. H. Pipberger (**right**).

Pipberger was born in Hamburg in 1920 and studied at the Rheinische Friedrich Wilhelms University in Bonn, Germany. He was an army doctor during World War II and was captured and imprisoned in France. He saved himself by telling stories in French, entertaining his captors [13].

A pioneer in the field of electrocardiology, he trained as a cardiologist and recognised the effectiveness of collaboration with electrical engineers, physicists, mathematicians, statisticians and computer programmers in problem solving and interdisciplinary research [14]. Dr. Pipberger's lab based its early analysis system on the three orthogonal lead ECG [15]. Analogue ECG recordings had to be converted into digital data using rather large equipment [11], which has to be compared with the current possibility of converting an ECG signal into a digital form for analysis within a wearable such as a wristwatch. Each diagnostic output from Pipberger's program had a probability attached on a scale of 0–1 and the sum of all outputs had to total 1. This could be confusing if a new diagnostic output was added, in a later ECG, to an existing abnormality which then had a reduced probability of being present.

In contrast, in 1959, Dr. Cesar Caceres (Figure 1) and his team in the National Institute of Health's Medical Systems Development Laboratory, also in Washington DC, based their approach on the analysis of the 12 lead ECG using conventional clinical ECG criteria, but initially by processing one lead at a time [16]. Caceres coined the term 'clinical engineering', putting engineering into the clinical world of medicine in order that the various disciplines could work hand in hand to improve healthcare in practice. He graduated from Georgetown University and specialised in Internal Medicine at Tufts and Boston Universities in Boston, Massachusetts. He received Cardiology specialisation and research training from George Washington University [17].

This early work led to the expectation of the technique playing a significant role in ECG interpretation.

3. The Glasgow Contribution

One of the authors (PWM) began work in Glasgow on Computer Assisted Reporting of Electrocardiograms (C.A.R.E) as a student with Professor T.D. Veitch Lawrie in the Department of Medical Cardiology, University of Glasgow, who had anticipated an expansion of the use of computers in ECG analysis. Early work determined that the conventional 12 lead ECG and the three orthogonal lead ECG could be used for computer interpretation of ECGs and so diagnostic criteria were developed for both lead systems [18]. Additional vectorcardiographic measurements were made and incorporated into the diagnostic criteria. To record and analyse the ECGs, a standard 3 channel VCG system was combined with 3 single-channel electrocardiographic amplifiers and a multi-channel analogue tape recorder linked to a small PDP8 digital computer with an analogue–digital converter, allowing ECGs to be replayed from the tape recorder to the computer. An ECG database was accumulated in both analogue and digital form. In the early 1970s, portable ECG recording units were assembled which could be transported to wards and clinics easily on a trolley. The modified axial lead system [19] was used and the analysis time was the order of one minute. What was thought to be the first hospital-based mini-computer system for routine

ECG interpretation was developed and introduced (Figure 2) in Glasgow Royal Infirmary around 1971 [20]. In the mid-70s, a hybrid lead system [21] was designed that combined the 12 lead and the three orthogonal lead ECG with the use of two additional electrodes (V5R and the neck), but there was little clinical acceptance.

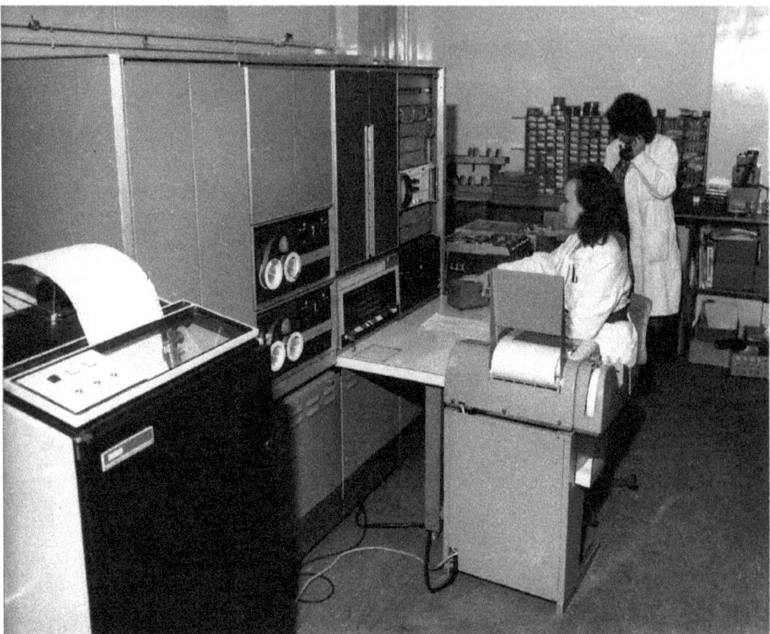

Figure 2. The first automated ECG interpretation system in operation in Glasgow Royal Infirmary around 1971. One technician controlled the tape recorder and listened to the patient details which were also recorded. The second technician monitored the three orthogonal lead ECG on the oscilloscope and started the analogue to digital conversion. The software was stored on the small digital tapes (DECtapes) and retrieved as necessary.

A major technological advance was the advent of the microprocessor and the arrival of automated ECG analysis at the bedside. In Glasgow, analogue to digital conversion at 500 samples per second was undertaken within an electrocardiograph designed and built by Dr. M. P. Watts, who had also introduced techniques for transmitting ECGs between a local hospital and the ECG lab for automated interpretation [22]. The analysis program was rewritten in Fortran and moved to a PDP11 series computer. New diagnostic criteria evolved for the 12 lead ECG including rhythm analysis and serial comparison [23]. Many clinical studies, some of which are described below, led to an enhanced program for automated ECG analysis, which was commercialised in the early 1980s.

4. Age, Sex and Racial Differences in the ECG
4.1. Neonatal and Paediatric ECG Analysis

The ECGs of the neonate, infant and child are completely different from the ECG of the adult. For this reason, special considerations apply to automated ECG interpretation of the neonatal and paediatric ECG.

The duration of the neonatal QRS complex is significantly shorter than that of the adult ECG, which implies a higher frequency content. This has often led to claims that the technology for recording the neonatal and paediatric ECG in general should be enhanced compared to that for recording the adult ECG. However, there has not been any study which has shown a clinically significant difference between a higher and a lower sampling rate when converting the ECG from analogue to digital form. Specifically, a major study by Rijnbeek [24] and colleagues from the Netherlands showed that reducing the sampling rate from 1000 samples per second to 500 samples per second had no impact on normal limits which they developed in infants and children.

The ECG of the newborn tends to have a QRS axis which is in the range of 90–180° which, for the adult, would be known as right axis deviation but for the baby is normal. This is simply a function of the path of circulation of blood through the foetus and the major role played by the right side of the heart at that stage in the development of the child. After the baby is born, the circulation changes and there is a gradual shift in emphasis of contraction, with the left ventricle becoming much more dominant than the right ventricle. It is probably not well appreciated that the ECG of the neonate therefore changes significantly, even over the first week of life. This was demonstrated by one of the authors (PWM) and colleagues [25]. Thus, interpretation of the neonatal ECG should ideally be based on a knowledge of the date of birth and date of recording so that changes from one day to another can be considered in an interpretation. Nowadays, new mothers are generally encouraged to leave hospital within 24 h or 48 h, and so trying to obtain a database of ECGs of neonates from birth to one week of life is currently extremely difficult.

Of course it goes without saying that, as the child grows, so does the heart and hence QRS duration, for example, linearly increases in duration from birth to adolescence. Allowance, therefore, has to be made for this in an interpretative program. Similarly, heart rate decreases shortly after birth, though not immediately, and again, simple equations can be used to set an upper limit of normal from the first week of life to adolescence [25].

It is suggested that with current advances, no matter how impressive, the use of machine learning will still prove challenging in the area of paediatric ECG interpretation.

4.2. Adult Age and Sex Differences in ECGs

There are many differences between adult male and female ECGs [26], and automated interpretation should be able to handle such variations with ease. In broad terms, QRS voltage is higher in younger compared to older persons, particularly in males, but this difference diminishes with increasing age. The same is true for ST amplitude, especially in the precordial leads, though it remains higher in males at all ages [27]. This latter publication eventually led to sex differences in ECG criteria being taken into account when reporting ST elevation myocardial infarction, as now acknowledged in the latest universal definition of myocardial infarction [28].

Mean QRS duration is higher in males than females though it is rare for this to be acknowledged in diagnostic criteria, with an exception being in 'true' left bundle branch block (LBBB) [29].

4.3. Racial Differences

It has been established that there is a clear ethnic variation in certain aspects of the ECG that should be acknowledged when making an interpretation. A number of studies have shown differences in normal limits of the ECG between Caucasians, Black people and Asian individuals [30,31] and diagnostic criteria should allow for this.

The availability of digital electrocardiographs and computers which can easily handle vast numbers of ECGs should allow for further work to be done on enhancing race-based diagnostic criteria. For example, it was noted that the mean ST segment amplitude is higher in Black people than in Caucasians and is higher in males than in females [30]. Rautaharju also showed that Black people had higher voltages than Caucasians in one of his studies [31]. From a historical perspective, Simonson pointed out racial differences in his 1961 treatise on the normal electrocardiogram [32]. His comparison was mainly between Caucasian and Japanese individuals, but differences were acknowledged at that time. One of the authors (PWM) also compared Caucasian with Indian, Nigerian and Chinese cohorts [30], showing a variety of differences, so it is important that race be acknowledged in ECG interpretation.

5. The European Contribution

In 1974, one of the authors (PWM) obtained a scholarship from the British Heart Foundation to spend one month in Europe visiting various centres which, by that time, had commenced work on some aspect of automated ECG interpretation. These included university departments in hospitals in Leuven in Belgium, Rotterdam in the Netherlands, Lyon in France, and Hannover in Germany. A report was compiled summarising the developments in progress and suggestions for collaboration were made. The net effect was that the European Economic Community, as it was known at the time, set up a project to further the technique of automated ECG analysis by supporting a joint project involving all those interested centres in various European countries. In due course, participation was opened to those from overseas, mainly the USA, who also had an interest in the topic. The North American delegates were mostly representatives from commercial companies developing products.

The project was entitled Common Standards for Quantitative Electrocardiography and it quickly became known as the CSE Project. A detailed summary of its main goals can be found elsewhere [33] and it arguably became the best-known project in automated electrocardiography. The project leader was Professor Jos Willems, who had spent time in Pipberger's laboratory in Washington DC in the early stages of the development of the technique. He led the project from 1976 through to the early 1990s, when unfortunately he was found to have a brain tumour and died shortly thereafter [34].

Early in the project, a steering committee was established, consisting of Professor Jos Willems (Chairman), Rosanna Degani (Padua, Italy), Christoph Zywietz (Hannover, Germany), Peter Macfarlane (Glasgow, UK), Jan van Bemmel (Rotterdam, The Netherlands) and Pierre Arnaud (Lyon, France). Later, Paul Rubel from Lyon replaced Pierre Arnaud (Figure 3). The steering committee met almost quarterly for over 10 years and there were biennial meetings of the full working group where multiple individuals from the same team could attend.

Figure 3. The CSE steering committee pictured in 1987 laughing at each other wearing glasses. The picture was taken by PWM. From left to right: (inset) Peter Macfarlane (1987); Christoph Zywietz; Jan van Bemmel; Rosanna Degani; Pierre Arnaud; Jos Willems; (inset) Paul Rubel (1992), who took over from Pierre Arnaud.

One of the biggest outcomes of the project was the establishment of databases, both of ECG waveforms for testing measurements and also of ECG interpretations from 1220 patients whose clinical condition was documented. These databases are still of importance to this day, over 30 years later.

The availability of the waveform database allowed the establishment of standards for ECG wave recognition which are still in use at the present time. The CSE diagnostic database resulted in a landmark publication [35] in 1991, where the accuracies of different diagnostic programs were assessed against both a clinical diagnosis and, separately, against the opinion of a group of eight cardiologists. Even today, companies wishing to submit data on performance of their software very often resort to using analysis based on the CSE diagnostic database. This requirement will continue in the new ISO/IEC standard for automated ECG interpretation which will be entitled '80601, Part 2-86: Particular requirements for the basic safety and essential performance of electrocardiographs, including diagnostic equipment, monitoring equipment, ambulatory equipment, electrodes, cables, and leadwires'. A summary of the expected contents based on a first draft already circulated can be found elsewhere in this issue [36].

The availability of a library of digitized ECGs made it possible for extensive recommendations on signal processing to be published as part of the CSE Study [37]. It should be noted that at that time (1985 and earlier), the majority of electrocardiographs produced recorded three leads simultaneously. Many of the recommendations therefore related to dealing with groups of three leads. Nevertheless, interval measures such as QT were recommended to be based on the longest QT interval measured in any single lead, though V1–V3 were suggested as giving the most accurate result. Other definitions, for example, included the exclusion of an isoelectric segment at the onset or termination of a QRS complex from the QRS duration in the lead in which it occurred. The overall QRS duration, nevertheless, was defined as the time from the earliest onset to the latest offset in the group of leads under consideration.

One of the interesting points to emerge from analysis of the CSE diagnostic database was that there were significant differences in sensitivity and specificity of diagnostic programs when the gold standard was, on the one hand, based on clinical data and, on the other, when it was based on the consensus interpretation of eight cardiologists. Some programs were developed on the basis of using clinical data while others were developed using cardiologist views as the gold standard, and that was reflected in the results of the study [35].

It is worthy of note that several programs developed in academic institutions in Europe were commercialised. These included the Glasgow program developed in the University of Glasgow [38], the HES program developed in the University of Hannover [39], and the MEANS program developed in the University of Rotterdam [40]. Software developed in the USA for commercial use was developed within industry, e.g., the Marquette—General Electric (GE) program, Hewlett Packard-Philips program, and the Mortara program.

It should also be noted that some programs used classical deterministic criteria, e.g., R amp in aVL > 1.5 mV, where others used a more statistical approach, involving probabilities. It was found that those programs which used classical criteria were more closely aligned with the gold standard based on cardiologist interpretations and conversely, those developed using probability theory were more closely aligned with the clinical data.

A typical example of this conundrum would be when the clinical diagnosis was left ventricular hypertrophy (LVH), which was based perhaps on a history of hypertension and an increased cardiothoracic ratio (1980s type of CSE criteria), but where the ECG was essentially within normal limits. The software developed according to a clinician's view would report a normal ECG and that would be in line with the cardiologist interpretation. There is therefore agreement in that case between the computer and the cardiologist. On the other hand, both the computer program and the cardiologist are wrong with respect to the clinical diagnosis of LVH. Conversely, the statistical program would be more likely to report LVH correctly in line with the clinical diagnosis but would be wrong with respect to the cardiologist interpretation. This example explains why different results are obtained with different software and different gold standards.

This still remains a problem to some extent even with the newer techniques of machine learning, because more often than not when a large dataset is used for training, it can be that the cardiologist over-read of an ECG is used as the gold standard, although not always. This point will be considered later.

A by-product of the project was the establishment of a Standard Communications Protocol (SCP) for electrocardiography [41]. This was designed with the aim of providing data from different manufacturers' ECG machines in a similar format which might for example contribute to a database or allow one vendor's system to analyse ECGs recorded on another's equipment. The SCP was strongly supported by Rubel who continued to regard it as of significant value, so much so that a new version has been released within the last year. Details can be found elsewhere in this issue [42].

In summary, the CSE Project has had a very significant influence over the field of automated ECG interpretation and still remains of great value at the present time.

6. The North American Contribution

In terms of software development for automated ECG analysis, the USA undoubtedly led the way with Caceres and Pipberger taking the lead as previously described. However, while the CSE project had a huge impact on standards for ECG wave recognition, other recommendations had previously been initiated by Pipberger [43]. These concentrated mainly on equipment for ECG and VCG recording but a recommendation for sampling analogue data for conversion to digital data of 500 samples per second was made. That recommendation is still followed by many systems. After the Caceres program fell out of favour through sampling one lead at a time [16], IBM with Ray Bonner at the helm, led the way commercially in developing a 12 lead ECG program [44] with three leads recorded simultaneously. Early players in the field included Telemed and Marquette, who were also prominent, with interpretative software predominantly developed by Dr. David Mortara. He later launched his own company which was many years later purchased by Hill-Rom. Marquette was subsequently taken over by GE. Hewlett-Packard also had a 12 lead ECG analysis program and this arm of the company was taken over by Philips.

From an academic standpoint, there was not the same proliferation of software produced by Universities in North America as had emerged from Universities in Europe. The principal exception was software [45] that was developed in Dalhousie University, Halifax, Nova Scotia, where Dr. Pentti Rautaharju had established the Epicare lab investigating various aspects of electrocardiography, including mathematical modelling, the local development of which he stimulated. He was very influential in the field which he had followed from 1960, when he was one of the authors of the Minnesota Code [46]. He moved from Halifax to Edmonton, Alberta and continued with a variety of studies and, together with his wife, published a book on the ECG in epidemiological studies and clinical trials [47]. His final academic move was to Winston-Salem. He died in 2018 [48].

Other North American contributions should not be overlooked. Early work in automated ECG analysis was undertaken by Dr. Ralph Smith and colleagues at the Mayo Clinic [49]. This was facilitated by an enormous number of ECGs being recorded annually in that establishment in the early 1970s, reportedly over 100,000 [50]. Smith introduced the concept of the ECG Interpretation Technician (EIT). In short, experienced ECG technicians received instruction in reviewing ECGs interpreted by an early IBM program and ultimately worked independently in over-reading the initial computer-based report [50].

Dr. Jim Bailey and colleagues at NIH produced a number of papers in the early 1970s relating to the assessment of automated ECG analysis programs. One very interesting and simple technique which he described was to take odd and even samples separately from data sampled at 1000 samples/s in order to create two ECGs sampled at 500 samples/s. The paired ECGs were then interpreted using two different programs and similar interpretations were found in only 49.8% and 79.7% after initial analogue filtering of the signal [51].

Bailey was also involved in preparing standards for ECG signal processing in 1990 [52]. This was followed in 2007 by an expanded set of recommendations by Dr. Paul Kligfield et al. [53] for the standardization and interpretation of the electrocardiogram. This was the first of six such scientific statements which appeared between 2007 and 2009 with the support of the American Heart Association, American College of Cardiology, Heart Rhythm Society and the endorsement of the International Society of Computerized Electro-cardiology.

It should also be noted that the International Electrotechnical Commission also became involved in creating standards and in 2003 produced guidelines such as IEC 60601-2-51 (subsequently replaced by IEC 60601-2-25) for ECG processing, including specifications on maximum tolerances on measuring ECG wave amplitudes and durations. These guidelines are discussed elsewhere in this issue [36].

7. The Technology

In parallel with developments in software for automated ECG analysis has been the miniaturisation of equipment. Initially, the early electrocardiographs which offered ECG interpretation on the spot were akin to the size of a washing machine on wheels (Figure 4). One advert proudly proclaimed that ECG interpretation was available within minutes. With hindsight, this was not necessarily the best way to advertise the product because no cardiologist would stand around at the bedside waiting minutes for a second opinion on an ECG interpretation. This has to be compared with the performance of current PCs, where 50 ECG analyses per second are commonplace.

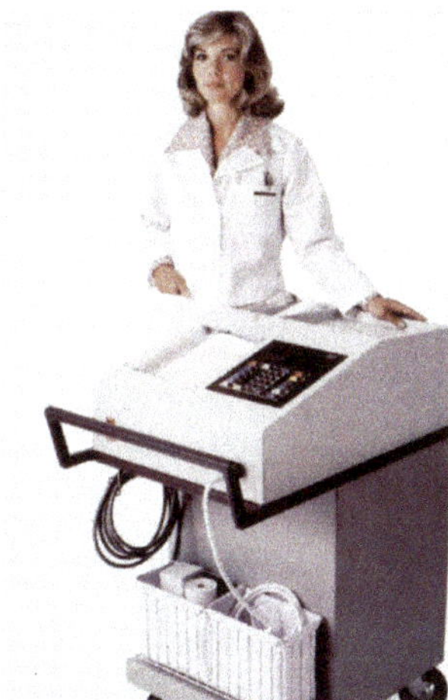

Figure 4. One of the earliest standalone electrocardiographs with automated ECG interpretation.

The reduction in size of equipment has continued through the possibility of having ECG acquisition and interpretation on a mobile phone with the actual ECG amplifiers being external to the mobile unit. On the other hand, wrist watches are now widely available which allow a single lead of the ECG to be recorded, displayed and a limited interpretation of rhythm offered. There are variations on the theme whereby a three electrode device can be used to record six limb leads via a mobile phone with transmission of the signals to a central facility for review if required. A review of some of these devices and techniques can be found elsewhere in this issue [54].

Although clinicians have always favoured the 12 lead ECG, current technology allows for the recording of many more leads, such as in body surface mapping, which is discussed elsewhere in this issue [55], to the other extreme of a single channel recording for ambulatory monitoring, also discussed elsewhere in this issue [54]. Because of the lack of redundancy, it is extremely important that the single channel ECG is of high quality when analysed. Most smart watches will use strong filtering in order to try to remove unwanted noise on the recording and provide a very stable baseline. This may be at the expense of a small loss of signal amplitude, but ultimately this is not of relevance in the interpretation of cardiac arrhythmias.

The initial internal report using the Apple Watch (versions 1–3), which used photoplethysmography (optical sensor)-based detection of the pulse to determine heart rhythm irregularity, showed extremely good sensitivity and specificity $\geq 98\%$ for detecting atrial fibrillation (AF) [56]. The Apple Heart Study, using the same type of watch, recruited 419,297 volunteers, and reported [57] that 0.52% received a notification of an irregular pulse. A total of 450 patients completed a follow up of long-term monitoring and only 34% were found to have AF. A more recent small study of 50 patients [58] using the Apple Watch 4, where the ECG is recorded using lead I (potential difference between left wrist and a finger on the right hand touching the crown of the watch), found 41% sensitivity and 100% specificity for AF determined from the watch-based interpretation, leading the authors from the Cleveland Clinic to conclude that 'physicians should exercise caution before undertaking action based on electrocardiographic diagnoses generated by this wrist-worn monitor'.

These results reflect the difficulty of providing high quality data to an algorithm for ECG signal processing and interpretation. A discussion on the use of 1 to 12 leads recorded from 10 s to 30 days, particularly for ambulatory monitoring with wearables, can be found elsewhere in this issue [54]. Various aspects of filtering/denoising the ECG in an attempt to provide a clean ECG are also presented in that article.

8. Machine Learning

The most recent development in the field of automated ECG analysis has been the use of artificial intelligence (AI), including a variety of machine learning techniques to aid interpretation. One of the authors (PWM) was involved in the use of neural networks in the early 1990s [59] but at that time, use of a simple neural network did not prove to be of any great advantage in ECG interpretation compared to the use of more basic, straightforward diagnostic criteria.

More recently, with advances in miniaturization but also very significant developments in software, the use of more advanced neural networks such as 'deep' convolutional neural networks has added greatly to the ability of software to undertake ECG interpretation without the need to develop diagnostic criteria. In addition, the easy availability of machine learning software has led many research groups to investigate the use of AI in this field.

There are essentially two approaches that can be used. In one case, the raw ECG data can be input to the software which 'detects' features, sometimes in an unknown manner, leading to a separation into different diagnostic groups. The other approach is to use actual ECG measurements such as wave amplitudes and durations and allow the software to pick out the features which will also lead to a separation of data. In some approaches, the ECG classification may be input to the training set and this is called supervised learning, whereas in other cases, there is no classification provided during training, i.e., this is unsupervised learning, and the system itself sorts out the different ECGs into a variety of classes.

Publications on the use of AI for 12 lead ECG interpretation are already appearing [60,61]. A recent study by Kashou et al. states that their AI-based approach 'outperforms an existing standard automated computer program' and also 'better approximates expert over-read for comprehensive 12 lead ECG interpretation' [62]. Rhythm analysis was included. Other studies dealing only with analysis of cardiac rhythm have been published [63,64].

One of the more interesting aspects of the use of machine learning, for example, is to use the ECG for the detection of abnormalities which are not in the ECG itself, e.g., in the contraction of the heart. Several papers have now been published on the detection of left ventricular diastolic dysfunction and reduced ejection fraction [65,66]. A recent meta-analysis of five such studies confirmed the ability of AI to identify heart failure from the 12 lead ECG [67].

This leads to the concept of certain packages based on AI being used in a very specific group of patients. A good example of this is the ability of an AI-based system to detect concealed long QT syndrome, where conventionally this is diagnosed when the QT interval exceeds a fixed threshold such as 500 ms Now, an AI-based system can report long QT syndrome when the QT interval is less than 450 ms [68], with confirmation being achieved via appropriate genetic testing. However, this approach would appear to be suited to use in a clinic where individuals suspected of having long QT or being screened for familial long QT are involved, but if applied to the general population, might result in a very high percentage of false positive reports of concealed long QT.

AI-based approaches have also been used for prediction, albeit retrospectively, based on large data bases. For example, mortality has been predicted from the 12 lead ECG in a large cohort, even in those with a normal 12 lead ECG [69]. Atrial fibrillation was predicted in patients in sinus rhythm by training a network with ECGs in sinus rhythm from patients who were known to have subsequently had an episode of atrial fibrillation [70]. The sensitivity was 79% and specificity was 79.5%.

A recent intriguing paper reported on the use of the 12 lead ECG for screening for SARS-CoV-2 [71]. The model could be adjusted to give varying sensitivity and specificity depending also on the prevalence of the virus. The sensitivity could be exceptionally high but with such a setting, the corresponding specificity was extremely low.

Critics of artificial intelligence techniques for ECG analysis will point to the fact that they do not give any indication of why a specific diagnosis has been made. However, there is an approach, termed saliency mapping, which purports to give an indication of the parts of the ECG waveform which contribute to reaching a specific interpretation [72]. The authors of this paper also point out that the technique allows users to find problems in their model which could be affecting performance and generalisability.

One of the advantages, or is it disadvantages, of some forms of machine learning is that thresholds can be chosen for deciding on the presence or absence of an abnormality. As yet, as far as is known, there is no commercial system available which allows the user to set such thresholds and hence 'control' the output. Furthermore, there can be a significant imbalance in groups used to test these newer techniques such that while sensitivity and specificity may seem reasonable, positive predictive value is simply not acceptable. For example, if there are 500 patients with an abnormality in a test population of 10,000 and the sensitivity and specificity of the algorithm for detecting the abnormality are each 80%, then the positive predictive value of the test is 17.4%.

It is generally recognised that the larger a training set can be for the development of AI-based techniques for ECG interpretation, the better will be the algorithm developed. This means that ECGs from a hospital database where cardiologists have verified (over-read) every ECG on the system can be used in a model based on supervised learning. Thus, in this situation, the AI-based approach is aimed at trying to perform as well as cardiologists would have done in interpreting an ECG. This comment relates only to those situations where the order of one million ECGs, for example, may be available for development of the newer techniques and clinical information may often be lacking, e.g., to substantiate an interpretation of left ventricular hypertrophy.

The foregoing suggests that an AI-based system for a complete interpretation of a 12 lead ECG cannot inherently improve on a 12 lead interpretation by cardiologists. Others might disagree with this view based, for example, on one recent report [62]. The same criticism applies to more conventional approaches to automated ECG analysis as shown in the CSE study [35]. Any form of automated ECG interpretation has the advantage of being able to apply the same thorough approach to interpretation 24/7, whereas a cardiologist will undoubtedly give different interpretations of some ECGs when seen several days, if not weeks, apart. The same could not be said of an AI-based system with identical data input to the logic to check on 'repeatability'. This may explain why cardiologists may not appear to give a performance which is equal to that of the AI-based approach [62].

It is becoming very fashionable to report performance in terms of the area under the curve (AUC) in the conventional format of plotting sensitivity versus (1-specificity). In addition, a different curve obtained by plotting precision recall (positive predictive value) versus sensitivity can also be obtained and the AUC similarly obtained. While the AUC may give an overall indication of the performance of the model, it still leaves the user with the problem of deciding at which point on the curve a threshold has to be chosen in order to set a desired sensitivity and specificity. Clearly, essentially by definition, the higher the sensitivity, the lower will be the specificity for the conventional AUC curve. Thus, it may be extremely difficult to gauge the potential day to day performance of a methodology from a knowledge of AUC alone, particularly in a site different from where it was developed.

Electrocardiography has seen many new ideas introduced through the years and essentially none has stood the test of time, e.g., vectorcardiography, QT dispersion, T wave alternans, late potentials, etc. AI is different, although it is aiming to improve diagnosis by finding hidden features in the ECG. Care will have to be taken that initial developments in one centre can be translated satisfactorily into routine practice in another. Much remains to be done for this to happen, including the ability of an AI-based approach to innately make use of age, sex, race, etc., as referenced in this review.

9. Supportive Organisations

There are two organisations which have played a major supporting role in the development of methods for computer analysis of ECGs, namely Computing in Cardiology (CinC) and the International Society for Computerized Electrocardiology (ISCE).

CinC was established in 1974 and has met annually ever since, either in the conventional manner or more recently as a hybrid in person and virtual conference. It is well attended and has a predominance of younger non-clinical researchers who participate in multiple parallel sessions and large poster sessions with the occasional plenary session. There are several competitions organised and, nowadays, conference proceedings are published online.

Of particular relevance is an annual Physionet/CinC Challenge where a problem relating to analysis of physiological signals is set and participants submit their own open source solutions. The challenge was initially organised by the Massachusetts Institute of Technology in Cambridge, MA. Dr Roger Mark and the late George Moody were principally involved in establishing this competition. More recently, Dr Gari Clifford, now at Emory University, Atlanta, has taken over the responsibility for the challenge. Further details can be found elsewhere [73].

A recent example of a challenge was the provision of several thousand single channel ECG recordings of at least 30 s duration obtained using the AliveCor KardiaMobile device. Competitors had to produce software that would determine whether or not atrial fibrillation was present. The interpretation of a large percentage of ECGs was provided for training purposes and the remainder was used for testing. This particular challenge attracted a very large number of participants.

This type of challenge is of current significance given the proliferation of wearables such as smart watches where only a single lead of ECG signal can be recorded. Thus, vendors can benefit perhaps by adopting some of the techniques used by competitors and even enter the competition themselves. Further discussion on analysis of ECGs in wearables can be found elsewhere in this issue [54].

As an aside, one of the most popular features of CinC is the social event where participants are divided into activists and passivists with the choice of being fully active, as in cycling or somewhat more sedate, perhaps undertaking a walking tour of the host city.

ISCE also started in the mid-70s, namely in 1975. It was initially organised by the Engineering Foundation based in New York. ISCE adopts a different style of a single session meeting and, traditionally, all delegates aim to attend all sessions. Each 15 min presentation is accompanied by a 15 min discussion period. ISCE advertises itself in the shape of the Einthoven Triangle with the vertices having Academia, Industry and the User as the three interlinked groups of participants. An average percentage of around 15% of delegates is medically qualified and a much higher percentage is from industry. ISCE also organises an annual Young Investigator Award, but is much more of a traditional conference with the major difference being afternoons free for socialising or engaging in business-like discussions or in potential scientific collaborations. Scientific presentations continue in the evening. Further details of the society can be found elsewhere [74].

An example of the value of ISCE can be given from an important study involving the developers of automated ECG analysis programs. Around 2012, it was suggested that there might be a comparative study of different commercially available software in respect of measuring common ECG intervals such as PR, QT and QRS duration. This was agreed and a database of 600 ECGs was assembled by the Cardiac Safety Research Consortium based at Duke University, North Carolina. Dr. Paul Kligfield from New York coordinated the exercise. Four participants, namely GE, Glasgow, Mortara, and Philips, participated.

Representatives from all groups met together on the occasion of an ISCE conference in Alabama and data was provided to all groups simultaneously gathered together in the same room (Figure 5). Each had their software running on a laptop and measurement data were provided, immediately after analyses were completed, to the study statistician, who was in the room. Results were later published [75] showing differences in measurements between all four programs. As there was no gold standard, there were no right or wrong answers. The study suggested that measurement differences between programs could lead to different interpretations of the same ECG, while normal limits developed by one program would differ from those of another. The study was later expanded to include seven programs [76] with the three new participants being AMPS-LLC, University of Rotterdam (MEANS program) and Schiller. A different database of ECGs was used.

Figure 5. Participants in the comparative study of ECG measurements gather together in 2012 in Birmingham, Alabama anxiously listening to Dr. Paul Kligfield on the extreme left explaining the rules. He is being assisted by Dr. Cindy Green and Dr. Fabio Badilini. Representatives (from centre to right) from Mortara Inc, Philips, GE and Glasgow make up the remainder of those present.

10. Conclusions

Automated electrocardiography has seen phenomenal advances from the 1960s to the 2020s. It is hard to predict where the technique will be even 10 years from now, but it would appear that the future of the 12 lead ECG remains secure, despite the fact that it has been under threat for many years. However, no matter how promising a new approach appears to be, individual cardiologists will still wish to make their own ECG interpretations and frequently claim that a particular automated report is incorrect. This has been true from the start of automated electrocardiography and will always be the case!

Author Contributions: Conceptualisation, P.W.M.; Investigation, J.K. and P.W.M.; Writing—original draft, P.W.M.; Writing—reviewing and editing, P.W.M. and J.K. All authors have read and agreed to the published version of the manuscript.

Funding: There was no financial support for the preparation of this review article.

Acknowledgments: The authors wish to thank all colleagues and collaborators past and present whose support is acknowledged via the references included.

Conflicts of Interest: There is no perceived conflict of interest in this review article other than that the work of the authors' laboratory is inevitably referenced.

References

1. Waller, A.D. A demonstration in man of electromotive changes accompanying the heart's beat. *J. Physiol.* **1887**, *8*, 229–234. [CrossRef]
2. McWilliam, J.A. Fibrillar contraction of the heart. *J. Physiol.* **1887**, *8*, 296–310. [CrossRef] [PubMed]
3. MacKenzie, J. *The Study of the Pulse, Arterial, Venous, and Hepatic, and of the Movements of the Heart*; Pentland: Edinburgh, UK, 1902.
4. Snellen, H.A. *Two Pioneers of Electrocardiography*; Donker: Rotterdam, The Netherlands, 1983; p. 17.
5. Macfarlane, P.W. Lead systems. In *Comprehensive Electrocardiology*, 2nd ed.; Macfarlane, P.W., van Oosterom, A., Pahlm, O., Kligfield, P., Janse, M., Camm, J., Eds.; Springer: London, UK, 2011; Volume 1, pp. 375–425.
6. Lux, R.L. Body surface potential mapping techniques. In *Comprehensive Electrocardiology*, 2nd ed.; Macfarlane, P.W., van Oosterom, A., Pahlm, O., Kligfield, P., Janse, M., Camm, J., Eds.; Springer: London, UK, 2011; Volume 3, pp. 1361–1374.
7. Ramanathan, C.; Ghanem, R.N.; Jia, P.; Ryu, K.; Rudy, Y. Electrocardiographic Imaging (ECGI): A Noninvasive Imaging Modality for Cardiac Electrophysiology and Arrhythmia. *Nat. Med.* **2004**, *10*, 422–428. [CrossRef] [PubMed]
8. Mann, H. A method of analysing the electrocardiogram. *Arch. Int. Med.* **1920**, *25*, 283–294. [CrossRef]
9. Frank, E. An accurate, clinically practical system for spatial vectorcardiography. *Circulation* **1956**, *13*, 737–749. [CrossRef] [PubMed]
10. Macfarlane, P.W. The coming of age of electrocardiology. In *Comprehensive Electrocardiology*, 2nd ed.; Macfarlane, P.W., van Oosterom, A., Pahlm, O., Kligfield, P., Janse, M., Camm, J., Eds.; Springer: London, UK, 2011; Volume 1, pp. 9–20.
11. Pipberger, H.V.; Freis, E.D.; Taback, L.; Mason, H.L. Preparation of electrocardiographic data for analysis by digital electronic computer. *Circulation* **1960**, *21*, 413–418. [CrossRef] [PubMed]
12. Rautaharju, P.M. The birth of computerized electrocardiography: Hubert V. *Pipberger. Cardiol. J.* **2007**, *14*, 420–421.
13. Bailey, J.J.; Dunn, R. Available online: http://ethw.org/Oral-History:James_J._Bailey_and_Rosalie_Dunn (accessed on 20 August 2021).
14. Berson, A. In Memoriam. Hubert V. Pipberger, MD 1920–1993. *Circulation* **1993**, *88*, 817–818.
15. Stallman, F.W.; Pipberger, H.V. Automatic recognition of electrocardiographic waves by digital computer. *Circ. Res.* **1961**, *9*, 1138–1143. [CrossRef]
16. Caceres, C.A.; Steinberg, C.A.; Abraham, S.; Carbery, W.J.; McBride, J.M.; Tolles, W.E.; Rikli, A.E. Computer extraction of electrocardiographic parameters. *Circulation* **1962**, *25*, 356–362. [CrossRef]
17. Cesar A. Caceres. Available online: https://bmet.fandom.com/wiki/Cesar_A._Caceres (accessed on 20 August 2021).
18. Macfarlane, P.W.; Lorimer, A.R.; Lawrie, T.D.V. 3 and 12 lead electrocardiogram interpretation by computer. A comparison on 1093 patients. *Brit. Heart J.* **1971**, *33*, 266–274. [CrossRef] [PubMed]
19. Macfarlane, P.W. Modified axial lead system for orthogonal lead electrocardiography. *Cardiovasc. Res.* **1969**, *3*, 510–515. [CrossRef] [PubMed]
20. Macfarlane, P.W.; Cawood, H.T.; Taylor, T.P.; Lawrie, T.D.V. Routine automated electrocardiogram interpretation. *Biomed. Eng.* **1972**, *7*, 176–180.
21. Macfarlane, P.W. A hybrid lead system for routine electrocardiography. In *Progress in Electrocardiology*; Macfarlane, P.W., Ed.; Pitman Medical: Tunbridge Wells, UK, 1979; pp. 1–5.
22. Watts, M.P.; Macfarlane, P.W. 3-lead electrocardiogram transmission over Post Office telephone lines. *Med. Biol. Eng. Comput.* **1977**, *15*, 311–318. [CrossRef] [PubMed]
23. Macfarlane, P.W.; Podolski, M. Serial comparison in the Glasgow ECG analysis program. In *Computerized Interpretation of the Electrocardiogram XI*; Bailey, J.J., Ed.; Engineering Foundation: New York, NY, USA, 1986; pp. 110–113.
24. Rijnbeek, P.R.; Witsenburg, M.; Schrama, E.; Hess, J.; Kors, J.A. New normal limits for the paediatric electrocardiogram. *Eur. Heart J.* **2001**, *22*, 702–711. [CrossRef] [PubMed]
25. Macfarlane, P.W.; Coleman, E.N.; Pomphrey, E.O.; McLaughlin, S.; Houston, A.; Aitchison, T.C. Normal limits of the high-fidelity ECG. Preliminary observations. *J. Electrocardiol.* **1989**, *22*, 162–168. [CrossRef]
26. Macfarlane, P.W. Influence of age and sex on the electrocardiogram. In *Sex Specific Analysis of Cardiac Function*; Kerkhof, P.L.M., Miller, V.M., Eds.; Springer: London, UK, 2018; pp. 93–106.
27. Macfarlane, P.W. Age, sex, and the ST amplitude in health and disease. *J. Electrocardiol.* **2001**, *34*, 234–241. [CrossRef] [PubMed]

28. Thygesen, K.; Alpert, J.S.; Jaffe, A.S.; Chaitman, B.R.; Bax, J.J.; Morrow, D.A.; White, H.D.; European Society of Cardiology (ESC); Scientific Document Group. Fourth universal definition of myocardial infarction (2018). *Eur. Heart J.* **2019**, *40*, 237–269. [CrossRef]
29. Strauss, D.G.; Wagner, G.S. Defining left bundle branch block in the era of cardiac resynchronization therapy. *Am. J. Cardiol.* **2011**, *107*, 927–934. [CrossRef] [PubMed]
30. Macfarlane, P.W.; Katibi, I.A.; Hamde, S.T.; Singh, D.; Clark, E.; Devine, B.; Franq, B.G.; Lloyd, S.; Kumar, V. Racial differences in the ECG. *J. Electrocardiol.* **2014**, *47*, 809–814. [CrossRef]
31. Rautaharju, P.M.; Park, L.P.; Gottdiener, J.S.; Siscovick, D.; Boineau, R.; Smith, V.; Powe, N.R. Race- and sex-specific models for left ventricular mass in older populations. Factors influencing overestimation of left ventricular hypertrophy prevalence by ECG criteria in African-Americans. *J. Electrocardiol.* **2000**, *33*, 205–218. [CrossRef] [PubMed]
32. Simonson, E. *Differentiation between Normal and Abnormal in Electrocardiography*; C.V. Mosby: St. Louis, MO, USA, 1961; pp. 126–129.
33. Willems, J.L.; Arnaud, P.; van Bemmel, J.H.; Degani, R.; Macfarlane, P.W.; Zywietz, C.; The CSE Working Party. Common Standards for Quantitative Electrocardiography: Goals and Main Results. *Methods Inf. Med.* **1990**, *29*, 263–271. [CrossRef]
34. Macfarlane, P.W. In memoriam: Jos Willems, MD, PhD (1939–1994). *J. Electrocardiol.* **1995**, *28*, 251–252. [CrossRef]
35. Willems, J.L.; Abreu-Lima, C.; Arnaud, P.; van Bemmel, J.H.; Brohet, C.; Degani, R.; Denis, B.; Gehring, J.; Graham, I.; van Herpen, G.; et al. The diagnostic performance of computer programs for the interpretation of electrocardiograms. *N. Engl. J. Med.* **1991**, *325*, 1767–1773. [CrossRef] [PubMed]
36. Young, B.; Schmid, J.J. The New ISO/IEC Standard for Automated ECG Interpretation. *Hearts* **2021**, *2*, 410–418. [CrossRef]
37. The CSE Working Party. Recommendations for measurement standards in quantitative electrocardiography. *Eur. Heart J.* **1985**, *6*, 815–825.
38. Macfarlane, P.W.; Devine, B.; Latif, S.; McLaughlin, S.; Shoat, D.B.; Watts, M.P. Methodology of ECG interpretation in the Glasgow program. *Methods Inf. Med.* **1990**, *29*, 354–361. [CrossRef]
39. Zywietz, C.; Borovsky, D.; Gotsch, G.; Joseph, G. Methodology of ECG interpretation in the Hannover program. *Methods Inf. Med.* **1990**, *29*, 375–385. [CrossRef]
40. Van Bemmel, J.H.; Kors, J.A.; Van Herpen, G. Methodology of the modular ECG analysis system MEANS. *Methods Inf. Med.* **1990**, *29*, 346–353. [CrossRef] [PubMed]
41. Willems, J.L.; Arnaud, P.; Rubel, P.; Degani, R.; Macfarlane, P.W.; van Bemmel, J.H. A standard communications protocol for computerized electrocardiography. *J. Electrocardiol.* **1991**, *24*, 173–178. [CrossRef]
42. Rubel, P.; Fayn, J.; Macfarlane, P.W.; Pani, D.; Schlögl, A.; Värri, A. The History and Challenges of SCP-ECG: The Standard Communication Protocol for Computer-Assisted Electrocardiography. *Hearts* **2021**, *2*, 384–409. [CrossRef]
43. Pipberger, H.V.; Arzbaecher, R.C.; Berson, A.S.; Briller, S.; Brody, D.A.; Flowers, N.C.; Geselowitz, D.B.; Lepeschkin, E.; Oliver, C.G.; Schmitt, O.H.; et al. Recommendations for standardization of leads and of specifications for instruments in electrocardiography and vectorcardiography. *Circulation* **1975**, *52*, 11–31.
44. Bonner, R.E.; Crevasse, L.; Ferrer, M.I.; Greenfield, J.C. A New Computer Program for Analysis of Scalar Electrocardiograms. *Comp. Biomed. Res.* **1972**, *5*, 629–653. [CrossRef]
45. Rautaharju, P.M.; MacInnis, P.J.; Warren, J.W.; Wolf, H.K.; Rykers, P.M.; Calhoun, H.P. Methodology of ECG interpretation in the Dalhousie program; NOVACODE ECG classification procedures for clinical trials and population health surveys. *Methods Inf. Med.* **1990**, *29*, 362–374. [CrossRef] [PubMed]
46. Blackburn, H.; Keys, A.; Simonson, E.; Rautaharju, P.; Punsar, S. The electrocardiogram in population studies. A classification system. *Circulation* **1960**, *21*, 1160–1175. [CrossRef] [PubMed]
47. Rautaharju, P.; Rautaharju, F. *Investigative Electrocardiography in Epidemiologic al Studies and Clinical Trials*; Springer: London, UK, 2007.
48. Macfarlane, P.W. Pentti Rautaharju (1932–2018). *J. Electrocardiol.* **2019**, *53*, I–II.
49. Hu, K.C.; Francis, D.B.; Gau, G.T.; Smith, R.E. Development and performance of Mayo-IBM Electrocardiographic Computer Analysis Programs (V70). *Mayo Clin. Proc.* **1973**, *48*, 260–268. [PubMed]
50. Hammill, S.C.; Andrist, E.M.; Thorkelson, L.A. Electrocardiogram interpreting technician: Training and role in a contemporary electrocardiogram practice. *J. Electrocardiol.* **2008**, *41*, 442–443. [CrossRef] [PubMed]
51. Bailey, J.J.; Horton, M.; Itscoitz, S.B. A Method for Evaluating Computer Programs for Electrocardiographic Interpretation. III. Reproducibility Testing and the Sources of Program Errors. *Circulation* **1974**, *50*, 88–93. [CrossRef]
52. Bailey, J.J.; Berson, A.S.; Garson, A.; Horan, L.G.; Macfarlane, P.W.; Mortara, D.W.; Zywietz, C. Recommendations for Standardization and Specifications in Automated Electrocardiography: Bandwidth and Digital Signal Processing. *Circulation* **1990**, *81*, 730–739. [CrossRef]
53. Kligfield, P.; Gettes, L.S.; Bailey, J.J.; Childers, R.; Deal, B.J.; Hancock, E.W.; van Herpen, G.; Kors, J.A.; Macfarlane, P.; Mirvis, D.M.; et al. Recommendations for the Standardization and Interpretation of the Electrocardiogram. Part I: The Electrocardiogram and Its Technology. *J. Am. Coll. Cardiol.* **2007**, *49*, 1109–1127. [CrossRef] [PubMed]
54. Xue, J.; Yu, L. Applications of Machine Learning in Ambulatory ECG. *Hearts* **2021**, in press.
55. Bergquist, J.; Rupp, L.; Zenger, B.; Brundage, J.; Busatto, A.; MacLeod, R.S. Body Surface Potential Mapping: Contemporary Applications and Future Perspectives. *Hearts* **2020**, in press.
56. Apple Inc. Using Apple Watch for Arrhythmia Detection. December 2020. Available online: https://www.apple.com/healthcare/docs/site/Apple_Watch_Arrhythmia_Detection.pdf (accessed on 10 September 2021).

57. Perez, M.V.; Mahaffey, K.W.; Hedlin, H.; Rumsfeld, J.S.; Garcia, A.; Ferris, T.; Balasubramanian, V.; Russo, A.M.; Rajmane, A.; Cheung, L.; et al. Apple Heart Study Investigators. Large-scale assessment of a smartwatch to identify atrial fibrillation. *N. Engl. J. Med.* **2019**, *381*, 1909–1917. [CrossRef]

58. Seshadri, D.R.; Bittel, B.; Browsky, D.; Houghtaling, P.; Drummond, C.K.; Desai, M.Y.; Gillinov, A.M. Accuracy of Apple Watch for Detection of Atrial Fibrillation. *Circulation* **2020**, *141*, 702–703. [CrossRef] [PubMed]

59. Edenbrandt, L.; Devine, B.; Macfarlane, P.W. Classification of electrocardiographic ST-T segments—human expert vs artificial neural network. *Eur. Heart J.* **1993**, *14*, 464–468. [CrossRef] [PubMed]

60. Kashou, A.H.; Ko, W.-Y.; Attia, Z.I.; Cohen, M.S.; Friedman, P.A.; Noseworthy, P.A. A comprehensive artificial intelligence enabled electrocardiogram interpretation program. *Cardiovasc. Digit. Health J.* **2020**, *1*, 62–70. [CrossRef]

61. Ribeiro, A.H.; Ribeiro, M.H.; Paixao, G.M.M.; Oliviera, D.M.; Gomes, P.R.; Canazart, J.A.; Ferreira, M.P.S.; Andersson, C.R.; Macfarlane, P.W.; Meira, W.; et al. Automatic diagnosis of the 12-lead ECG using a deep neural network. *Nat. Commun.* **2020**, *11*, 1760. [CrossRef]

62. Kashou, A.H.; Mulpuru, S.K.; Deshmukh, A.J.; Ko, W.-Y.; Attia, Z.I.; Carter, R.E.; Friedman, P.A.; Noseworthy, P.A. An artificial intelligence-enabled ECG algorithm for comprehensive ECG interpretation: Can it pass the 'Turing test'? *Cardiovasc. Digit. Health J.* **2021**, *2*, 164–170. [CrossRef]

63. Hannun, A.Y.; Rajpurkar, P.; Haghpanahi, M.; Tison, G.H.; Bourn, C.; Turakhia, M.P.; Ng, A.Y. Cardiologist-level arrhythmia detection and classification in ambulatory electrocardiograms using a deep neural network. *Nat. Med.* **2019**, *25*, 65–69. [CrossRef]

64. Zhu, H.; Cheng, C.; Yin, H.; Li, X.; Zou, P.; Ding, J.; Lin, F.; Wang, J.; Zhou, B.; Li, Y.; et al. Automatic multilabel electrocardiogram diagnosis of heart rhythm or conduction abnormalities with deep learning: A cohort study. *Lancet Digit. Health* **2020**, *2*, e348–e357. [CrossRef]

65. Attia, Z.I.; Kapa, S.; Lopez-Jimenez, F.; McKie, P.M.; Ladewig, D.J.; Satam, G.; Pellika, P.A.; Enriquez-Sarano, M.; Noseworthy, P.A.; Munger, T.M.; et al. Screening for cardiac contractile dysfunction using an artificial intelligence-enabled electrocardiogram. *Nat. Med.* **2019**, *25*, 7074. [CrossRef]

66. Kagiyama, N.; Piccirilli, M.; Yanamala, N.; Shrestha, S.; Farjo, P.D.; Casaclang Verzosa, G.; Tarhuni, W.M.; Nezarat, N.; Budolf, M.J.; Narula, I.; et al. Machine learning assessment of left ventricular diastolic function based on electrocardiographic features. *J. Am. Coll. Cardiol.* **2020**, *76*, 930–941. [CrossRef]

67. Grun, D.; Rudolph, F.; Gumpfer, N.; Hannig, J.; Elsner, L.K.; von Jeinsen, B.; Hamm, C.W.; Rieth, A.; Guckert, M.; Keller, T. Identifying Heart Failure in ECG data with artificial intelligence—A meta-analysis. *Front. Digit. Health* **2021**, *2*, 67. [CrossRef]

68. Bos, J.M.; Attia, Z.I.; Albert, D.E.; Noseworthy, P.A.; Friedman, P.A.; Ackerman, M.J. Use of artificial intelligence and deep neural networks in evaluation of patients with electrocardiographically concealed long QT syndrome from the surface 12-lead electrocardiogram. *JAMA Cardiol.* **2021**, *6*, 532–538. [CrossRef] [PubMed]

69. Raghunath, S.; Cerna, A.E.; Jing, L.; van Maanen, D.P.; Stough, J.; Hartzel, D.N.; Leader, J.B.; Kirchner, H.L.; Stumpe, M.C.; Hafez, A.; et al. Prediction of mortality from 12-lead electrocardiogram voltage data using a deep neural network. *Nat. Med.* **2020**, *26*, 886–891. [CrossRef] [PubMed]

70. Attia, Z.I.; Lopez-Jimenez, F.; Asirvatham, S.; Deshmukh, B.; Gersh, B.J.; Carter, R.E.; Yao, X.; Rabinstein, A.A.; Erickson, B.J.; Kapa, S.; et al. An artificial intelligence-enabled ECG algorithm for the identification of patients with atrial fibrillation during sinus rhythm: A retrospective analysis of outcome prediction. *Lancet* **2019**, *394*, 861–867. [CrossRef]

71. Attia, Z.I.; Kapa, S.; Dugan, J.; Pereira, N.; Noseworthy, P.A.; Lopez-Jimenez, F.; Cruz, J.; Halamka, J.; Gari, N.R.C.; Madathala, R.S.; et al. Rapid Exclusion of COVID Infection with the Artificial Intelligence ECG. *Mayo Clin. Proc.* **2021**, *96*, 2081–2094. [CrossRef] [PubMed]

72. Jones, Y.; Deligianni, F.; Dalton, J. Improving ECG classification interpretability using saliency maps. In Proceedings of the 20th IEEE International Conference of BioInformatics and BioEngineering (BIBE), Cincinnati, OH, USA, 26–28 October 2020; pp. 675–682. [CrossRef]

73. The Physionet/CINC Challenge. Available online: https://www.cinc.org/physionet-cinc-challenge-awards/ (accessed on 10 September 2021).

74. International Society of Computerized Electrocardiology. Available online: https://www.isce.org/ (accessed on 10 September 2021).

75. Kligfield, P.; Badalini, F.; Rowlandson, I.; Xue, J.; Clark, E.; Devine, B.; Macfarlane, P.; de Bie, J.; Mortara, D.; Babaeizadeh, S.; et al. Comparison of automated measurements of electrocardiographic intervals and durations by computer-based algorithms of digital electrocardiographs. *Am. Heart J.* **2014**, *167*, 150–159.e1. [CrossRef] [PubMed]

76. Kligfield, P.; Badalini, F.; Denjoy, I.; Babaeizadeh, S.; Clark, E.; de Bie, J.; Devine, B.; Extramiana, F.; Generali, G.; Gregg, R.; et al. Comparison of automated interval measurements by widely used algorithms in digital electrocardiographs. *Am. Heart J.* **2018**, *200*, 1–10. [CrossRef]

 *hearts*

Review

The History and Challenges of SCP-ECG: The Standard Communication Protocol for Computer-Assisted Electrocardiography

Paul Rubel [1,*], Jocelyne Fayn [2], Peter W. Macfarlane [3], Danilo Pani [4], Alois Schlögl [5] and Alpo Värri [6]

1   INSA-Lyon, University of Lyon, 69100 Villeurbanne, France
2   INSERM US7 SFR Santé Lyon-Est: eTechSanté, University of Lyon, 69008 Lyon, France; jfayn.mtics@gmail.com
3   Institute of Health and Wellbeing, Electrocardiology Section, Royal Infirmary, Glasgow G31 2R, Scotland, UK; Peter.Macfarlane@glasgow.ac.uk
4   Electrical and Electronic Engineering Department, University of Cagliari, 09123 Cagliari, Italy; danilo.pani@unica.it
5   Institute of Science and Technology Austria, 3400 Klosterneuburg, Austria; alois.schloegl@ist.ac.at
6   Faculty of Medicine and Health Technology, Tampere University, 33720 Tampere, Finland; Alpo.Varri@elisanet.fi
*   Correspondence: prubel.lyon@gmail.com; Tel.: +33-609-266-429

Citation: Rubel, P.; Fayn, J.; Macfarlane, P.W.; Pani, D.; Schlögl, A.; Värri, A. The History and Challenges of SCP-ECG: The Standard Communication Protocol for Computer-Assisted Electrocardiography. *Hearts* **2021**, 2, 384–409. https://doi.org/10.3390/hearts2030031

Academic Editor: Lukas J. Motloch

Received: 22 July 2021
Accepted: 16 August 2021
Published: 24 August 2021

Abstract: Ever since the first publication of the standard communication protocol for computer-assisted electrocardiography (SCP-ECG), prENV 1064, in 1993, by the European Committee for Standardization (CEN), SCP-ECG has become a leading example in health informatics, enabling open, secure, and well-documented digital data exchange at a low cost, for quick and efficient cardiovascular disease detection and management. Based on the experiences gained, since the 1970s, in computerized electrocardiology, and on the results achieved by the pioneering, international cooperative research on common standards for quantitative electrocardiography (CSE), SCP-ECG was designed, from the beginning, to empower personalized medicine, thanks to serial ECG analysis. The fundamental concept behind SCP-ECG is to convey the necessary information for ECG re-analysis, serial comparison, and interpretation, and to structure the ECG data and metadata in sections that are mostly optional in order to fit all use cases. SCP-ECG is open to the storage of the ECG signal and ECG measurement data, whatever the ECG recording modality or computation method, and can store the over-reading trails and ECG annotations, as well as any computerized or medical interpretation reports. Only the encoding syntax and the semantics of the ECG descriptors and of the diagnosis codes are standardized. We present all of the landmarks in the development and publication of SCP-ECG, from the early 1990s to the 2009 International Organization for Standardization (ISO) SCP-ECG standards, including the latest version published by CEN in 2020, which now encompasses rest and stress ECGs, Holter recordings, and protocol-based trials.

Keywords: standardization; computerized ECG; personalized medicine; telemedicine; digital ECG data interchange protocol; eHealth

1. Introduction

The electrocardiogram was the very first biosignal ever processed by computers, pioneered by Cesar Caceres at the National Institute of Health (NIH) and Hubert Pipberger from the Veterans Administration (VA) [1]. In the early 1960s, the problem with digital ECG processing was the lack of commercial amplifiers and frequency modulation (FM) recorders for multichannel recordings of the 12-lead ECGs. The standard 12-lead ECG signals were transmitted—lead-by-lead or 3-leads at a time—in analog form by means of FM modems to central computing facilities, and the interpretation results were returned in a textual format by means of teletypes.

Significant changes took place in computer-based ECG analysis and interpretation in the late 1970s and in the 1980s, thanks to the introduction of minicomputers and PCs and the development of microcomputer-based ECG interpreting carts [1–3]. Machine intelligence could be brought at the bedside of the patient.

Early evaluation studies of computer-based ECG measurements and interpretations demonstrated the need for the development of reference databases to assess the performance of ECG measurements and interpretation programs [4–9], and the use of ECG-independent evidence as the gold standard to assess the diagnosis accuracy of ECG interpretation programs [6–8,10]. There was a lack of agreement concerning the definitions of waves, common measurements, standardized criteria for classification, and common terminology for reporting. This created a situation whereby large differences in measurements by different computer programs hampered not only the exchange of diagnostic criteria, but also of ECG measurements and their uses for decision support and clinical studies [4–9]. To solve these issues, a group of several European investigators, led by the late Jos L Willems (J.L.W.), proposed, in 1978, a concerted action to different advisory councils of the European Union (at that time, called European Community, abbreviated EC), aimed at establishing "Common Standards for Quantitative Electrocardiography" (abbreviated CSE) [11]. The proposal was officially approved in 1980 with three other European projects, in the second research and development program for medical and public health research of the EC, and extended during the two following European research and development programs (up until 1990). The objectives were to standardize the ECG measurement procedures in quantitative terms, and to develop methodologies and reference databases to assess the accuracy of ECG measurement programs and the diagnostic performances of computer interpretation programs against the combined wave delineation and interpretative results of a panel of highly skilled cardiologists (and against the clinical truth).

The (approximate) 100 person-years pioneering work is a leading example of how to assess medical information processing techniques [10–22]. CSE databases and measurement libraries are now widely used to assess the quality of ECG measurements and interpretation programs as required by some clauses of the International Electrotechnical Commission (IEC) 60621-2-25 and IEC 60621-2-51 standards [23,24], and are the starting points of new research programs to improve ECG processing and interpretation methodologies.

One interesting result (among the lessons learned during CSE) is that, as is also the case for panel reviews, computer interpretation could be further improved by combining the interpretation results of individual programs [10,17,20,21], a direct result of the application of the Central Limit Theorem. However, it was also shown that, when analyzing a single standard 10 s rest ECG, at the very best, a diagnostic accuracy of about 80 to 82% can be obtained [10,17,20,21], and that further improving the diagnostic performance of the ECG can only be obtained by analyzing, as recommended by several international cardiology societies [25,26], the trend of serial ECG changes, with reference to baseline recordings [25–29], and/or by designing advanced, data fusion-based artificial intelligence (AI) decision support systems, merging ECG signal data (or measurements) and demographic and clinical data featured in the patient's electronic health record (EHR) [30,31]. However, the lack of open communication and interlinking of various types of ECG equipment and software, and the difficulty of overcoming the proprietary solutions proposed by the manufacturers for the exchange of ECG data, had to be solved.

To open this field, J.L.W., Jan van Bemmel, P.W.M., P.R. and the late Rosanna Degani and Christoph Zywietz (C.Z.) launched, in 1989, an international collaboration project called SCP-ECG, aimed at developing a "Standard Communications Protocol for computerized electrocardiography", consisting of standards for the interchange, encoding, and storage of digital ECG data [32–36]. The development was achieved under the aegis of the preliminary Advanced Medical Informatics (AIM) program of the EC, in close collaboration with representatives from academia and industry.

In the following, we first present the main achievements of the AIM A1015 SCP-ECG project. We then recall the different steps of the development of the first two main versions, V1.x and V2.x, of European Norm (EN) 1064 and the International Organization for Standardization (ISO) 11073-91064 "standard communication protocol–computer-assisted electrocardiography" standards [37,38]. Finally, we present the main content of the latest version SCP-ECG V3.0 of the standard, approved by the European Committee for Standardization, or, in French, Comité Européen de Normalisation (CEN) as EN 1064:2020 22 June 2020 [39].

2. Main Achievements of the AIM A1015 SCP-ECG Project

The basic aim of the SCP-ECG project was to arrive at a standard communications protocol for computerized electrocardiography, to make possible the interconnection and exchange of ECG data between different computerized ECG acquisition and analysis devices of various manufacturers, preserve the quality of the original signals, to provide the data and metadata that are relevant for decision support and/or for reanalysis and interpretation by another device or system, and to facilitate the integration of ECG signals and data with other information systems for departmental, hospital, or ambulatory care.

To reach this goal, the project was divided into three closely related work packages, respectively dealing with the exchange of digital ECG data (Work Package 1), encoding and compression (Work Package 2), and storage (Work Package 3) [32,33]. The three work packages were simultaneously worked out and their issues extensively discussed during three working conferences held in Leuven (Belgium) in 1989 and 1990. The conferences were attended by 45 experts in the field, from 11 countries (USA, Japan, and Europe), including representatives of 13 leading manufacturers, responsible, at that time, for the production of over 80% of the world market share of computerized electrocardiographs.

Extensive proceedings were published for the first conference. During the last SCP-ECG conference, consensus was obtained among the different participants with respect to the final SCP-ECG project report [40], which provided a set of recommendations and a first draft of the agreed SCP-ECG standard communication protocol.

In the following, we briefly present the different achievements performed by the three work packages WP1–WP3. In Chapter 3, we present the different steps performed to obtain the draft SCP-ECG communication protocol approved by CEN and, further, by the US American National Standards Institute (ANSI) and ISO.

2.1. WP1: Standards for Digital ECG Data Interchange

The main goal of this work package was to develop a universal protocol, acting as a functional standard, and making recommendations as to how (and when) a given standard should be used for two-way digital ECG data transmission and communication between heterogeneous ECG computer systems. Three specific objectives were addressed:

1. Definition of the data content and format of the ECG records to be exchanged, including the ECG signal data, demographic and acquisition data, as well as measurement and interpretation results;
2. Definition of specific query and control messages, to initiate and control the flow of digital ECG data between different devices or users;
3. Selection of the transport communication protocol and application services (A profiles) for the transfer of digital ECG data, such as File Transfer, Access and Management (FTAM), the Message Handling Services of X.400, Electronic Data Interchange For Administration, Commerce and Transport (EDIFACT), or others.

Various existing protocols from different research groups and manufacturers of computerized electrocardiographs were first examined. The VA transmission protocol [41] was then used as a starting point to define the content and data formats for the interchange. After several discussions and amendments, agreement was achieved on structuring the digital ECG data and metadata into the following 11 data sections, corresponding to the different steps of data acquisition and analysis, preceded by a header called Section 0, and containing pointers to Sections 1 to 11 in the record:

- Section 0: pointers to data areas in the record;
- Section 1: header information—patient data/ECG acquisition data;
- Section 2: Huffman tables used in encoding of ECG data, if used;
- Section 3: ECG lead definition section;
- Section 4: QRS locations, if medians are encoded;
- Section 5: encoded median data, if medians are stored;
- Section 6: encoded rhythm data if no medians are stored, or "error signal" after median subtraction, if medians are stored;
- Section 7: global measurements, wave onsets and offsets as well as global intervals;
- Section 8: textual diagnosis from the "interpretative" device;
- Section 9: manufacturer specific diagnostic and overreading data from the "interpretative" device;
- Section 10: lead measurement results, including the duration and amplitudes of major ECG waves (P+,P-,Q,R,S,R',S',J,T+,T-);
- Section 11: universal interpretative statement codes.

The messaging part of the agreed draft SCP-ECG standard describes the type of information that could be requested and transmitted, and the format of the message headers. Standard query request functions for networked or remotely connected ECG machines, terminals, workstations, ECG management systems, and hospital information systems were also defined, but in these early times, it was only required that every system, at minimum, be able to produce on demand the final interpretative report and the final measurement matrix. Future systems should provide means for interactive communication sessions, with requests that will automatically be transformed into Structured Query Language (SQL) queries.

2.2. WP2: Standards for Digital ECG Data Encoding

Although to a lesser degree than medical imaging, electrocardiography results in large amounts of digital data compared to other medical data, such as patient history, diagnostic codes, and biomedical laboratory data. Thus, data reduction was (and still is) highly desirable, especially in these early times, when transmission was performed over the normal telephone network. This required the use of efficient encoding and, eventually, encryption techniques.

The SCP-ECG project has opened up this field by soliciting publication and comparison of various algorithms. These experiments have demonstrated that, at the end, compression ratios of more than 20 could be obtained without significant loss of information for the so-called residual rhythm record, and with no loss at all for the representative ECG cycle (Figure 1). Based on this analysis, recommendations on minimum performance requirements were proposed. Manufacturers were free to apply their own methods. However, each manufacturer had to clearly define the degree to which signals could be reproduced after decompression, and to classify their systems into one of the following three categories:

Category A: systems that could provide a fully reproducible ECG record, so that each sample can be reconstructed, as it has been obtained from the analog to digital converter with its original specifications.

Category B: systems that could provide a fully reproducible representative cycle and a minimally distorted so-called residual record. Consensus was obtained on the minimum requirements for data loss and on a general approach to ECG compression [42].

Category C: systems that perform data compression, but do not specify the amplitude/time distortion of their algorithms.

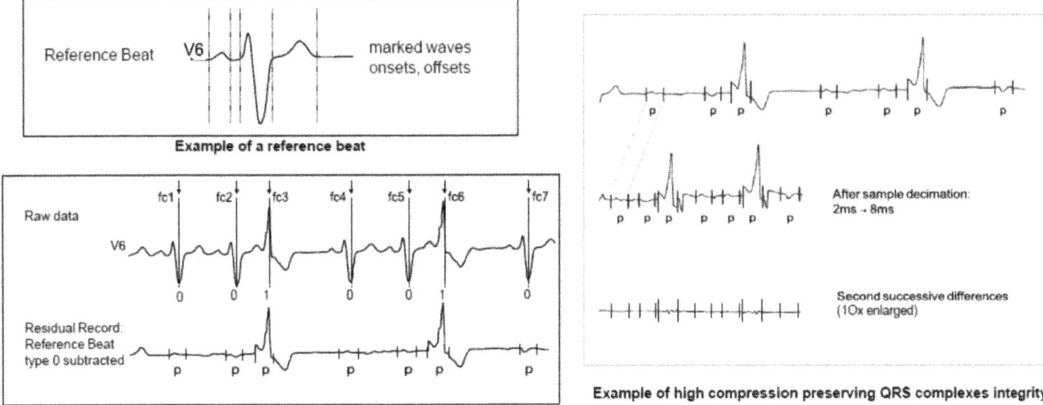

Figure 1. Example of the high compression obtained by subtracting a reference beat from all complexes (including extrasystoles), filtering and sample decimation of the non-protected areas, and second difference calculations [42]. These types of lossy compression schemes, e.g., filtering, sample decimation, and beat subtraction, are no longer allowed in SCP-ECG V3.0 (See Chapter 4 and [39]). Figure adapted from EN 1064:2020 [39].

2.3. WP3: Conceptual Reference Model for Digital ECG Data Storage

The main objective of this work package was to develop a conceptual reference model for digital ECG storage and retrieval, according to different scenarios of use, e.g., for serial comparison, display on a remote workstation, integration in the patient's EHR, different managerial and research purposes, such as statistics for management, evaluation of program performance and, eventually, re-analysis by a modified or totally different ECG program.

The conceptual reference model developed during the SCP-ECG project [43,44] was then further refined throughout the Open European Data Interchange and Processing for computerized Electrocardiography (OEDIPE) project, to fully comply with SCP-ECG V1.0, to improve its performance, in terms of implementation and use, and to support the setting up of generic, dynamic, and patient-specific management and control strategies of serial analysis processes [27,45–48].

The overall core model was designed around the concept of a relational database, which allows various data objects to be linked together via defined relations, based on the J. Martins entity–relationship formalism [48]. It was split into six sub-models, each centered on a major processing task that an ECG database should be able to support. Such a modular architecture was proven to ease the set-up of the different databases implemented during the OEDIPE project, whatever the clinical applications and the scenarios of use. The complete conceptual core data model was then obtained by merging the six sub-models into one model that was composed of about 50 tables and 200 attributes. It allowed the storage of multiple analysis and interpretation results of a given ECG, as well as their traceability. A snapshot of the ECG data acquisition model is displayed in Figure 2.

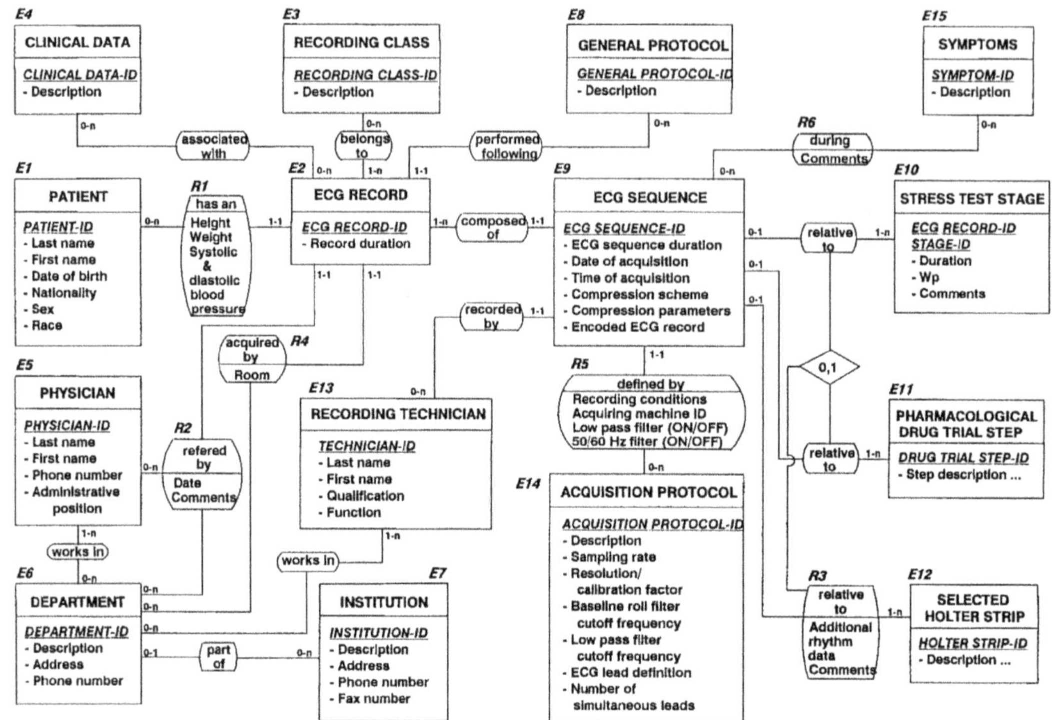

Figure 2. Snapshot of the conceptual ECG data acquisition reference model developed during the AIM A1015 SCP-ECG project and fully implemented during the OEDIPE project [40,43,48–51].

Application area include rest, stress testing and Holter ECG, and the management of ECGs during pharmacological drug trials [44].

3. Development of SCP-ECG Versions V1.x and V2.x and Finalization of the First Versions of the Official EN 1064, ANSI EC71, and ISO 11073-91064 Standards

The final document of the recommendations and agreed transmission protocol developed from 1 June 1989 to 31 December 1990 during the AIM A1015 project was then further worked out with the support of DG XIII/F (Directorate-General XIII for Telecommunications, Information Industries, and Innovation, renamed DG CNECT: Directorate-General for Communications Networks, Content and Technology in 2012) of the European Commission, until February 1993, when it was approved by CEN as an official pre-standard, known as SCP-ECG V1.0 and identified by CEN as prENV 1064:1993.

Preparation of the final draft of this first version of the official SCP-ECG standard was undertaken by project team PT007 (coordinated by J.LW., C.Z. and P.R.) of the newly created CEN TC 251, the so-called Technical Committee (TC) for Medical Informatics established by the technical board (BT) of the European Standardization Committee CEN in March 1990. CEN TC 251 was later renamed Technical Committee on Health Informatics, and is nowadays the counterpart of ISO TC 215, which was created in 1998.

Concomitantly to the finalization of the SCP-ECG V1.0 draft—different research teams, led by J.F., P.R., J.L.W., and C.Z., and some associated industrial partners fully implemented the proposed SCP-ECG standard and carried out different experimentations within the frame of the AIM A2026 OEDIPE project (1992–1995) [27,45–59], and J.L.W. and C.Z. set-up Conformance Testing Services (CTS) within the frame of the CTS-ECG project [60].

In the following, we briefly present the main SCP-ECG related achievements of the OEDIPE and CTS-ECG projects. We then briefly describe the content and the differences between SCP-ECG versions V1.x and V2.x, and we summarize the objectives and accomplishments of the OpenECG project [61].

3.1. Objectives and Main Achievements of the AIM A2026 OEDIPE Project

The ultimate goal of the OEDIPE Project (project leader P.R.) was to develop demonstration systems for the follow-up of cardiac-diseased patients, integrating serial analysis, decision support, open databases, and communication protocols [27,46,47]. OEDIPE successfully implemented and fully tested the SCP-ECG standard communications protocol and the related compression algorithms on stand-alone, industry standard microprocessor-based ECG carts from two different manufacturers [52]. Another successful step towards the electronic data interchange of digital ECGs was the implementation and the operational testing of a low level, Xmodem-based ECG file transfer protocol and an Enhanced Xmodem file transfer protocol [53], along with implementation and testing of the previously mentioned conceptual ECG data storage reference model, and the development of a general scheme, called "file to database", which allowed the extraction of the data contained in an SCP-ECG file and the automation of their storage in different fields of a relational database (and vice versa) [49–51,62], as illustrated by Figure 3.

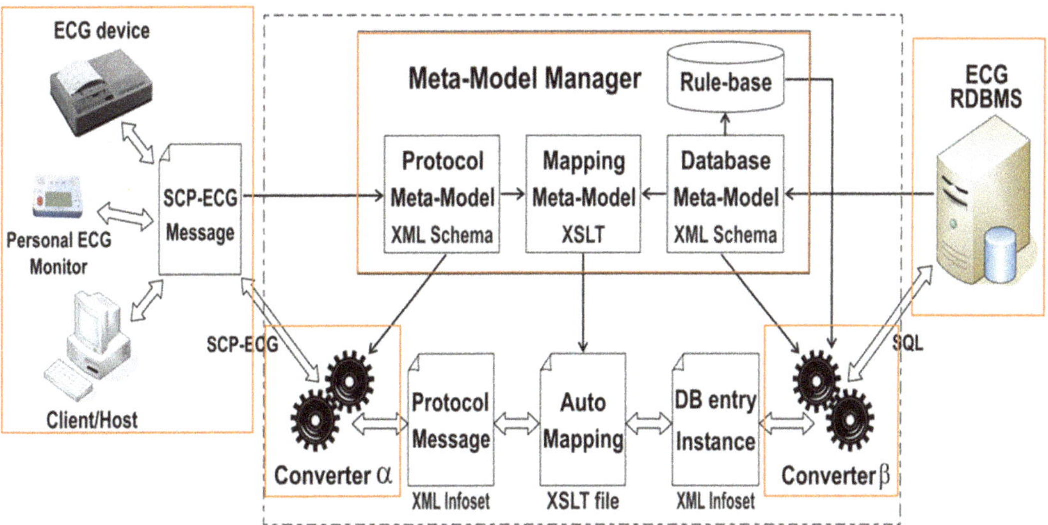

Figure 3. Generic bi-directional SCP-ECG message to database interface schema. This is an updated version of the original "File to Database" schema developed by the OEDIPE project [49], where eXtensible Markup Language (XML) and eXtensible Stylesheet Language Transformations (XSLT) tools were based on Abstract Syntax Notation 1 (ASN.1) [50,51]. The interface updates the database with electrocardiographic information coming from the messages and gives the message handler data retrieved from the database. The solution contains generic software modules independent of the database and SCP-ECG protocol layout. It accesses a descriptive data dictionary containing the database structure, the data format layout, and the mapping between both. The design involves issues related to structure description and standard query language generation and allows automating the development of SCP-ECG Vx.i to Vy.j converters. For more details, see [50,62,63].

Operational testing of the implemented SCP-ECG protocol also included (i) the follow-up of patients who suffered from acute myocardial infarction during transportation in the emergency car or after admission to a coronary care unit; (ii) the follow-up of patients during experimental drug studies; and (iii) the follow-up of heart transplant patients.

OEDIPE was nominated as one of the two best Advanced Informatics in Medicine projects and selected for demonstration during the showcase "Building the Information Society for the Citizens of the World", organized at the initiative of (then) U.S. Vice President Al Gore, within the frame of the G7 Ministerial Conference on Global Information Society held in Brussels, 25–26 February 1995. The objective of the showcase, entitled "Cardiology Network Berlin, Lyon", was to demonstrate a comprehensive healthcare scenario covering several phases of treatment of patients suffering from a chronic coronary disease. An OEDIPE cardiology workstation was used to retrieve and display the patient's health records (clinical data, ECGs, images, etc.), to transmit an ECG from a real ambulance, using an SCP-ECG compliant electrocardiograph via a global system for mobile (GSM) communication-based digital service (note that general packet radio services (GPRSs) were not yet operational), to use the patient's smart card to identify and access health records, retrieve the patient's previous ECG records from the SCP-ECG conceptual reference model-compliant Lyon Cardiology Hospital ECG database management system (DBMS) in near-real time, by means of an 128 Kbps Integrated Services Digital Network (ISDN) link; and retrieve some of the patient's cardiac ultrasound and angiography images from the Deutsches HerzZentrum Picture Archiving and Communication System (PACS) in Berlin, by means of a broadband 35 Mbps asynchronous transfer mode (ATM) network.

A similar demonstrator, using an OEDIPE workstation, an SCP-ECG compliant interpreting electrocardiograph, and an International Maritime Satellite organization (INMARSAT-C) satellite link was set-up to conduct remote follow-up of a subject with cardiac disease (a former French Minister) during his transatlantic race in November 1995.

3.2. Objectives and Main Achievements of the CTS-ECG Project

Concomitantly to the SCP-ECG project, J.L.W. and C.Z. worked out a proposal for the Conformance Testing Services (CTS) program of DG XIII, of the European Commission [60]. The objective was to promote harmonized European testing services by setting-up at least two laboratories in two different countries that would be capable of implementing, executing and, if necessary, further refining IEC specifications, with regard to the performance and safety of computer-based electrocardiographs, and the minimum requirements worked out by other international groups or standards (e.g., the American Heart Association (AHA), CSE, Association for the Advancement of Medical Instrumentation (AAMI), SCP-ECG, EN, ISO, Consultative Committee on International Telephone and Telegraph, now International Telecommunication Union-Technical (CCITT)), along various technical steps, from signal acquisition to the measurement, and the transmission and communication of ECG data.

The 4-year CTS-ECG project (July 1989–June 1993) successfully implemented two pilot test centers, at the University of Leuven (Belgium) and at the Medical School in Hanover (Germany), and developed a PC-based Conformance Testing Services test bench, as well as a set of tools, including a database of test and calibration signals to test both analogue-based conventional electrocardiographs and electrocardiographs with integrated signal processing capabilities, e.g., filtering, ECG measurement and interpretation, signal compression and communication, according to the SCP-ECG standard [64–66].

The CTS-ECG database includes 16 calibration and three analytical ECGs, five type of noise signals, and a few ECGs, to be used for testing the conformance with the minimum requirements for ECG compression and decompression of the SCP-ECG standard. A subset of this database and the corresponding reference values, also known as CTS ECG Test Atlas, containing 14 calibration and three analytical signals, have been included in the IEC 60601-2-25:2011 standard [23], and are currently used to test the essential performance and accuracy of some of the characteristics and measurement functions of electrocardiographs.

3.3. Outline of prENV 1064:1993 SCP-ECG V1.0

prENV 1064:1993 consists of a 145-page document organized in eight chapters (called Clauses) and five annexes. It relates to the conventional recording of the electrocardiogram, i.e., the so-called standard 12-lead electrocardiogram and the vectorcardiogram (VCG). The clinical need to transmit specialized recordings such as body surface maps, Holter and long-term ECG recordings was very low in these early times, and was therefore not considered in this SCP-ECG V1.0 version of the official SCP-ECG standard.

SCP-ECG V1.0 defines the common conventions required for digital transmission of patient-specific data (demographic, recording metadata, etc.), ECG signal data, ECG measurement, and ECG interpretation results between digital electrocardiographs (ECG carts), computer ECG management systems (ECG DBMS), and any other type of computer system (hosts).

The standard specifies the content and structure of the information, which is to be interchanged, provides minimum requirements for encoding and compression of digital ECG data, and defines a set of specific query and control messages that may be used to initiate and control the flow of data for cart-to-host and host-to-host communication between different users.

The various data that may be transmitted by means of the standard ECG communications protocol are defined in Clause 5 of SCP-ECG V1.0. The data are organized in 11 sections as defined during the SCP-ECG A1015 project and enumerated in 2.1. For the sake of usability of the standard, only sections 0 and 1 are mandatory, all other sections are optional.

Although a large number of parameters of Section 1 may be transmitted, most devices will only send a subset of that number. The formatting of data in Section 1 was, thus, made flexible by introducing tagged fields (illustrating the use of tagged fields can be found hereafter in Figure 4), with the provision that, at minimum, the following set of patient demographics and ECG acquisition data must be present: Patient ID number; patient name; sex and birthdate; identification of acquiring cardiograph; date and time of ECG recording; identification of the analyzing device. Accurate patient identification of ECG data and reports is, of course, of utmost importance for data queries and replies.

Minimum requirements for data encoding and compression are defined in Clause 6. A minimum set of control and query messages for cart-to-cart and cart-to-host interchange are defined in Clause 7. A low-level transport protocol for the exchange of data between an ECG cart and a host based on an enhanced X-Modem protocol is defined in Clause 8.

Annex A provides specifications for encoding alphanumeric ECG data in a multilingual environment, so that the standard can be applied worldwide (the nowadays widely used UTF-8, universal coded character set transformation format–8-bit, did not exist yet). Annex B presents a detailed description of the universal ECG interpretation statement codes (to be used for coding Section 11) that were developed during the SCP-ECG project and extended under the PT007 mandate. Annex C (informative) provides a detailed description of the recommended signal compression methodology and outlines the minimum conformance requirements for ECG data compression. Annex D specifies how to encode the level of compatibility of the devices and systems with the SCP-ECG standard, according to four levels of compliance. The last three annexes (informative) provide a glossary of specific terms used in SCP-ECG (Annex E), a list of additional references (Annex F), and a subject index (Annex G).

3.4. From SCP-ECG V1.0 to V1.3

In 1997, AAMI set-up a working group with the mission to reformat and update the contents of ENV 1064:1993 and to adopt it as an ANSI/AAMI standard. Several updated proposals were drafted by the AAMI working group in close cooperation with CEN TC 251 WG4, and then extensively discussed during seven AAMI SCP-ECG WG meetings, also attended by experts having developed ENV 1064:1993.

The final draft, dated 11 September 1999, and identified as SCP-ECG V1.3, was then approved on 11 May 2001 by ANSI as an official ANSI standard catalogued ANSI/AAMI EC71:2001. The ANSI/AAMI standard was revised in 2007 (ANSI/AAMI EC71:2001/(R)2007) and reaffirmed by ANSI 10 September 2013 (ANSI/AAMI EC71:2001/(R)2013).

The main differences between SCP-ECG V1.0 and V1.3 (other than the renumbering of a few clauses and annexes and the introduction of some additional explanatory sentences for the purpose of clarification) are listed hereafter:

- Deprecation of the acquiring device identification encoding scheme, which is now identified by a text string;
- Extension of the language support encoding scheme;
- Amendment of the ECG device capabilities encoding scheme;
- Introduction of two new tags to support electrode configuration and date time zone encoding, and a tagged field to store the patient's medical history in free text;
- Extension of the number of supported ECG leads from 65 to 85;
- Extension of the content of Section 7 global ECG measurements and provision of means to store the QRS type for each detected QRS, the type and source of each detected pacemaker spike (if any), and some additional global measurements in tagged fields (see example of tagged global ECG measurements data field in Figure 4);
- Extension of the confirmation status encoding possibilities in sections 8 and 11;
- Extension of the number of supported per-lead ST measurements;
- Introduction of a set of state diagrams in Clause 7 "Definition of a minimum set of control and query messages for the interchange of ECG data", describing the process by which ID messages are exchanged by cart and host devices;
- Amendment and substantial extension of former Annex D "Definition of compliance with the SCP-ECG standard" (normative), which becomes Annex B (normative).

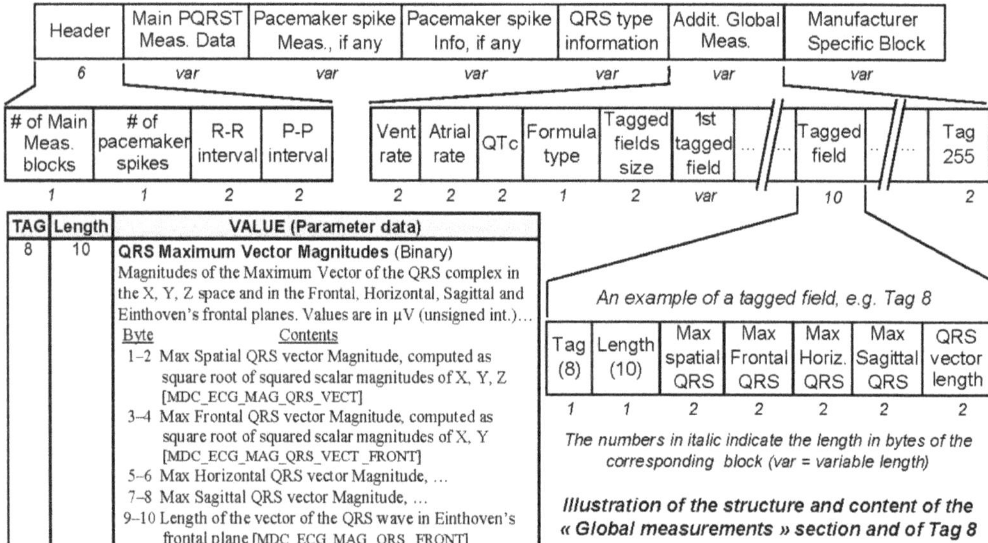

Figure 4. Snapshot of the data part of Section 7 (global measurements), highlighting the structure of the additional global measurements data block and of one of the optional tagged fields, e.g., Tag 8, "QRS Maximum Vector Magnitudes". SCP-ECG V3.0 defines 17 tagged global ECG measurement data fields, numbered from 0 to 16. The structure and content of tag 8 are detailed in the bottom left (tag, length, value) table. The number of tagged fields actually stored may vary from one SCP-ECG record to another.

3.5. From ENV 1064:1993 to EN 1064:2005+A1:2007 and ISO 11073-91064:2009

On 17 December 2004, CEN approved a slightly reformatted version of ANSI/AAMI EC71:2001 as EN 1064:2005, also known as SCP-ECG V2.1. The main differences between SCP-ECG V1.3 and SCP-ECG V2.1 are the move of Clause 7 "Definition of a minimum set of control and query messages for the interchange of ECG data" and Clause 8 "Standard low-level ECG-Cart to host protocol" (both normative) to annexes D and E (both informative).

In 2006, CEN TC 251 Working Group 4 (WG4) further worked out SCP-ECG to take account of new requirements. This updated version identified as SCP-ECG V2.3 was then approved by CEN on 15 January 2007, as standard EN 1064:2005+A1:2007 and by ISO in May 2009 as standard ISO 11073-91064:2009. The latter was reconfirmed by ISO in 2017.

The main differences between SCP-ECG V2.1 and SCP-ECG V2.3 are:

- Extension of the number of supported ECG leads from 85 to 184;
- Amendment of the ECG leads descriptions in order to cross reference the SCP-ECG electrode names and codes with the MDC_ECG_LEAD REFIDs (nomenclature code REFerence IDentifier) from the newly adopted ISO/IEEE 11073-10101:2004 standard;
- Reduction from four to two, concerning the number of data format categories (specified in Annex B "Definition of compliance with the SCP-ECG standard"). The latter are used to encode the type of SCP-ECG related features and information content provided by a specific device.

3.6. Objectives and Main Achievements of the OpenECG Project

Most smaller companies have successfully implemented the different versions of SCP-ECG, but not the largest companies, which preferred to protect their market shares by selling quite expensive integrated departmental ECG analysis and management systems uniquely based on their old proprietary interchange protocols. Increased need for interoperability between medical devices and systems and new requirements coming from new applications, e.g., in the context of image processing using Digital Imaging and Communications in Medicine (DICOM), clinical studies, drug approval by the USA Food and Drugs Administration (FDA), telemedicine and home care, and the EHR, triggered the set-up of the two-year (July 2002–June 2004) European Information Society Technologies (IST)–Thematic Networks Program IST-2001-37711 OpenECG project (project leader Catherine Chronacki) [61,67,68]. The objective was to promote the consistent use of existing interchange format and communication standards for computer-assisted resting electrocardiography, to consolidate expertise, assist integration, and support correct implementations of the ECG interoperability standards, to pave the way towards developing similar standards for exercise and Holter ECGs [61,67–71].

To reach these goals, OpenECG set up an internet portal (www.OpenECG.net, which is unfortunately no longer active since the original OpenECG domain has expired in 2014) [70] to support information exchanges between a steadily growing network of registered OpenECG members (682 members from 56 countries in September 2006) [71] and to grant the OpenECG community access to data and tools for SCP-ECG conformance testing, a database containing annotated samples of real SCP-ECG files, Open Source SCP-ECG viewers [72], and two way converters between SCP-ECG and other standards such as HL7 (Health Level Seven) aECG [73] and DICOM 3.0 (supplement 30) [74], etc.

OpenECG also organized two workshops, in October 2002 in Crete (Greece) and in April 2004 in Berlin (Germany), bringing together about 60 users, manufacturers, healthcare providers, and standardization experts from CEN, AAMI, IEEE, ISO, and IEC, with the objective to consolidate experience and provide practical feedback on user experiences and manufacturer thoughts, ECG interchange formats and viewers, medical devices interoperability with the EHR, harmonization with the different ECG formats from other application areas, related R&D initiatives and trends in Europe, and future challenges [68,69].

The experience and feedback gained during the OpenECG project was taken into account for the development of SCP-ECG V2.3 and V3.0.

4. Development of SCP-ECG V3.0 and EN 1064:2020

According to the policy of CEN, each standard needs to be revised every 5 years. To this end, an EN 1064 SCP-ECG revision kickoff meeting was organized by A.V., convenor of CEN/TC 251/WG IV, in the premises of the CEN and the European Committee for Electrotechnical Standardization (CENELEC) meeting Center in Brussels (Belgium), 25 November 2013. An SCP-ECG revision project team lead by P.R. and A.S. has been nominated, with the mission of keeping useful and up-to-date parts of the standard more or less intact, removing or revising the outdated parts and, if relevant, extending the standard by including new rest ECG measurements and additional recording modalities with related metadata, measurements, and annotations.

The SCP-ECG revision project team first performed a "strengths, weaknesses, opportunities, and threats" (SWOT) analysis and an in depth review of the literature on the SCP-ECG standard [75–79] and complementary standards such as the newly approved ISO/IEEE 11073-10102 "Nomenclature—Annotated ECG" standard [80], the under revision version of the ISO/IEEE 11073-10101 "Nomenclature" standard [81], ANSI/HL7 V3 ECG [82–84], DICOM [85], and the medical waveform description format encoding rules (MFER) series of standards [86–89], as well as the recommendations from leading scientific societies, e.g., AHA, American College of Cardiology (ACC), European Society of Cardiology (ESC), etc. [90–94]. SCP-ECG updates and new section proposals were then drafted by and circulated within the project team members, and some stakeholders in the field of quantitative electrocardiology, and have been amply discussed over 30 WebEx meetings and during a special session on "standardization" held during the Staff 2015 meeting in Vence (France), with different representatives from the industry.

SCP-ECG V3.0 was approved by CEN, on 22 June 2020, as EN 1064:2020 standard [39]. Institutional contacts have been taken by CEN, asking ISO TC 215 to amend some of the terms used in ISO/IEEE/FDIS 11073-10101:2020 [81] and to confirm SCP-ECG V3.0 as an ISO standard that would supersede ISO 11073-91064:2009 [38].

In the following, we summarize the changes that were performed on the already existing sections in SCP-ECG versions 1.x and 2.x, briefly present the new sections that were included in version 3.0, and conclude by recalling the medical challenges of SCP-ECG and why it is a unique standard that is different from other signal-related standards.

4.1. Outline of SCP-ECG V3.0

EN 1064:2020 consists of a 240-page document that comprises a comprehensive, five-page introduction, describing the scene and the application areas, five clauses, eight annexes, and the bibliography. It specifies the means to be used to encode and exchange standard and medium- to long-term electrocardiogram waveforms, and the related metadata acquired in physiological laboratories, hospital wards, clinics, and during primary care medical check-ups, ambulatory, and homecare. It covers electrocardiograms, such as 12-lead, 15-lead, 18-lead, Cabrera lead, Nehb lead, Frank lead, XYZ lead, Holter ECGs, and exercise ECGs that were recorded, measured, and analyzed by equipment, such as electrocardiographs, patient monitors, and wearable devices. It also covers intracardiac electrograms recorded by implantable devices, as well as the analysis results produced by ECG analysis and interpretation systems and software that are compatible with SCP-ECG.

ECG waveforms and data that are not in the scope of this technical specification include real-time ECG waveform encoding and transmission (used for physiological monitors), which are covered by other standards, and intra-cardiac or extra-cardiac ECG mappings.

The various ECG data and related metadata and formats addressed by EN 1064:2020 are, as for the previous versions, specified in Clause 5. They are now structured into 18 sections that are summarized in Table 1. Sections 0, 1, 3, and at least one of the signal

Sections 6, 12, or 14 are required. Section 13 is required if Section 14 is present. All other sections are optional.

Table 1. List of SCP-ECG V3.0 data sections. Sections 0, 1, 3, and at least one of the signal sections 6, 12, or 14 are required. Section 13 is required if Section 14 is present. Section 4 was deprecated and will no longer be used. All other sections are optional.

Section Id	Type	Content
0	Required	Pointers to data areas in the record
1	Required	Header information—patient data/ECG metadata
2	Optional	Huffman tables used in encoding of ECG data (if used)
3	Required	ECG leads definition
4	Reserved	Reserved for legacy SCP-ECG versions
5	Optional	Encoded type 0 reference beat data (if reference beat is stored)
6	Optional *	Short-term ECG rhythm data
7	Optional	Global ECG measurements
8	Optional	Textual diagnosis from the "interpretive" device
9	Optional	Manufacturer specific diagnostic and over-reading data from the "interpretive" device
10	Optional	Per-lead ECG measurements
11	Optional	Universal statement codes resulting from the interpretation
12	Optional *	Long-term ECG rhythm data
13	Optional *	Stress tests, drug trials and protocol-based ECG recordings metadata
14	Optional *	Selected ECG sequences repository
15	Optional	Beat-by-beat ECG measurements and annotations
16	Optional	Selected ECG beat measurements and annotations
17	Optional	Pacemaker spike measurements and annotations
18	Optional	Additional ECG annotations

* See Table 1 legend.

Annex A (normative) provides a set of supplementary information for the encoding of drugs, of the functionalities of implanted cardiac devices, the filtering methods used for ECG processing, and the physical units and/or the type of measurements and annotations.

Annex B "Universal ECG interpretation statements codes" provides, as in the previous versions of SCP-ECG, a "pragmatic" approach to the problem of mapping computer interpretation statements onto a common and understandable lexicon. It was substantially amended to comply with the latest cardiology societies recommendations, and was extended by cross-referencing semantically equivalent statement codes and acronyms from AHA recommendations and related nomenclature standards.

Annex C provides a set of SCP-ECG compliance definitions and specifications, and outlines a testing procedure that may be followed by manufacturers who intend to state SCP-ECG compliance for their devices and/or systems or software.

Annex D recalls the recommended ECG compression techniques that were used in the previous versions of SCP-ECG. This annex was previously normative and is now only informative. It was kept for educational reasons, to support understanding of Huffman encoding and decoding, and conversion of high compressed ECGs from legacy SCP-ECG version 1.x and 2.x files into SCP-ECG version V3.0 files.

Annex E provides some cross-references to the coding schemes used by other ECG-related standards namely AHA recommendations [92], CDISC (Clinical Data Interchange Standards Consortium) C-Code [95], ISO/IEEE 11073-10102 aECG [80], and DICOM code values [85].

Annexes F, G, and H, respectively, provide some implementation recommendations, a short glossary of the terms used in EN 1064:2020 that are beyond the usual backgrounds (in electrocardiography and in computer science of the potential readers), and a revision history, since the initial release of the EN 1064 standard in 1993.

Sections 0 to 11 already existed in the previous versions of SCP-ECG and, although they have been significantly updated, they remain almost backwards compatible with SCP-ECG V1.x and V2.x, except for non-UTF-8 text strings encoding and beat subtraction or bimodal compression schemes, which are no longer supported. Starting with SCP-ECG V3.0, only lossless compression (difference and Huffman encoding) of the long-term rhythm data (Section 6) and of the reference beat type 0 data (Section 5) are now allowed. In addition, to simplify encoding, the present standard recommends storing all ECG signal data uncompressed as a series of fixed length, signed integers, and to reserve difference data calculations and Huffman encoding for mobile and/or wearable devices, when they are intended to be used in poorly served areas with limited wireless connectivity, such as GPRS, where lossless data compression strategies are still relevant. Converting legacy SCP-ECG V1.x and V2.x files into SCP-ECG V3.0 compliant files would, thus, only require (i) converting non-UTF-8 text strings into UTF-8; (ii) store ECG signal data, if any, uncompressed. Sections 12 to 18, which are new, were introduced to support the storage of continuous, long-term ECG recordings, of selected sequences of stress tests, drug trials, and protocol-based ECG recordings, and the related metadata, measurements, and annotations.

All over the document, emphasis was put on cross-referencing, whenever possible, all SCP-ECG terms, measurements, annotations, diagnosis statements, and metadata with the REFIDs, specified by the ISO/IEEE 11073-10102 annotated ECG (aECG) [80] and 11073-10101 nomenclature (vital signs) [81] standards, and on removing the ambiguities and inaccuracies of some of these, other than SCP-ECG standards [96].

Additional changes to Sections 0 to 11 are shortlisted in 4.2 and new Sections 12 to 18 are described in 4.3.

4.2. Main Updates to Sections 0 to 11 Existing in SCP-ECG V2.3

The revisions to sections 0 to 11 of SCP-ECG V2.3 are the following:

- Update and harmonization of the description of the various terms, measurements, annotations, diagnosis statements, and metadata to be compliant with the existing health informatics norms (e.g., ISO/IEEE 11073, CDISC, DICOM, HL7, IEC, etc.), the recommendations from cardiology societies (e.g., AHA, ACC, ESC, etc.), and the scientific literature;
- Precise definition of the semantics inherent to the various terms, measurements, annotations, diagnosis statements, and metadata [96];
- Renaming of Section 1: header information—patient data/ECG metadata;
- Update of SCP-ECG drug codes and move of the drug codes lists to new Annex A;
- Update of the medical history codes;
- Amendment of the definitions of the electrode configuration codes;
- Introduction of two new tags to support the description of implanted cardiac devices and the storage of the patient's drugs, according to the World Health Organization (WHO) Anatomical Therapeutic Chemical (ATC) classification system [97];
- Introduction of a new "Global, virtual lead" in Section 2, to support the coding of per-lead measurements with suffix _LEAD_CONFIG;
- Deprecation of Section 4 (now reserved for legacy SCP-ECG versions);
- Introduction of three new fields in the header of Section 5: a Huffman-encoding specifier (HES), the number of samples per lead, and the location of a fiducial point (QRS trigger point) in the reference beat type 0. These changes are intended to simplify

the implementation of the SCP-ECG protocol in case no (or only) default Huffman tables are used;

- Deprecation of the bimodal compression scheme in Sections 5 and 6;
- Substantial amendment and extension of the content of Section 7 "Global ECG measurements" and Section 10 "Per-lead ECG measurements" in order to be compliant with ISO/IEEE 11073-10101 and 11073-10102, and with the latest recommendations of the scientific and medical societies [96];
- Introduction of the possibility to store the local time zone in sections 8 and 11;
- Extension of the content of Section 11 [96] by introducing the possibility to store interpretive statements, encoded according to the categorized AHA statement codes [92] and CDISC code specifications [95].

4.3. New Sections in SCP-ECG V3.0

4.3.1. Long-Term and Protocol-Based ECG Recordings—Sections 12 to 14

Starting with version V3.0, in addition to the short duration resting ECG (Section 6) and the corresponding type 0 reference beat (Section 5), the standard now provides means for storing long-term ECG rhythm data in Section 12, e.g., up to 40 days of continuous recording of 3-lead ECG signals sampled at 200 samples/sec, with a 16 bit resolution, in Section 14, several selected short- to medium-duration ECG sequences, and, in Section 13, the related metadata and reference beats (or pointers to selected reference beats). These two additional sections were included to support protocol-based ECG recordings, namely stress tests and drug trials procedures.

Section 14 may also be used to store high resolution, large bandwidth ECG recordings, with sampling rates beyond 150,000 samples/s for improved pacemaker spikes detection and analysis [98,99], while simultaneously storing a standard ECG in Sections 6 or 12.

The format of Section 12 is very similar to the International Society for Holter and Noninvasive Electrocardiology (ISHNE) format [100,101]. In order to preserve random access to the record's segments, no compression or encoding is allowed in this section.

4.3.2. Beat-by-Beat ECG Measurements and Annotations—Section 15

In addition to the full set of global measurements (Section 7), and the per-lead measurements (Section 10) of the type 0 reference beat, starting with version V3.0, the standard now allows storing, in this section, several pre-defined global and per-lead beat measurements and annotations, for the reference beats stored in or pinpointed by Section 13, and for all (or for only some) selected beats of the long-term and/or short-term ECGs stored in Sections 12 and 14 and/or in Section 6. The beats may have been selected one-by-one by a physician or by a beat typing algorithm (reference beats of different types, etc.), or may refer to the entire set of beats from one or more selected time windows within the long-term ECG stored in Section 12 or in the long-term ECGs stored in Sections 6 or 14. The data format is designed to support a large number of use cases, such as selecting and daily analyzing a set of 10 min duration time windows from a continuous long-term ECG recording, for example, for time windows starting at 2 a.m. and 2 p.m., and then storing (P-on, P-off, QRS-on, QRS-off, T-off), and some additional useful annotations to assess day and night differences and day-to-day variabilities of the selected measurements.

In another scenario, e.g., performing thorough QT studies, one may select and store the measurements and annotations for K preselected, not necessarily consecutive beats, in as much measurement blocks (MB) as there are selected beats. To facilitate the comparison of reference beat measurements, the standard also allows saving, in separate MBs, the measurements and annotations performed on the reference beats stored in Sections 5 and 13.

4.3.3. Additional ECG Beat and Spike Measurements, and ECG Annotations—Sections 16–18

Although Section 15 already provides means for storing several pre-defined global and per-lead beat measurements and annotations for different subsets of computed or selected (reference) beats of the analyzed signals, there are various scenarios which, for example, require storage of a few measurements and annotations for all beats of the rhythm signals, and a larger set of measurements and annotations for a much smaller number of beats, i.e., for some selected or computed reference beats. One solution would be to extend the number of (optional) additional measurements in Section 15 in order to include the additional measurements used to quantify the selected or computed reference beats, but this could introduce huge overheads, as all measurement and annotation arrays in Section 15 do have the same MB length, which would require the storage of void measurement values for the non-selected beats, even if not computed.

Section 16 thus provides a solution for storing a different set of measurements and annotations than those stored in Section 15 and is therefore complementary to Section 15. Its structure and format are much the same as for Section 15, except that there is no provision for specifying analysis time windows and that there are no reserved fields for systematically storing the PP and RR intervals (the latter can nevertheless be stored, if need be, as optional additional measurements).

Section 16 is the preferred section for storing selected ECG beat measurements and annotations, if no beat-by-beat measurements and annotations are required (i.e., if Section 15 is not present).

Section 17 was designed to include support for pre-defining and storing (much like the way used for storing beat-by-beat ECG measurements in Section 15) large sets of global, and/or per-lead spike measurements and annotations, spike-by-spike, in one or more spike measurement array(s), one measurement array per analyzed ECG sequence (full long-term ECG record, selected ECG sequence), or reference beat.

The main objective of Section 18 "Additional ECG annotations" is to provide a solution for storing any type of manually or automatically produced annotation, which has not been stored in a systematic way in Sections 7, 8, 10, 11, and 15 to 17, such as the onset (and end) of a bigeminal rhythm or atrial fibrillation, the identification of a pacemaker spike that was not listed in Section 17, measurements that were not foreseen in Sections 15 and 16 (or a few measurements, such as QT intervals in drug studies, in case neither Section 15 nor Section 16 were implemented), manual annotation of complex cases with different types of aberrant QRS complexes (LBBB aberrancy, etc.), and P waves (AV dissociation, etc.), noise annotations in a given lead, etc.

5. Discussion

The ECG, non-invasive, easy-to-use (anywhere, anytime), and inexpensive, has been (and still is) the main source of information for the early detection of cardiovascular diseases and rhythm disorders, which are the leading causes of death in the world, excluding pandemics. The ECG, the first physiological signal ever to be analyzed by computer, benefited from unprecedented international cooperative research, bringing together clinicians, engineers, and scientists from academia and industry.

Digital ECG analyses, in different modalities at rest, during exercise, and on an outpatient basis, have been the subject of intensive and pioneering methodological research for the improvement of the quality of the signals, thanks to advanced and specific filtering techniques, for the recognition, identification, characterization, and delineation of waveforms, and for medical decision support. All known theories in signal processing and data analysis have been applied, compared, and evaluated on the ECG. Although deep learning was not yet mature some decades ago (however, in 1970, neuron weights were already adjusted by means of backpropagation algorithms, based on linear programming [102]), it is on the ECG that the most sophisticated wavelet and artificial intelligence techniques, for example, have been tested at a large-scale, thanks to the advent of databases. This proliferation of results on the ECG was a pioneering asset in the medical field, and explains why (and how) the CSE concerted research action was initiated, in order to objectively, independently, and consensually compare the results of the different methods used. One piece of fundamental knowledge produced and proven by the CSE concerted action from the end of the 1970s, and then during the 1980s, was the importance of inter-subject, inter-observer, as well as intra-observer variability in digital and diagnostic medicine. Medicine is not an exact science as laymen in the field have just discovered in these times of the COVID-19 pandemic.

Once again, in a pioneering way, and since the 1970s, the main world actors in digital and quantitative electrocardiology research developed the concept of personalized medicine to improve the performance of ECG analysis and interpretation methods, by creating the so-called serial analysis approach. Since the patient's ECG is like a fingerprint and the patient himself becomes his own reference, any suspicion of abnormal changes in the ECG descriptors, over time, is a cardiac risk factor.

Thus, while the CSE concerted action was almost over, J.L.W. launched, in 1989, a new collaborative research project aimed at bringing together the entire R&D community in the field of quantitative electrocardiology at the international level, and to propose a standard for the exchange of digital ECG signals and data, in order to retrieve (for any patient) a previous self-documenting and re-analyzable reference ECG, and to compare it with any of the just-recorded ECGs by a so-called serial analysis.

SCP-ECG was born, bringing together more than 80% of the electrocardiograph manufacturers from all over the world and the main European research teams in the field of quantitative electrocardiology, thanks to the support of DG XIII of the European Commission.

Millions of ECGs are recorded every day, and in most cases, they need to be transmitted and/or archived for reviewing and/or for serial ECG analysis. However, although all ECGs today are digitally acquired, most so-called rest ECGs are immediately printed and the printouts are either stored as a paper record or are scanned and sent as a pdf file to be included in the patient's EHR. Very often, and for different aims, these hard copies were also sent by fax, which was considered, until the mid-2010s, to be "more secure"! According to Badilini et al. [103], one reason for this situation is because many companies have never widely implemented a universal ECG storage and exchange standard format such as SCP-ECG, and still use proprietary formats, mainly to protect intellectual property and technology; thus, inciting users to adopt printed/scanned solutions for ECG transmission and archival.

Paradoxically, while we are already in the digital era, in several use cases, these "ECG images" need to be processed shortly after being printed, to recover the "voltage-versus-time" nature of the signals [104–106] (this was the subject of P.R.'s PhD thesis in the late 1960s!), either for research purposes, for submitting annotated digital ECGs to the FDA for drug approval [104], or for computing new ECG measurements. Some authors claim that by using this "reverse-printing" process they can achieve (only) up to 97% digitization accuracy [105], a figure that does not guarantee that the reconstructed signal will allow detecting small Q waves as defined in [13]. Other AI researchers further argue that the original ECG signals are not really needed as AI-based ECG interpretation may be performed by directly processing the "ECG images", just as any medical image record [103,106,107]. However, these new approaches have not yet been assessed against well-documented databases such as the CSE Diagnostic database. Knowing that, because of the high slew rate of the QRS wave, especially in pediatric ECGs, the digitalization of a standard ECG with a 5 µV resolution requires that the uncertainty in the sampling time (i.e., the maximum allowable sample-to-sample variation in the encoding process) of the acquiring device is less than 12.5 µs (which represents less than 0.015 pixels for an 25 mm/s ECG printout on a 1200 dpi printer); it is difficult to imagine that any "reverse printing" process would ever achieve this performance. It is P.R.'s belief that, because the electrocardiogram is a spatiotemporal image, e.g., (8D+time) for the standard 12-lead ECG, none of the methods cited above will allow recovery, with enough accuracy, of one of the most important features of the original digital ECG signals, the phase relationship between the different leads [28]. Image processing researchers are working on pixel-based images, but that is the only thing they have. Why should researchers in quantitative electrocardiology and cardiologists turn their ECG signals into pixels and lose spatiotemporal information (intra-beat, beat-to-beat, intra-day, day-to-day, month-to-month, etc.) when information age technologies already allow retrieval of the original signals, anywhere, anytime, using a universal ECG storage and exchange format such as SCP-ECG?

Several other arguments are also in favor of SCP-ECG, which allows, if need be, retrieval of the original, unfiltered, high fidelity multilead ECG signals for reprocessing and/or interpretation by another ECG analysis program, for serial comparison, for pacemaker spike analysis, clustering, and identification [98,99], for operator-guided retrospective relabeling, or sometimes just for displaying the ECG in a different way than the one used on the standard printout. Among the different types of presentations used, the most common options are: filtered vs. unfiltered ECG, 25 vs. 50 mm/sec, panoramic display according to the Cabrera sequence [91], 24-lead ECG display for enhanced recognition of STEMI-equivalent patterns in the 12-lead ECG [108], display of the VCG computed by using an inverse Dower-like reconstruction method [109], etc. SCP-ECG also provides easy access to additional measurements, clinical information, and ECG metadata already embedded in the SCP-ECG record—information that is usually not printed on the paper copy, but that is very important for decision-making.

SCP-ECG V3.0 was designed to accommodate almost any use case, from the simplest to the most complex, e.g., storing in Section 17 spike measurements performed with a standard spike detection program, which analyzed the standard 12-lead rest ECG stored in Section 6 or in Section 12, and then storing the spike measurements performed with a high-resolution ECG analysis program on a simultaneously-recorded high-bandwidth ECG stored in Section 14. The expected benefit is to empower the monitoring of the functioning of implanted pacing devices, improve computerized interpretation or human over-reading of paced rhythms, and allow correlations with clinical patient outcomes.

Various authors have recently compared and provided extensive reviews on complementary standards dealing with ECG formats (DICOM, EDF, HL7 aECG, MFER, PDF-ECG, etc.) and on converters between different ECG formats, e.g., SCP-ECG to DICOM, HL7 aECG, MFER, and vice-versa [73–75,77–79,103,110–112]. European data format (EDF) and MFER are intended to only store the raw signals. HL7 aECG is intensively used for drug trials annotations, but to overcome the problem of excessive message size inherent to the use of XML, the HL7 aECG standard recommends that the ECG signals are stored in a separate file, which may be an issue for integrity preservation, whereas SCP-ECG V3.0 may store the HL7 aECG annotations in Section 18. DICOM 3.0 (supplement 30) is mainly used to store the ECG signals that have been used for synchronizing cardiovascular magnetic resonance imaging (MRI) image acquisition with heart motion. PDF-ECG is another approach that allows encapsulating, in the same PDF/A-3u file, both the standard graphical report of a rest ECG in the pdf format and the ECG signal data in any of the existing formats/standards, such as HL7 or DICOM, and possibly SCP-ECG [103], but it needs ad-hoc tools to extract the embedded files, and it is not guaranteed that the standard graphical report can be displayed by any PDF viewer, whereas SCP-ECG V3.0 could store PDF reports as single annotations in Section 18.

SCP-ECG V3.0 is the latest, most comprehensive, and universal ECG storage and ECG exchange standard today. It has been significantly amended, with the objective to provide means to support the storage and interchange of almost all existing ECG recording modalities, processing results, annotations and diagnoses, and to alleviate some implementation burdens for manufacturers by removing the little-used lossy compression methods that were supported in previous versions of the standard, and that are no longer allowed in version 3.0, as the need for compression is significantly lower compared to

30 years ago. All terms, measurements, and metadata are precisely defined, enabling the harmonization with other standards in health informatics.

The information conveyed by SCP-ECG is structured in a coherent way with regard to the different fundamental stages of the processing of the ECG signal, and has been selected to supply the knowledge useful for the exploitation and interpretation of the transmitted data. However, the option chosen in SCP-ECG is not to transmit another complete patient medical record encapsulating information, which is redundant with other standards, but rather to provide an ECG-centric data and metadata record. The size of an SCP-ECG file is thus reduced to its minimum, and allows its transmission via a smartphone in case of emergency. Binary encoding of the stored information not only reduces the size of the SCP-ECG files, but also secures the exchange of medical data. In addition, the exploitation of SCP-ECG files does not require expensive computer equipment, but only basic PCs and it is expected that the use of the SCP-ECG standard will be all the more widespread, as automatic decoding tools and viewers of SCP-ECG files will be inexpensive or open access.

Hence, only the syntax of the language used for the exchanges and the semantics of the conveyed data are standardized, and not the logic for obtaining these data nor the intelligence embedded in the devices, which must nevertheless be documented to be exploitable. Indeed, like universal thought, which can only result from the application of a rule known as majority voting, especially in medicine, and because no method of reasoning can be privileged and standardized, only the fusion of the results of these methods allows to approach the supposed truth, in agreement with the Central Limit Theorem. This is why SCP-ECG was designed to be a self-documented standard and to be open to all numerical methods used to compute, analyze, and interpret ECG descriptors, and to the different modalities of ECG signal acquisition techniques. Except for the patient characteristics that are essential to link the ECG data to the persons on whom the stored ECGs have been recorded, and the specification of the ECG recording modalities that are useful for the spatiotemporal exploitation of the conveyed signal data, all other information is optional, thus facilitating the opening of the ECG devices market.

SCP-ECG V3.0 seems to be complex, but in reality, it is not. Only Sections 0, 1, and one of the signal sections (e.g., Section 6) are compulsory, all other sections are optional. Moreover, only some information stored in Section 1 is mandatory, most tagged fields are optional. An original SCP-ECG record, storing only the raw signals, may then be completed step-by-step, either automatically by a (remote) ECG device/software and/or annotated by an over-reading physician. All amendments to the SCP-ECG record are time-stamped. Most sections are self-contained, i.e., if, for example, one wants to exploit the data content of one of the measurement data sections, and only exchange lead-by-lead, beat-by-beat, or spike-by-spike measurements and annotations, without exchanging signals, no other information than the patient data stored in Section 1 is needed.

SCP-ECG is an open, efficient, and flexible standard, with a small footprint that may be easily extended to accommodate other types of vital signs and related context-aware and patient-centric data, as recently demonstrated by Mandellos et al. [76,113], who used sections 200 to 212 to store saturation of peripheral oxygen (SPO2), accelerometer, gyroscopic and magnetometer signals, and metadata not yet covered by other standards.

Should SCP-ECG be further extended to several other signal modalities? This was already experimented with in the mid-1990s within the frame of the European AIM A2050 Biosignal Representation, Integration, and Telecommunication services in Rehabilitation (BRITER), 1993–1996 project, which attempted to define a general communication protocol based on ASN.1, including different signals, such as ECG, EEG, EMG, EOG, movement analysis signals, MRI, etc. [59]. The project was technically successful, but evidenced that a "universal" protocol, which encompasses so many different signal modalities would be very difficult to maintain, and that the only solution is to have and maintain interchange protocols that are optimized to the clinical application domain. However, the building concepts of SCP-ECG may be reused as a reference for other areas of medical applications; thus, easing data interoperability. If need be, data and signal fusion (used for designing integrated clinical workstations, storage in databases, epidemiology studies, etc.) could then be achieved by converting each of the signal specific data formats into an intermediate-, ASN.1-, or XML-based format, which would be part of a mediation meta model [62,63].

Henceforth, one future challenge of the SCP-ECG standard is the widespread use of its powerful capabilities, to empower clinical applications and the development of novel approaches, e.g., self-improving ECG interpretation programs [114] and ambient intelligence solutions [29,30,115], for the enhancement of ECG-based quality of care and the early prevention of cardiovascular diseases.

6. Conclusions

The consensual objective of the initiators of SCP-ECG, which makes SCP-ECG an exemplary standard, was to exploit the wealth of knowledge and experience acquired in the field of quantitative electrocardiography, in an open manner, and independent of particular interests, in order to allow secure and low-cost data exchange, to improve timely medical decision-making and for the benefit of citizens. This approach was followed throughout the years, allowing the various versions (up to the most recent one, SCP-ECG V3.0) to continuously improve.

The SCP-ECG standard was designed to support practically any use case, in clinical practice, research and evaluation trials, of raw and computerized digital ECG data storage in a stand-alone file, for electronic data interchange and archiving, for display, reviewing, reprocessing, converting into another format, or for distributed storage within a medical information system.

Numerous researchers, engineers, and stakeholders from the scientific and medical computerized ECG community and industry, have made significant contributions toward systematic progress in the development of the SCP-ECG standard. We should all be grateful to them for their accomplished efforts in developing such an exemplary standard, and highly encourage all cardiologists and general practitioners to promote its use in order to enhance ECG diagnosis accuracy, for the benefit of patients, and to reduce societal costs.

Author Contributions: P.R. and J.F. wrote the original draft. All of the authors are major contributors to the development of the SCP-ECG standard—J.F., P.W.M., and P.R. (since 1989), A.S. and A.V. (since the late 1990s), and D.P. (since 2013). All authors have read and agreed to the published version of the manuscript.

Funding: This research received no external funding.

Acknowledgments: The authors thank the large number of researchers, engineers, cardiologists, and clinicians from academia, industry, and normalization organizations who contributed to the development and testing of the SCP-ECG standards.

Conflicts of Interest: The authors declare no conflict of interest.

References

1. Rautaharju, P.M. Eyewitness to history: Landmarks in the development of computerized electrocardiography. *J. Electrocardiol.* **2016**, *49*, 1–6. [CrossRef]
2. Rubel, P.; Bailly, G.; Giard, M.H. Microprocessor based biomedical instrumentation. A modular approach. In *Digest of Papers, Proceedings of the 11th International Conference on Medical and Biological Engineering, Ottawa, ON, Canada, 2–6 August 1976*; Durie, N.D., Swail, E.I., Eds.; National Research Council: Ottawa, ON, Canada, 1976; pp. 560–561.
3. Rubel, P.; Varrot, M.; Morlet, D.; Arnaud, P.; Forlini, M.C.; Bailly, G. Architecture and performance of digital ECG acquisition systems. In *Optimization of Computer ECG Processing*; Wolf, H.K., Macfarlane, P.W., Eds.; North Holland Publ. Comp.: Amsterdam, The Netherlands, 1980; pp. 41–53.
4. Bailey, J.J.; Horton, M.; Itscoitz, S.B. A Method for Evaluating Computer Programs for Electrocardiographic Interpretation. III. Reproducibility Testing and the Sources of Program Errors. *Circulation* **1974**, *50*, 88–93. [CrossRef]
5. Willems, J.L.; Paerdens, J. Differences in measurement results obtained by four different ECG computer programs. In *Proceedings of the Computers in Cardiology, Rotterdam, The Netherlands, 29 September–1 October 1977*; Ostrow, H.G., Ripley, K.L., Eds.; IEEE: Long Beach, CA, USA, 1977; pp. 115–121.

6. Rautaharju, P.M.; Ariet, M.; Pryor, T.A.; Arzbaecher, R.C.; Bailey, J.J.; Bonner, R.; Goetowski, C.R.; Hooper, J.K.; Klein, V.; Millar, C.K.; et al. The quest for optimal electrocardiography. Task Force III: Computers in diagnostic electrocardiography. *Am. J. Cardiol.* **1978**, *41*, 158–170. [CrossRef]
7. Willems, J.L. A plea for common for common standards in computer aided analysis. *Comput. Biomed. Res.* **1980**, *13*, 120–131. [CrossRef]
8. Willems, J.L. A review of computer ECG analysis: Time to evaluate and standardize. *CRC Crit. Rev. Med. Inform.* **1986**, *1*, 165–207.
9. Brohet, C.; Willems, J.L.; Derwael, D.; De Schreye, D.; Fesler, R.; Pardaens, J. Impact of measurement precision on diagnostic ECG computer classification. In *Computer ECG Analysis: Towards Standardization*; Willems, J.L., van Bemmel, J.H., Zywietz, C., Eds.; North-Holland Publ.: Amsterdam, The Netherlands, 1986; pp. 171–176.
10. Willems, J.L.; Abreu-Lima, C.; Arnaud, P.; Brohet, C.R.; Denis, B.; Gehring, J.; Graham, I.; van Herpen, G.; Machado, H.; Michaelis, J.; et al. Evaluation of ECG Interpretation Results Obtained by Computer and Cardiologists. *Methods Inf. Med.* **1990**, *4*, 308–316.
11. The CSE European Working Party; Willems, J.L.; Arnaud, P.; van Bemmel, J.H.; Bourdillon, P.J.; Brohet, C.; Dalla Volta, S.; Degani, R.; Denis, B.; Demeester, M.; et al. Common Standards for Quantitative Electrocardiography: The CSE Pilot Study. In *Medical Informatics Europe 81. Lecture Notes in Medical Informatics*; Grémy, F., Degoulet, P., Barber, B., Salamon, R., Eds.; Springer: Berlin/Heidelberg, Germany, 1981; Volume 11, pp. 319–326. [CrossRef]
12. Willems, J.L. Common standards for quantitative electrocardiography. *J. Med. Eng. Technol.* **1985**, *9*, 209–217. [CrossRef] [PubMed]
13. The CSE Working Party. Recommendations for measurement standards in quantitative electrocardiography. *Eur. Heart J.* **1985**, *6*, 815–825.
14. Willems, J.L.; Arnaud, P.; van Bemmel, J.H.; Bourdillon, P.; Degani, R.; Denis, B.; Graham, I.; Harms, F.M.A.; Macfarlane, P.W.; Mazzocca, G.; et al. A reference data base for multi-lead electrocardiographic computer measurement programs. *J. Am. Coll. Cardiol.* **1987**, *10*, 1313–1321. [CrossRef]
15. Morlet, D.; Rubel, P.; Arnaud, P.; Willems, J.L. An improved method to evaluate the precision of computer ECG measurement programs. *Int. J. Bio-Med. Comput.* **1988**, *22*, 199–216. [CrossRef]
16. Morlet, D.; Rubel, P.; Willems, J.L. Value of Scatter-Graphs for the Assessment of ECG Computer Measurement Results. *Methods Inf. Med.* **1990**, *29*, 413–423. [CrossRef]
17. Willems, J.L.; Abreu-Lima, C.; Arnaud, P.; van Bemmel, J.H.; Brohet, C.; Degani, R.; Denis, B.; Graham, I.; van Herpen, G.; Macfarlane, P.W. Effect of combining electrocardiographic interpretation results on diagnostic accuracy. *Eur. Heart J.* **1988**, *9*, 1348–1355. [CrossRef]
18. Van Bemmel, J.H.; Willems, J.L. Standardization and validation of medical support-systems: The CSE Project. *Methods Inf. Med.* **1990**, *29*, 261–262. [CrossRef] [PubMed]
19. Willems, J.; Arnaud, P.; Van Bemmel, J.; Degani, R.; Macfarlane, P.; Zywietz, C. Common standards for quantitative electrocardiography: Goals and main results. *Methods Inf. Med.* **1990**, *29*, 263–271.
20. Willems, J.L.; Abreu-Lima, C.; Arnaud, P.; van Bemmel, J.H.; Brohet, C.; Degani, R.; Denis, B.; Gehring, J.; Graham, I.; van Herpen, G.; et al. The diagnostic performance of computer programs for the interpretation of the electrocardiogram. *N. Engl. J. Med.* **1991**, *325*, 1767–1773. [CrossRef]
21. Willems, J.L. Assessment of diagnostic ECG results using information and decision theory: Results from the CSE diagnostic study. *J. Electrocardiol.* **1992**, *25*, 120–125. [CrossRef]
22. Fayn, J.; Rubel, P.; Macfarlane, P.W. Can the lessons learned from the assessment of automated electrocardiogram analysis in the Common Standards for quantitative Electrocardiography study benefit measurement of delayed contrast-enhanced magnetic resonance images? *J. Electrocardiol.* **2007**, *40*, 246–250. [CrossRef]
23. International Electrotechnical Commission. *IEC 60601-2-25:2011, Medical Electrical Equipment—Part 2-25: Particular Requirements for the Basic Safety and Essential Performance of Electrocardiographs*, 2nd ed.; IEC: Geneva, Switzerland, 19 October 2011; pp. 1–196, ISBN 978-2-88912-719-1.
24. International Electrotechnical Commission. *IEC 60601-2-51:2003, Medical Electrical Equipment—Part 2-51: Particular Requirements for Safety, Including Essential Performance, of Recording and Analysing Single Channel and Multichannel Electrocardiographs*, 1st ed.; IEC: Geneva, Switzerland, 27 February 2003; pp. 1–175.
25. Schlant, R.C.; Adolph, R.J.; DiMarco, J.P.; Dreifus, L.S.; Dunn, M.I.; Fisch, C.; Garson, A.; Haywood, L.J.; Levine, H.J.; Murray, J.A.; et al. Guidelines for electrocardiography. A report of the American College of Cardiology/American Heart Association Task Force on Assessment of Diagnostic and Therapeutic Cardiovascular Procedures (Committee on Electrocardiography). *Circulation* **1992**, *85*, 1221–1228. [CrossRef]
26. Kadish, A.H.; Buxton, A.E.; Kennedy, H.L.; Knight, B.P.; Mason, J.W.; Schuger, C.D.; Tracy, C.M.; Winters, W.L.; Boone, A.W.; Elnicki, M.; et al. ACC/AHA clinical competence statement on electrocardiography and ambulatory electrocardiography: A report of the ACC/AHA/ACP-ASIM task force on clinical competence (ACC/AHA Committee to develop a clinical competence statement on electrocardiography and ambulatory electrocardiography) endorsed by the International Society for Holter and noninvasive electrocardiology. *Circulation* **2001**, *104*, 3169–3178.
27. Rubel, P.; Willems, J.L.; Zywietz, C.; Fayn, J. Towards open data interchange and processing of serial ECGs. In Proceedings of the Computers in Cardiology, Durham, NC, USA, 11–14 October 1992; Murray, A., Arzbaecher, R., Eds.; IEEE Computer Society Press: Los Alamitos, CA, USA, 1992; pp. 79–82. [CrossRef]

28. Fayn, J.; Rubel, P.; Pahlm, O.; Wagner, G.S. Improvement of the detection of myocardial ischemia thanks to information technologies. *Int. J. Cardiol.* **2007**, *120*, 172–180. [CrossRef]
29. Fayn, J. A Classification Tree Approach for Cardiac Ischemia Detection Using Spatiotemporal Information from Three Standard ECG Leads. *IEEE Trans. Biomed. Eng.* **2011**, *58*, 95–102. [CrossRef]
30. Fayn, J.; Rubel, P. Toward a personal health society in cardiology. *IEEE Trans. Inf. Technol. Biomed.* **2010**, *14*, 401–409. [CrossRef]
31. Atoui, H.; Fayn, J.; Gueyffier, F.; Rubel, P. Cardiovascular Risk Stratification in Decision Support Systems: A Probabilistic Approach. Application to pHealth. In Proceedings of the Computers in Cardiology, Valencia, Spain, 17–20 September 2006; Murray, A., Ed.; IEEE: Piscataway, NJ, USA, 2006; pp. 281–284. Available online: https://www.cinc.org/archives/2006/pdf/0281.pdf (accessed on 15 August 2021).
32. Willems, J.L.; Zywietz, C.; Rubel, P.; Degani, R.; Macfarlane, P.W.; van Bemmel, J.H. A standard communications protocol for computerized electrocardiography. *J. Electrocardiol.* **1991**, *24*, 173–178. [CrossRef]
33. Willems, J.L.; Zywietz, C.; Rubel, P.; Degani, R.; Macfarlane, P.W.; van Bemmel, J.H. Development of a standard communications protocol for computerized electrocardiography. In *Telematics in Medicine*; Duisterhout, J.S., Hasman, A., Salamon, R., Eds.; Elsevier Science Publ., North Holland: Amsterdam, The Netherlands, 1991; pp. 299–311.
34. Rubel, P.; Willems, J.L.; Zywietz, C.; Degani, R.; Macfarlane, P.W.; van Bemmel, J.H. Development of a standard communications protocol for the exchange and the storage of digital ECGs. In Proceedings of the 13th Annual International Conference of the IEEE Eng. Med. Biol. Soc., Orlando, FL, USA, 31 October–3 November 1991; Nagel, J.H., Smith, W.M., Eds.; IEEE Press: New York, NY, USA, 1991; pp. 572–573. [CrossRef]
35. Willems, J.L.; Zywietz, C.; Rubel, P.; Degani, R.; Macfarlane, P.W.; van Bemmel, J.H. SCP-ECG: A standard communications protocol for computerized electrocardiography. In *Advances in Medical Informatics. Results for the AIM Exploratory Action*; Noothoven van Goor, J., Christensen, J.P., Eds.; IOS Press: Amsterdam, The Netherlands, 1992; pp. 325–332. [CrossRef]
36. Willems, J.L.; Rubel, P.; Zywietz, C. Standard Interchange for Computerized Electrocardiography. In *Progress in Standardization in Health Care Informatics*; De Moor, G.J.E., McDonald, C.L., Noothoven van Goor, J., Eds.; IOS Publisher: Amsterdam, The Netherlands, 1993; pp. 185–194. [CrossRef]
37. European Committee for Standardization. *EN 1064:2005+A1:2007, Health Informatics—Standard Communication Protocol—Computer-Assisted Electrocardiography*; CEN: Brussels, Belgium, 2007; pp. 1–188.
38. International Organization for Standardization. *ISO 11073 91064:2009, Health Informatics—Standard Communication Protocol—Part 91064: Computer-Assisted Electrocardiography*; ISO: Geneva, Switzerland, 2009; pp. 1–159.
39. European Committee for Standardization. *EN 1064:2020, Health Informatics—Standard Communication Protocol—Computer-Assisted Electrocardiography*; CEN: Brussels, Belgium, 2020; pp. 1–240.
40. Willems, J.L. *Standard Communications Protocol for Computerized Electrocardiography: Final Specifications and Recommendations*; ACCO Publ.: Leuven, Belgium, 1991; ISBN 90-73402-01-7.
41. Marquette Electronics, Inc. *Internal Document: Definition of the Data Contents within a Universal ECG Protocol*; GE Healthcare: Milwaukee, WI, USA, 1987; pp. 1–24.
42. Zywietz, C.; Joseph, G.; Fischer, R.; Degani, R.; Willems, J.L. Compression and encoding of ECG data within the European standard communications protocol. In Proceedings of the Computers in Cardiology, Venice, Italy, 23–26 September 1991; Murray, A., Arzbaecher, R., Eds.; IEEE Computer Society Press: Los Alamitos, CA, USA, 1991; pp. 105–108. [CrossRef]
43. Rubel, P.; Fayn, J.; Macfarlane, P.W.; Willems, J.L. Development of a conceptual reference model for digital ECG data storage. In Proceedings of the Computers in Cardiology, Venice, Italy, 23–26 September 1991; Murray, A., Arzbaecher, R., Eds.; IEEE Computer Society Press: Los Alamitos, CA, USA, 1992; pp. 109–112. [CrossRef]
44. Fayn, J.; Rubel, P.; Girard, P.; Ravel, C.; Forlini, M.C.; Boissel, J.P. CAVIAR, a Serial ECG Processing and Management System for Pharmacological Drug Trials. In Proceedings of the Computers in Cardiology, Venice, Italy, 23–26 September 1991; Murray, A., Arzbaecher, R., Eds.; IEEE Computer Society Press: Los Alamitos, CA, USA, 1992; pp. 301–304. [CrossRef]
45. Fayn, J.; Rubel, P.; Willems, J.L. Management of serial ECGs and control strategies for the comparison process. *Meth. Inform. Med.* **1994**, *33*, 148–152. [PubMed]
46. Rubel, P.; Willems, J.L.; Zywietz, C.; Fayn, J.; Assanelli, D.; Malossi, C.; Todd, S. OEDIPE: Open European Data Interchange and Processing for Electrocardiography. In *Health in the New Communications Age*; Laires, M.F., Ladeira, M.J., Christensen, J.P., Eds.; IOS Press: Amsterdam, The Netherlands, 1995; pp. 313–321.
47. Rubel, P.; Fayn, J.; Zywietz, C.; Willems, J.L.; Assanelli, D.; Malossi, C.; Florin, C. Open European electrocardiological data interchange. In *Health in the Information Society, Proceedings of the Health Telematics '95, Laco Ameno Ischia, Naples, Italy, 2–6 July 1995*; Bracale, M., Denoth, F., Eds.; CNR: Pisa, Italy, 1995; pp. 381–386, ISBN 88-7958-003-5.
48. Fayn, J.; Rubel, P.; Willems, J.L.; Conti, L.; Reniers, R. Design and implementation strategies of a core database model for the storage and retrieval of serial ECG data. In Proceedings of the Computers in Cardiology, Bethesda, MD, USA, 25–28 September 1994; Murray, A., Arzbaecher, R., Eds.; IEEE Computer Society Press: Los Alamitos, CA, USA, 1994; pp. 109–112. [CrossRef]
49. Rubel, P.; Fayn, J.; Willems, J.L.; Zywietz, C. New trends in serial ECG analysis. *J. Electrocardiol.* **1993**, *26*, 122–128.
50. Pfirsch, F.; Reniers, R.; Rubel, P.; Willems, J.L. A Generic Interface between Communication Protocols and Relational Databases. In *Communication and Integration in Healthcare Informatics, Proceedings of the Medical Informatics Conference M.I.C. '93, Gent, Belgium, 11–12 November 1993*; De Moor, G.J.E., Ed.; Acco Publ.: Leuven, Belgium, 1993; pp. 23–32.

51. Al-Ahmad, W.; Rubel, P.; Willems, J.L. OSI standards for ECG data interchange: Use of ASN.1 in the OEDIPE project. In *Communication and Integration in Healthcare Informatics, Proceedings of the Medical Informatics Conference M.I.C. '93, Gent, Belgium, 11–12 November 1993*; De Moor, G.J.E., Ed.; Acco Publ.: Leuven, Belgium, 1993; pp. 115–122.
52. Mertins, V.; Fischer, R.; Joseph, G.; Zywietz, C.; Malossi, C.; Wack, K. Implementation of the SCP-ECG Protocol into stand-alone Electrocardiographs. In Proceedings of the Computers in Cardiology, Durham, NC, USA, 11–14 October 1992; Murray, A., Arzbaecher, R., Eds.; IEEE Computer Society Press: Los Alamitos, CA, USA, 1992; pp. 83–86. [CrossRef]
53. Zywietz, C.; Mertins, V.; Assanelli, D.; Malossi, C. Digital ECG Transmission from Ambulance Cars with Application of the European Standard Communications Protocol SCP-ECG. In Proceedings of the Computers in Cardiology, Bethesda, MD, USA, 25–28 September 1994; Murray, A., Arzbaecher, R., Eds.; IEEE Computer Society Press: Los Alamitos, CA, USA, 1994; pp. 341–344. [CrossRef]
54. Al-Ahmad, W.; Willems, J.L.; Rubel, P. ECG Data Interchange within the Framework of the SCP-ECG and the OEDIPE Projects. In Proceedings of the Computers in Cardiology, Bethesda, MD, USA, 25–28 September 1994; Murray, A., Arzbaecher, R., Eds.; IEEE Computer Society Press: Los Alamitos, CA, USA, 1994; pp. 337–340. [CrossRef]
55. Badr, A.; Al-Ahmad, W.; Rubel, P.; Willems, J.L. Protection of Data Privacy for Open ECG Data Interchange. In *Trends in Medical Informatics, Proceedings of the Medical Informatics Conference M.I.C. '94, Velthoven, The Netherlands, 25–26 November 1994*; De Moor, G.J.E., Ed.; Acco Publ.: Leuven, Belgium, 1994; pp. 145–153.
56. Ron, J.L.; Fayn, J.; Rubel, P. Towards a generic domain information model in cardiology. Experiences from OEDIPE. In Proceedings of the Computers in Cardiology, Indianapolis, IN, USA, 8–11 September 1996; Murray, A., Arzbaecher, R., Eds.; IEEE: Piscataway, NJ, USA, 1996; pp. 301–304. [CrossRef]
57. Fayn, J.; Conti, L.; Fareh, S.; Maison-Blanche, P.; Nony, P.; Rubel, P. Interactive and Dynamic ECG Analysis. Is it Just an IDEA or a clinically relevant approach? *J. Electrocardiol.* **1996**, *29* (Suppl. S1), 21–25. [CrossRef]
58. Ferrari, G.; Zywietz, C.; Willems, J.L.; Fayn, J.; Poeta, M.; Assanelli, D. User interface database for digital SCP-ECG and epidemiological information. In Proceedings of the Computers in Cardiology, Bethesda, MD, USA, 25–28 September 1994; Murray, A., Arzbaecher, R., Eds.; IEEE Computer Society Press: Los Alamitos, CA, USA, 1994; pp. 177–180. [CrossRef]
59. Loukil, A.; Fayn, J.; Conti, L.; Bortone, S.; Fioretti, S.; Bedini, R.; Nentidis, G.; Leo, T.; Rubel, P.; Talbot, A. A system for accessing and handling heterogeneous sources of patient data. Application to the rehabilitation domain. In *Proceedings of the 18th Annual International Conference of the IEEE Engineering in Medicine and Biology Society, Amsterdam, The Netherlands, 31 October–3 November 1996*; IEEE: Piscataway, NJ, USA, 1996; Volume 18, pp. 1240–1241. [CrossRef]
60. Willems, J.L.; Zywietz, C. Conformance Testing for Computerized Electrocardiography. In *Proceedings of the European Conference on Conformance Testing and Certification in Information Technology and Telecommunications, Brussels, Belgium, 13–15 June 1990*; CEN/CENELEC&ETSI, Ed.; IOS Press: Amsterdam, The Netherlands, 1991; pp. 307–319, ISBN 9051990383.
61. Chronaki, C.E.; Chiarugi, F.; Lees, P.J.; Bruun-Rasmussen, M.; Conforti, F.; Ruiz Fernandez, R.; Zywietz, C. Open ECG: A European Project to Promote the SCP-ECG Standard, a Further Step Towards Interoperability in Electrocardiography. In Proceedings of the Computers in Cardiology, Memphis, TN, USA, 22–25 September 2002; Murray, A., Ed.; IEEE: Piscataway, NJ, USA, 2002; pp. 285–288. [CrossRef]
62. Jumaa, H.; Fayn, J.; Rubel, P. XML based mediation for automating the storage of SCP-ECG data into relational databases. In *Proceedings of the Computers in Cardiology, Bologna, Italy, 14–17 September 2008*; Murray, A., Ed.; IEEE: Piscataway, NJ, USA, 2008; pp. 445–448. [CrossRef]
63. Jumaa, H.A. XML and Relational Databases Mediation Automation. Ph.D. Thesis, National Institute of Applied Sciences of Lyon (INSA-Lyon), Lyon, France, 16 December 2010. Available online: http://theses.insa-lyon.fr/publication/2010ISAL0120/these.pdf (accessed on 12 July 2021).
64. Zywietz, C.; Willems, J.L. European Conformance Testing Services for computerized electrocardiography. New procedures and standards. *J. Electrocardiol.* **1993**, *26*, 137–146.
65. Zywietz, C.; Alraun, W.; Mertins, V. Quality Assurance in Electrocardiography: Inappropriate Performance Standards, ECG Characteristics and a New Test Database. In Proceedings of the Computers in Cardiology, Bethesda, MD, USA, 25–28 September 1994; Murray, A., Arzbaecher, R., Eds.; IEEE Computer Society Press: Los Alamitos, CA, USA, 1994; pp. 333–336. [CrossRef]
66. Zywietz, C.; Alraun, W.; Fischer, R. Quality assurance in biosignal processing—Procedures and recommendations for evaluation for electrocardiological analysis systems. In Proceedings of the Computers in Cardiology, Rotterdam, The Netherlands, 23–26 September 2001; Murray, A., Ed.; IEEE: Piscataway, NJ, USA, 2001; pp. 201–204. [CrossRef]
67. Chronaki, C.E.; Chiarugi, F.; Lees, P.J.; Macerata, A.; Conforti, F.; Bruun-Rasmussen, M.; Ruiz Fernandez, R.; Zywietz, C. A Year in the Life of the OpenECG Network. In Proceedings of the Computers in Cardiology, Thessaloniki, Greece, 21–24 September 2003; Murray, A., Ed.; IEEE: Piscataway, NJ, USA, 2003; pp. 17–20. [CrossRef]
68. Lees, P.J.; Chronaki, C.E.; Chiarugi, F.; The OpenECG Consortium. Standards and Interoperability in Digital Electrocardiography. The OpenECG Project. *Hell. J. Cardiol.* **2004**, *45*, 364–369.
69. Chronaki, C.E.; Chiarugi, F.; Macerata, A.; Conforti, F.; Voss, H.; Johansen, I.; Ruiz-Fernandez, R.; Zywietz, C. Interoperability in digital electrocardiography after the OpenECG project. In Proceedings of the Computers in Cardiology, Chicago, IL, USA, 19–22 September 2004; Murray, A., Ed.; IEEE: Piscataway, NJ, USA, 2004; pp. 49–52. [CrossRef]

70. Chronaki, C.E.; Chiarugi, F.; Sfakianakis, S.; Zywietz, C. A Web Service for Conformance Testing of ECG Records to the SCP-ECG Standard. In Proceedings of the Computers in Cardiology, Lyon, France, 25–28 September 2005; Murray, A., Ed.; IEEE: Piscataway, NJ, USA, 2005; pp. 961–964. [CrossRef]
71. Chronaki, C.E.; Chiarugi, F.; Fischer, R. OpenECG: Medical Device Interoperability as a Quality Label for eHealth Services. In Proceedings of the Computers in Cardiology, Valencia, Spain, 17–20 September 2006; Murray, A., Ed.; IEEE: Piscataway, NJ, USA, 2006; pp. 465–468.
72. Chiarugi, F.; Spanakis, M.; Lees, P.J.; Chronaki, C.E.; Tsiknakis, M.; Orphanoudakis, S.C. ECG in Your Hands: A Multi-Vendor ECG Viewer for Personal Digital Assistants. In Proceedings of the Computers in Cardiology, Thessaloniki, Greece, 21–24 September 2003; Murray, A., Ed.; IEEE: Piscataway, NJ, USA, 2003; pp. 359–362. [CrossRef]
73. Schloegl, A.; Chiarugi, F.; Cervesato, E.; Apostolopoulos, E.; Chronaki, C.E. Two-way converter between the HL7 aECG and SCP-ECG data formats using BioSig. In Proceedings of the Computers in Cardiology, Durham, NC, USA, 30 September–3 October 2007; Murray, A., Ed.; IEEE: Piscataway, NJ, USA, 2007; pp. 253–256. [CrossRef]
74. Sakkalis, V.; Chiarugi, F.; Kostomanolakis, S.; Chronaki, C.E.; Tsiknakis, M.; Orphanoudakis, S.C. A gateway between the SCP-ECG and the DICOM supplement 30 waveform standard. In Proceedings of the Computers in Cardiology, Thessaloniki, Greece, 21–24 September 2003; Murray, A., Ed.; IEEE: Piscataway, NJ, USA, 2003; pp. 25–28. [CrossRef]
75. Schlögl, A. An overview on data formats for biomedical signals. In Proceedings of the World Congress on Medical Physics and Biomedical Engineering, Munich, Germany, 7–12 September 2009; Dössel, O., Schlegel, W.C., Eds.; Springer: Berlin/Heidelberg, Germany, 2009; Volume 25/4, pp. 1557–1560. [CrossRef]
76. Mandellos, G.J.; Koukias, M.N.; Styliadis, I.S.; Lymberopoulos, D.K. e-SCP-ECG+ Protocol: An Expansion on SCP-ECG Protocol for Health Telemonitoring—Pilot Implementation. *Int. J. Telemed. Appl.* **2010**, *2010*, 137201. [CrossRef]
77. Gonçalves, B.; Guizzardi, G.; Pereira Filho, J.G. Using an ECG reference ontology for semantic interoperability of ECG data. *J. Biomed. Inf.* **2011**, *44*, 126–136. [CrossRef] [PubMed]
78. Bond, R.R.; Finlay, D.D.; Nugent, C.D.; Moore, G. A review of ECG storage formats. *Int. J. Med. Inform.* **2011**, *80*, 681–697. [CrossRef] [PubMed]
79. Trigo, J.D.; Alesanco, A.; Martinez, I.; Garcia, J. A Review on Digital ECG Formats and the Relationships Between Them. *IEEE Trans. Inf. Technol. Biomed.* **2012**, *16*, 432–444. [CrossRef]
80. International Organization for Standardization. *ISO/IEEE 11073-10102:2014, Health Informatics—Point-of-Care Medical Device Communication—Part 10102: Nomenclature—Annotated ECG*, 1st ed.; ISO: Geneva, Switzerland, 2014; pp. 1–177.
81. International Organization for Standardization. *ISO/IEEE 11073-10101:2020, Health informatics—Point-of-Care Medical Device Communication—Part 10101: Nomenclature*, 2nd ed.; ISO: Geneva, Switzerland, 2020; pp. 1–1040.
82. Healh Level Seven. *ANSI/HL7 V3 ECG, R1-2004 (R2019), HL7 Version 3 Standard: Regulated Studies—Annotated ECG, Release 1*; HL7 International: Ann Arbor, MI, USA. Available online: https://www.hl7.org/implement/standards/product_brief.cfm?product_id=70 (accessed on 15 August 2021).
83. Healh Level Seven. *HL7 V3 IG ECGR1 R2, HL7 Version 3 Implementation Guide: Regulated Studies; Annotated ECG R1, Release 2*; HL7 International: Ann Arbor, MI, USA, 2015; pp. 1–111. Available online: https://www.hl7.org/implement/standards/product_brief.cfm?product_id=102 (accessed on 15 August 2021).
84. Healh Level Seven. *HL7 Annotated ECG Implementation Guide. Continuous Waveforms Supplement*; Brown, B., Ed.; HL7 International: Ann Arbor, MI, USA, 2014; pp. 1–25. Available online: https://www.hl7.org/documentcenter/public/wg/healthcaredevices/HL7_aECG_Continuous_ECG_Update%20-%20Brown.zip (accessed on 15 August 2021).
85. DICOM Standards Committee, Working Group 1-Cardiac and Vascular Information. *Digital Imaging and Communications in Medicine (DICOM) Supplement 30: Waveform Interchange*; DICOM Standard Committee: Arlington, VA, USA, 2000; pp. 1–77. Available online: https://www.dicomstandard.org/News-dir/ftsup/docs/sups/sup30.pdf (accessed on 15 August 2021).
86. International Organization for Standardization. *ISO 22077-1:2015, Health Informatics—Medical Waveform Format—Part 1: Encoding Rules*, 1st ed.; ISO: Geneva, Switzerland, 2015; pp. 1–42.
87. International Organization for Standardization. *ISO 22077-2:2015, Health Informatics—Medical Waveform Format—Part 2: Electrocardiography*, 1st ed.; ISO: Geneva, Switzerland, 2015; pp. 1–38.
88. International Organization for Standardization. *ISO/TS 22077-3:2015, Health Informatics—Medical Waveform Format—Part 3: Long Term Electrocardiography*, 1st ed.; ISO: Geneva, Switzerland, 2015; pp. 1–29.
89. International Organization for Standardization. *ISO/TS 22077-4:2019, Health Informatics—Medical Waveform Format—Part 4: Stress Test Electrocardiography*, 1st ed.; ISO: Geneva, Switzerland, 2019; pp. 1–30.
90. Bailey, J.J.; Berson, A.S.; Garson, A.; Horan, L.G.; Macfarlane, P.W.; Mortara, D.W.; Zywietz, C. Recommendations for standardization and specifications in automated electrocardiography: Bandwidth and digital signal processing. A report for health professionals by an ad hoc writing group of the Committee on Electrocardiography and Cardiac Electrophysiology of the Council on Clinical Cardiology, American Heart Association. *Circulation* **1990**, *81*, 730–739. [CrossRef] [PubMed]
91. Kligfield, P.; Gettes, L.S.; Bailey, J.J.; Childers, R.; Deal, B.J.; Hancock, E.W.; van Herpen, G.; Kors, J.A.; Macfarlane, P.; Mirvis, D.M.; et al. Recommendations for the standardization and interpretation of the electrocardiogram. Part I: The electrocardiogram and its technology. A Scientific Statement from the American Heart Association Electrocardiography and Arrhythmias Committee, Council on Clinical Cardiology, the American College of Cardiology Foundation, and the Heart Rhythm Society. *J. Am. Coll. Cardiol.* **2007**, *49*, 1109–1127. [CrossRef]

92. Mason, J.W.; Hancock, E.W.; Gettes, L.S.; Bailey, J.J.; Childers, R.; Deal, B.J.; Josephson, M.; Kligfield, P.; Kors, J.A.; Macfarlane, P.; et al. Recommendations for the standardization and interpretation of the electrocardiogram. Part II: Electrocardiography diagnostic statement list. *J. Am. Coll. Cardiol.* **2007**, *49*, 1128–1135. [CrossRef]
93. Rautaharju, P.M.; Surawicz, B.; Gettes, L.S. AHA/ACCF/HRS Recommendations for the standardization and interpretation of the electrocardiogram. Part IV: The ST segment, T and U waves, and the QT interval. A Scientific Statement from the American Heart Association Electrocardiography and Arrhythmias Committee, Council on Clinical Cardiology; the American College of Cardiology Foundation; and the Heart Rhythm Society: Endorsed by the International Society for Computerized Electrocardiology. *J. Am. Coll. Cardiol.* **2009**, *53*, 982–991. [CrossRef] [PubMed]
94. Wagner, G.S.; Macfarlane, P.; Wellens, H.; Josephson, M.; Gorgels, A.; Mirvis, D.M.; Pahlm, O.; Surawicz, B.; Kligfield, P.; Childers, R.; et al. AHA/ACCF/HRS recommendations for the standardization and interpretation of the electrocardiogram. Part VI: Acute ischemia/infarction: A scientific statement from the American Heart Association Electrocardiography and Arrhythmias Committee, Council on Clinical Cardiology; the American College of Cardiology Foundation; and the Heart Rhythm Society. Endorsed by the International Society for Computerized Electrocardiology. *J. Am. Coll. Cardiol.* **2009**, *53*, 1003–1111. [CrossRef] [PubMed]
95. CDISC Controlled Terminology. Available online: http://www.cdisc.org/terminology (accessed on 15 August 2021).
96. Rubel, P.; Pani, D.; Schloegl, A.; Fayn, J.; Badilini, F.; Macfarlane, P.W.; Varri, A. SCP-ECG V3.0: An enhanced standard communication protocol for computer-assisted electrocardiography. In Proceedings of the Computing in Cardiology Conference (CinC), Vancouver, BC, Canada, 11–14 September 2016; Murray, A., Ed.; IEEE: Piscataway, NJ, USA, 2016; pp. 309–312.
97. The Anatomical Therapeutic Chemical (ATC) Classification System and the Defined Daily Dose (DDD). WHO Collaborating Centre for Drug Statistics Methodology. Available online: http://www.whocc.no/ (accessed on 11 July 2021).
98. Ricke, A.D.; Swiryn, S.; Bauernfeind, R.A.; Conner, J.A.; Young, B.; Rowlandson, G.I. Improved pacemaker pulse detection: Clinical evaluation of a new high-bandwidth electrocardiographic system. *J. Electrocardiol.* **2011**, *44*, 265–274. [CrossRef]
99. Luo, S.; Johnston, P.; Macfarlane, P.W. Implanted cardiac pacemaker pulses as recorded from the body surface. *J. Electrocardiol.* **2012**, *45*, 663–669. [CrossRef] [PubMed]
100. Zareba, W.; Locati, E.H.; Maison Blanche, P. The ISHNE Holter Standard Output File Format: A Step Toward Compatibility of Holter Systems. *Ann. Noninvasive Electrocardiol.* **1998**, *3*, 261–262. [CrossRef]
101. Badilini, F. The ISHNE Holter Standard Output File Format. *Ann. Noninvasive Electrocardiol.* **1998**, *3*, 263–266. [CrossRef]
102. Ibaraki, T.; Muroga, S. Adaptive Linear Classifier by Linear Programming. *IEEE Trans. Syst. Sci. Cybern.* **1970**, *6*, 53–62. [CrossRef]
103. Badilini, F.; Young, B.; Brown, B.; Vaglio, M. Archiving and exchange of digital ECGs: A review of existing data formats. *J. Electrocardiol.* **2018**, *51*, S113–S115. [CrossRef]
104. Badilini, F.; Erdem, T.; Zareba, W.; Moss, A.J. ECGScan: A method for conversion of paper electrocardiographic printouts to digital electrocardiographic files. *J. Electrocardiol.* **2005**, *38*, 310–318. [CrossRef]
105. Li, Y.; Qu, Q.; Wang, M.; Yu, L.; Wang, J.; Shen, L.; He, K. Deep learning for digitizing highly noisy paper-based ECG records. *Comput. Biol. Med.* **2020**, *127*, 104077. [CrossRef] [PubMed]
106. Mishra, S.; Khatwani, G.; Patil, R.; Sapariya, D.; Shah, V.; Parmar, D.; Dinesh, S.; Daphal, P. ECG Paper Record Digitization and Diagnosis Using Deep Learning. *J. Med. Biol. Eng.* **2021**, *41*, 422–432. [CrossRef]
107. Loresco, P.J.M.; Africa, A.D. ECG Print-Out Features Extraction Using Spatial-Oriented Image Processing Techniques. *JTEC* **2018**, *10*, 15–20.
108. Pahlm, U.; Pahlm, O.; Wagner, G.S. The 24-lead ECG display for enhanced recognition of STEMI-equivalent patterns in the 12-lead ECG. *J. Electrocardiol.* **2014**, *47*, 425–429. [CrossRef]
109. Fayn, J.; Rubel, P. An ECG web services portal for standard and serial ECG analysis with enhanced 3D graphical capabilities. In Proceedings of the Computing in Cardiology, Rennes, France, 24–27 September 2017; IEEE: Piscataway, NJ, USA, 2017; pp. 1–4. [CrossRef]
110. Husain, K.; Zahid, M.S.M.; Ul Hassan, S.; Hasbullah, S.; Mandala, S. Advances of ECG Sensors from Hardware, Software and Format Interoperability Perspectives. *Electronics* **2021**, *10*, 105. [CrossRef]
111. Zentara, T.; Murawski, K. ECG signal coding methods in digital systems. In *Communication Papers of the 2018 Federated Conference on Computer Science and Information Systems FedCSIS, Poznań, Poland, 9–12 September 2018*; Ganzha, M., Maciaszek, L., Paprzycki, M., Eds.; ACSIS: New Jersey, NJ, USA, 2018; Volume 17, pp. 95–102. [CrossRef]
112. Trigo, J.D.; Chiarugi, F.; Alesanco, A.; Martinez-Espronceda, M.; Serrano, L.; Chronaki, C.E.; Escayola, J.; Martinez, I.; Garcia, J. Interoperability in Digital Electrocardiography: Harmonization of ISO/IEEE x73-PHD and SCP-ECG. *IEEE Trans. Inf. Technol. Biomed.* **2010**, *14*, 1303–1317. [CrossRef] [PubMed]
113. Mandellos, G.J.; Papaioannou, M.; Panagiotakopoulos, T.; Lymberopoulos, D.K. e-SCP-ECG+v2 Protocol: Expanding the e-SCP-ECG+ Protocol. In *Broadband Communications, Networks, and Systems BROADNETS 2018, Lecture Notes of the Institute for Computer Sciences, Social Informatics and Telecommunications Engineering*; Sucasas, V., Mantas, G., Althunibat, S., Eds.; Springer: Cham, Switzerland, 2019; Volume 263, pp. 125–135. [CrossRef]

114. Behlouli, H.; Miquel, M.; Fayn, J.; Rubel, P. Towards Self-improving NN based ECG Classifiers. In Proceedings of the 18th Annual International Conference of the IEEE Engineering in Medicine and Biology Society, Amsterdam, The Netherlands, 31 October–3 November 1996; IEEE: Piscataway, NJ, USA, 1996; Volume 18, pp. 927–928. [CrossRef]

115. Fayn, J.; Rubel, P. False Alarm Reduction in Self-Care by Personalized Automatic Detection of ECG Electrode Cable Interchanges. *Int. J. Telemed. Appl.* **2020**, 9175673. [CrossRef] [PubMed]

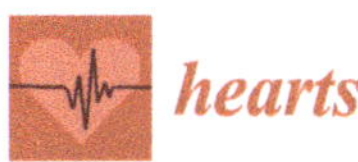

Review

The New ISO/IEC Standard for Automated ECG Interpretation

Brian Young [1,*] and Johann-Jakob Schmid [2]

[1] GE Healthcare, Wauwatosa, WI 53226, USA
[2] Schiller AG, 6341 Baar, Switzerland; JJ.Schmid@schiller.ch
* Correspondence: brian.young@med.ge.com; Tel.: +1-414-721-2454

Abstract: Updates to industry consensus standards for ECG equipment is a work-in-progress by the ISO/IEC Joint Work Group 22. This work will result in an overhaul of existing industry standards that apply to ECG electromedical equipment and will result in a new single international industry, namely 80601-2-86. The new standard will be entitled "80601, Part 2-86: Particular requirements for the basic safety and essential performance of electrocardiographs, including diagnostic equipment, monitoring equipment, ambulatory equipment, electrodes, cables, and leadwires". This paper will provide a high-level overview of the work in progress and, in particular, will describe the impact it will have on requirements and testing methods for computerized ECG interpretation algorithms. The conclusion of this work is that manufacturers should continue working with clinical ECG experts to make clinically meaningful improvements to automated ECG interpretation, and the clinical validation of ECG analysis algorithms should be disclosed to guide appropriate clinical use. More cooperation is needed between industry, clinical ECG experts and regulatory agencies to develop new data sets that can be made available for use by industry standards for algorithm performance evaluation.

Keywords: ECG equipment; computerized electrocardiograph; ECG analysis algorithms; computerized ECG interpretation

Citation: Young, B.; Schmid, J.-J. The New ISO/IEC Standard for Automated ECG Interpretation. *Hearts* **2021**, *2*, 410–418. https://doi.org/10.3390/hearts2030032

Academic Editors: Peter W. Macfarlane and Matthias Thielmann

Received: 7 July 2021
Accepted: 22 August 2021
Published: 27 August 2021

Publisher's Note: MDPI stays neutral with regard to jurisdictional claims in published maps and institutional affiliations.

1. Introduction

Industry standards are published for particular types of electromedical equipment by the International Electrotechnical Commission (IEC) and the International Organization for Standardization (ISO). These industry standards are updated on a regular basis by ISO/IEC workgroups. Work that is in progress by the Joint Workgroup 22 (JWG22) under the ISO/IEC 62D Electromedical Equipment Subcommittee will result in the publication of a new standard for ECG devices and systems with the designation of ISO/IEC 80601-2-86, which will be entitled "80601, Part 2-86: Particular requirements for the basic safety and essential performance of electrocardiographs, including diagnostic equipment, monitoring equipment, ambulatory equipment, electrodes, cables, and leadwires" [1]. JWG22 is a joint workgroup formed between the maintenance team that oversees the ECG particular standards and liaisons from other standard workgroups. This new standard is currently in draft form and constitutes a significant overhaul of current ECG equipment standards and, in effect, combines the three current ECG particular standards published by the IEC for diagnostic electrocardiographs [2], ECG patient monitors [3], and ambulatory ECG equipment [4]. The standard will additionally incorporate three ECG-related standards published by the Association for the Advancement of Medical Instrumentation (AAMI), which is the national standard development organization in the United States for health technology. The three additional AAMI standards will have safety and performance requirements for disposable electrodes (AAMI EC12 [5]), ECG cables and leadwires (AAMI EC53 [6]), and arrhythmia analysis performance reporting (AAMI EC57 [7]). Finally, 80601-2-86 will restore requirements that were omitted from a previously deprecated IEC diagnostic ECG particular standard that addressed the performance of computerized ECG analysis [8]. The enormous effort required for the development of the new 80601-2-86 standard represents a formidable task with far reaching implications, such that a comprehensive discussion of changes is beyond the scope of this paper. Therefore, the intention of this paper is narrowed in focus to give the reader a cursory level of understanding of the work in progress and a more detailed discussion about the impact it will have, specifically on performance requirements for computerized diagnostic ECG analysis algorithms, which is also commonly referred to by other terms, such as automated ECG interpretation or computerized ECG interpretation.

2. Background

Industry standards for ECG equipment have existed since the 1980s and were historically developed by separate workgroups for different types of ECG equipment and across different standard organizations, such as the IEC or AAMI. This has resulted in a complex landscape of industry standards, which are recognized at different levels for compliance by different regulatory agencies. Moreover, specific standards were developed for different types of ECG devices (namely diagnostic electrocardiographs, ambulatory ECG equipment, and ECG patient monitors). It is now common that contemporary ECG equipment does not clearly fit into any of these historical definitions of specific types of ECG devices, which has led to confusion and resulted in the inconsistent use of these standards. This is a challenge for manufacturers, clinicians, regulatory agencies, and testing facilities alike. Table 1 contains the list of the different historical standards that have been applied to ECG equipment. Despite the efforts that have been made over the years by the AAMI and IEC organizations to harmonize redundant ECG standards [9], challenges continue to exist because the current standards were developed decades ago and very few changes have been made over the years to requirements or conformance testing methods to update the standards with advancements made in either ECG technology or the clinical use of the ECG. Consequently, there is a tremendous opportunity recognized by the standard developers in JWG22 to accomplish three primary goals: combine current standards into a single standard that will encompass all types of ECG medical equipment, harmonize requirements and testing methods, and make updates to bring requirements and conformance testing methods into alignment with the current state of the art for ECG equipment and clinical use of ECG. These goals, albeit a monumental effort, have been adopted for current work in progress by JWG22 and will result in the new single ISO/IEC 80601-2-86 ECG equipment standard.

The first committee draft of 80601-2-86 [1] has been published and the second committee draft is in progress at the time of preparing this paper. Draft 80601-2-86 is organized as a set of general basic safety and performance requirements that will apply to all ECG devices, together with additional clauses that retain specific requirements for ECG equipment included within the definitions of diagnostic electrocardiographs, ECG patient monitors, and ambulatory ECG equipment. The definitions for these different types of ECG equipment are based on the intended use for the ECG equipment claimed by the manufacturer. A fundamental goal of the 80601-2-86 standard is to provide updated definitions for these specific types of ECG equipment linked to intended use. Explanation of these intended use definitions have been updated with guidance intended to make it easier to understand how current types of devices as well as emerging novel devices fit into defined categories of ECG equipment. These updates should provide better clarity and understanding regarding how the new standard should be applied to existing types of ECG devices as well as future innovations.

Table 1. List of historical industry standards that apply to ECG equipment.

Standard Designation	Title
AAMI EC11 [10]	Diagnostic electrocardiographic devices
AAMI EC12 [5]	Disposable ECG Electrodes
AAMI EC13 [11]	Cardiac monitors, heart rate meters, and alarms
AAMI EC38 [12]	Ambulatory electrocardiographs
AAMI EC53 [6]	ECG cables and leadwires
AAMI EC57 [7]	Testing and reporting performance results of cardiac rhythm and ST segment measurement algorithms
IEC 60601-2-25 [2]	Medical electrical equipment—Part 2-25: Particular requirements for the basic safety and essential performance of electrocardiographs
IEC 60601-2-27 [3]	Medical electrical equipment—Part 2-27: Particular requirements for the basic safety and essential performance of electrocardiographic monitoring equipment
IEC 60601-2-47 [4]	Medical electrical equipment—Part 2-47: Particular requirements for the basic safety and essential performance of ambulatory electrocardiographic systems
IEC 60601-2-51 [8]	Medical electrical equipment—Part 2-51: Particular requirements for the basic safety and essential performance of recording and analyzing single channel and multichannel electrocardiographs

3. Impact of 801601-2-86 on Automated ECG Interpretation

One of the key aspects impacted by this work is the update to requirements for computerized analysis of ECG signals. Great efforts have been made in this new standard to combine the different requirements and testing methods and clarify how they should be applied to different types of ECG analysis algorithms in a single standard. There are currently different sets of requirements, testing methods, and test data sets defined in existing standards [2–4,7]. While it may seem to be a simple task to combine the algorithm testing requirements, methods and data sets from the existing standards, and to add clarifications and rationale, 80601-2-86 addresses a long-standing challenge to understand the scope and purpose of the different algorithm testing requirements, as well as how to apply them across different types of ECG equipment. Historical ECG device standards have each had clauses that apply to ECG analysis algorithms [2–4,7,8,10–13] along with corresponding definitions of the types of ECG devices to which they apply. Unfortunately, the definitions focus more on the type of device containing the algorithm rather than the intended clinical use of the algorithm. Moreover, there was no guidance for manufacturers regarding how to apply these standards to ECG analysis algorithms that were contained in ECG equipment that did not meet these specific ECG device definitions.

When the new 80601-2-86 standard is introduced, ECG algorithm testing requirements, testing methods and data sets will be applied based on the intended use of the algorithm and not just the type of ECG device, which contains the algorithm. The requirements in the existing draft are structured with two different clauses, namely 201.12.4.1 Algorithm testing for Diagnostic 12 Lead and 201.12.4.2 requirements for testing computerized arrhythmia analysis algorithms [1]. Requirements in each of these clauses are applied to the specific types of ECG equipment for which they were originally defined in historical standards. In addition, these requirements are also applied to other computerized ECG analysis algorithms based on the intended use of the computerized ECG analysis output rather than the definition of the equipment itself.

In general, the diagnostic 12 lead algorithm runs on a static recording of an ECG snapshot to generate measurements, rhythm interpretation and may include interpretive statements for conduction, and morphologic patterns [1] for abnormalities that may include a wide range of diseases such as hypertrophic disease, ischemic disease, acute myocardial infarction, primary and secondary repolarization abnormalities, etc. Algorithms that are intended to provide a diagnostic 12 lead ECG interpretation using data that are derived from a non-standard reduced lead set, such as the EASI system [14] or from a reduced precordial lead [15], also meet the description of a diagnostic 12 lead algorithm, even though the devices which may contain these algorithms do not meet the definition of a diagnostic electrocardiograph ("DIAGNOSTIC ECG ME EQUIPMENT" [1]) in 80601-2-86.

In contrast, arrhythmia analysis algorithms are intended to analyze data in a more continuous nature and may analyze long term data, such as those from Holter or ECG patch devices, may analyze continuous and/or real time data, such as those from ECG patient monitors, or may analyze short term ECG data, such as those from ECG event recorders or mobile cardiac telemetry (MCT) type devices. The intended purpose of arrhythmia algorithms is to detect and classify QRS complexes and detect arrhythmic events [1]. Arrhythmia analysis algorithms may also perform ECG measurements for the purpose of trending measurement or detecting events, such as ischemic episodes.

There is some overlap between the outputs of these two types of ECG analysis algorithms, but they have different intended uses, and, therefore, the requirements, testing methods, and testing data sets are different for each of these two types of algorithms. The following discussion will focus on the impact of 80601-2-86 on performance testing for diagnostic 12 lead ECG analysis algorithms, which are also referred to by other descriptions, such as "automated ECG interpretation". The statistical metrics, limitations of testing and underlying principles for automated ECG interpretation also apply to arrhythmia analysis algorithms as well but will not be discussed in this paper.

Most current diagnostic electrocardiographs now have the ability for computer automated ECG interpretation and, by 2006, it was estimated that 100 million ECGs were being interpreted by computerized algorithms in the United States and a similar number in Europe and in the rest of the world [16]. The performance of these computerized ECG analyses has reached a point where the algorithms can make routine ECG measurements accurately and provide useful clinical benefits, yet also having well studied limitations when compared to humans over reading [17]. Because of the widespread use of computer-automated ECG interpretation and the impact it can have on clinical decision making, it is imperative to include the performance-testing requirements that provide as comprehensive a characterization of the algorithm performance as possible. This goal has been a cornerstone part of industry standards for electrocardiographic equipment and is maintained in the 80601-2-86 standard. It is based on requirements and testing methods developed by the Common Standards for Quantitative Electrocardiography (CSE) project [18,19] and the European Conformance Testing Services (CTS) project [20].

Current requirements for testing diagnostic ECG interpretation algorithms only require testing the accuracy of amplitude measurements and interval measurements on CTS and CSE data using calibration, analytic and biologic waveforms [2]. CTS analytic and calibration ECG waveforms are both simulated ECG-like waveforms with a range of characteristics. Calibration ECG waveforms are artificial in nature and designed to test both the hardware response of a device, as well as the automated ECG measurement performance used for automatic diagnostic ECG interpretation programs. Analytic ECG waveforms are more physiologically realistic in nature and designed to measure the accuracy of 12 lead diagnostic ECG interpretative programs in the detection and measurement of the ECG features. Figure 1 shows examples of CTS calibration and analytic waveforms. Biologic waveforms consist of a small set of actual physiologic ECG recordings that have been annotated by human over readers.

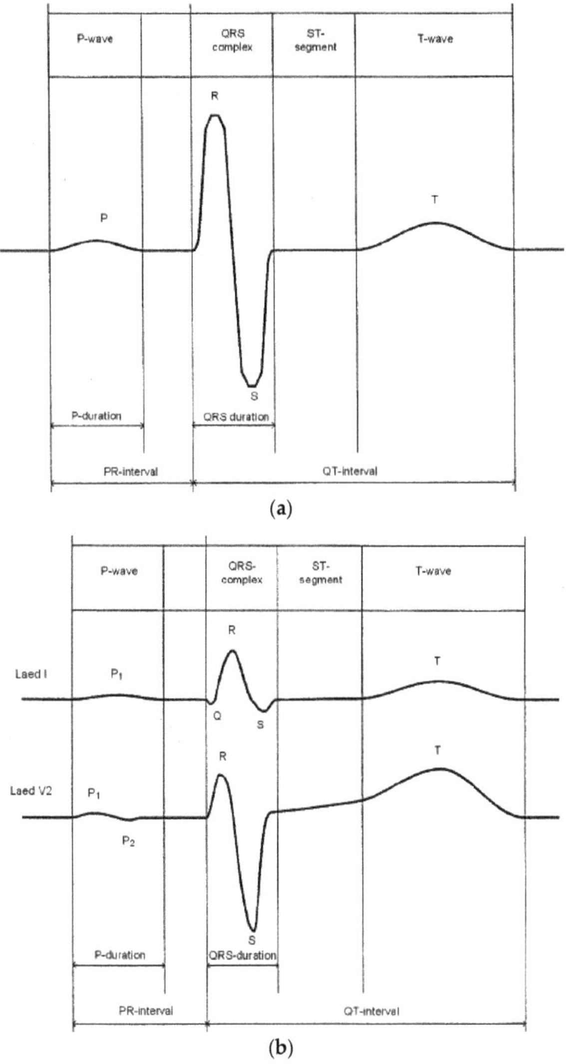

Figure 1. Examples of simulated waveforms from the CTS database used for algorithm performance testing in current ECG standards [2]. (**a**) Example of a calibration ECG waveform with ECG waves and interval nomenclature indicated; (**b**) example of an analytic ECG waveform with ECG waves and interval nomenclature indicated.

Diagnostic statements for automated ECG interpretation algorithms are the fundamental output, and the accuracy of these statements should be well characterized by algorithm testing, including both ECG contour and ECG rhythm diagnostic statements. The methods for measuring accuracy have been consistently and well defined, and previously included in ECG standards [8]. However, the current industry ECG standards have omitted these historical requirements to test the accuracy of diagnostic statements [2], which is a gap that is being addressed in 80601-2-86. This testing is particularly important with the emergence of new types of algorithms, such as machine learning, for which little guidance is available regarding the validation of clinical accuracy. The challenge with making even further improvements to this situation is that advancement requires better test datasets and, at this time, there are no new available databases with appropriate types of data including properly adjudicated reference annotations, and which are publicly accessible for inclusion in an industry standard. Consequently, improvements that are being made in 80601-2-86 are limited in nature.

The content of 80601-2-86 combines and harmonizes the safety and performance requirements from historical ECG standards. This includes requirements that address the technical aspects of signal acquisition and signal conditioning to ensure that ECG equipment will operate safely and acquire signals that are appropriate for the intended use of the equipment, which may include both human interpretation and/or the computer analysis of the ECG [21]. These technical requirements address necessary performance specifications to ensure that resulting signals are appropriate for their intended use and include specifications, such as filtering, bandwidth, common mode rejection, and system noise [22]. It is beyond the scope of this paper to discuss the effects of inadequate signal acquisition and signal conditioning on the effects of computerized ECG interpretation. It is assumed that, if ECG equipment is compliant with conformance testing for ECG signal acquisition, then the output ECG signals will be appropriate for both human and computerized interpretation, based on the intended use claimed by the manufacturer.

There are strong data to support the proposition that computerized ECG interpretation programs provide an important clinical adjunct to the physician that may even enhance physician overreading [23,24], but it is also clearly understood that the outputs of all computerized ECG interpretation algorithms have limitations [25] and require physician overread [21,26]. The historical requirements, testing methods and testing data sets have changed little over the years. Methods for measuring automated ECG interpretation have been consistently applied over the years by current [2] and past [8] industry standards. However, the data used for testing can heavily influence the measurement of accuracy and, at this time, there are no new additional databases that are appropriate for use as an industry standard, although some new efforts are ongoing [27]. Consequently, little progress has been made in improving the current quality of performance testing for algorithms in 80601-2-86. The work required to create better reference data sets for algorithm testing is particularly daunting and the improvements that can be made to the current performance testing are limited until better data sets are available for use within the context of an industry standard.

Developers will continue to improve the accuracy of automated ECG interpretation programs and individual manufacturers will continue to validate algorithm performance with private data sets. Furthermore, the emerging use of machine learning and artificial intelligence algorithms for ECG interpretation will add new complexities to the problem of understanding and characterizing algorithm safety and performance. This is also challenging regulatory agencies to expand their considerations for algorithm development and validation to address these new complexities [28]. Nevertheless, because of the ubiquitous presence of automated ECG interpretation software and because of the impact it can have on clinical diagnosis and decision making, it remains critical for the performance evaluation of algorithms to be a compulsory element of industry standards for ECG equipment.

It has been debated through the years as to the value and validity of some of testing requirements that have been included in ECG equipment standards. In fact, when the IEC 60601-2-51 standard was combined with the IEC 60601-2-25 standard, two areas of algorithm testing were not included in the update, namely (1) the testing requirements that pertain to the evaluation of diagnostic ECG measurements in the presence of noise and (2) reporting for interpretive 12 lead diagnostic statements [2]. At the time when these two standards were combined, the consensus of the workgroup was that they had limited value. However, the consensus of the JWG22 workgroup has changed based on constituency feedback and now acknowledges that these deprecated algorithm-testing requirements are important elements of computerized ECG interpretation and should be mandatory to improve algorithm-testing requirements. Reviews of the use of the CSE and

CTS data sets in 80601-2-86 clearly indicate that these test data sets have limitations [29], and consequently manufacturers often use proprietary data sets for validating clinical performance of automatic ECG interpretation. Manufacturers must be cognizant that measuring diagnostic accuracy depends on the quality of the data composition and should use data that are representative of the intended clinical environment and consider that the predictive merits of performance evaluation must be examined in relation to sample size, patient populations and the prevalence values for each diagnostic category.

In general, it is expected that each diagnostic category should be validated by an adequate number of clinical cases, and the use of enriched datasets may be acceptable if supplemented by supporting analysis. Noise is a common occurrence in clinical environments and the variation of errors will increase with degraded signal quality [30] and the impact of noise on the measurement accuracy of diagnostic ECG measurement algorithms is an important characteristic to evaluate [31]. In general, a manufacturer should assess whether the databases and methods defined by the standard are fully representative of the device under test as well as its use and determine when deviations or additional testing may be needed (e.g., additional device-specific data, numerical transformation, additional noise patterns, etc.). While methods included in 80601-2-86 were originally developed for tradition rule-based algorithms (i.e., those that implement classification rules based on clinical consensus), the general concepts for testing and reporting may also be applied to algorithms based on machine learning/artificial intelligence, although larger testing datasets and additional analysis may be needed to ensure a robust validation.

It is important to note that 80601-2-86 does not specify pass–fail criteria for automated ECG analysis performance. This obviously does not mean that any performance is acceptable; instead, it is a recognition that the performance of an automated ECG analysis algorithm should be evaluated in the context of the device's intended use, to ensure that the device performs sufficiently well in clinical practice.

4. Discussion

At the time of preparing this paper, the first committee draft of 80601-2-86 had been published and circulated for comments by national standard organizations members of the IEC JWG22. The second committee draft is in preparation for circulation to the national committees for a second call for comments. The current state of 80601-2-86 combines several existing standards that apply to ECG equipment into a single standard that will include all ECG equipment within its scope and will also contain specific requirements for particular types of ECG equipment based on intended use claimed by the manufacturer. This will include requirements and conformance testing methods for computerized ECG analysis algorithms, which are defined in two broad categories, namely diagnostic 12 lead ECG interpretative algorithms and arrhythmia analysis algorithms. The quantification of performance and testing data sets have been in existence for decades. The goal of the new 80601-2-86 standard is to update the rationales and guidance contained in the informative annexes in such a way that it is more clearly understood how to apply the standard to the range of contemporary computerized ECG analysis algorithms based on the intended use of the ECG equipment in which they are used.

In particular, the requirements for measurement and analysis algorithms for diagnostic ECG interpretation restore some historical performance testing requirements and conformance testing methods that had been previously deprecated from current standards. Although the limitations of the conformance testing data sets have been well recognized and published, they still provide the only method of uniformly and consistently benchmarking algorithm performance. This is especially important because of the ubiquitous use of automated ECG interpretation by the clinical community and the important influence it can have on physician over reading.

Furthermore, the profound influence that automated ECG interpretation programs can have on physician ECG interpretation and clinical decision making has been well published by experts in electrocardiography and the importance of developing and evaluating the performance of these algorithms with scientific rigor is critical to ensuring that the appropriate use of computerized ECG interpretation programs is well understood and benefits patient care.

While the 80601-2-86 standard applies to the vast majority of ECG devices, the requirements for automated ECG analysis and interpretation are mostly relevant for traditional device types (e.g., rule-based analysis of resting 12-lead ECG and traditional Holter ECGs). However, the same concepts can be applied to novel technologies (e.g., machine learning/AI-based algorithms, non-standard lead technology/lead configuration) by using additional datasets relevant for the device's intended use. Manufacturers should pay particular attention to factors that impact the quality and appearance of data sets for both algorithm development and testing, in particular, establishing appropriate sample sizes, patient population representation, and disease prevalence/representation to accurately reflect the clinical environment and intended use for which the algorithm is designed.

5. Summary

The introduction of 80601-2-86 is a significant overhaul of existing industry standards and will result in a single international standard that can be applied to all ECG equipment. The goals are to combine, update, and harmonize the safety and performance requirements from the multiple existing industry standards so that the new standard can be applied to all types of ECG and be appropriate for current ECG technology and clinical use.

Although the CSE and CTS test data sets have well known limitations, no other data sets have been accepted for inclusion in 80601-2-86. Manufacturers should continue to work together with clinical ECG experts to continue clinically meaningful improvements to computerized ECG analysis and should disclose the clinical validation of algorithm improvements to guide appropriate clinical use. More cooperation is needed between industry, clinical ECG experts and regulatory agencies to develop new data sets that can be made available for use by industry standards for algorithm performance evaluation.

Author Contributions: All authors have been directly involved with work to draft the ISO/IEC 80601-2-86 particular standard as active members of JWG22. J.-J.S. is the current convenor for JWG22. B.Y. also serves as a Co-Chair of the AAMI ECG Committee. All authors have read and agreed to the published version of the manuscript.

Funding: This research received no external funding.

Institutional Review Board Statement: Not applicable.

Informed Consent Statement: Not applicable.

Data Availability Statement: Not applicable.

Conflicts of Interest: The author declares no conflict of interest.

References

1. ISO/IEC. *80601, Part 2-86: Particular Requirements for the Basic Safety and Essential Performance of Electrocardiographs, Including Diagnostic Equipment, Monitoring Equipment, Ambulatory Equipment, Electrodes, Cables, and Leadwires*; Committee Draft; ISO: Geneva, Switzerland, 2018.
2. ANSI/AAMI/IEC. *ANSI/AAMI/IEC 60601-2-25: 2011/(R)2016 Medical Electrical Equipment—Part 2-25: Particular Requirements for the Basic Safety and Essential Performance of Electrocardiographs*; AAMI: Arlington, VA, USA, 2016.
3. ANSI/AAMI/IEC. *ANSI/AAMI/IEC 60601-2-27: 2011/(R)2016 Medical Electrical Equipment—Part 2-27: Particular Requirements for the Basic Safety and Essential Performance of Electrocardiographic Monitoring Equipment*; AAMI: Arlington, VA, USA, 2016.
4. ANSI/AAMI/IEC. *ANSI/AAMI/IEC 60601-2-47: 2012/(R)2016 Medical Electrical Equipment—Part 2-47: Particular Requirements for the Basic Safety and Essential Performance of Ambulatory Electrocardiographic Systems*; AAMI: Arlington, VA, USA, 2016.
5. ANSI/AAMI. *ANSI/AAMI EC12: 2000/(R)2015 Disposable ECG Electrodes*; ANSI/AAMI: Arlington, VA, USA, 2015.
6. ANSI/AAMI. *ANSI/AAMI EC53: 2013 ECG Trunk Cables and Patient Leadwires*; ANSI/AAMI: Arlington, VA, USA, 2013.
7. ANSI/AAMI. *ANSI/AAMI EC57-2012, Testing and Reporting Performance Results of Cardiac Rhythm and ST Segment Measurement Algorithms*; ANSI/AAMI: Arlington, VA, USA, 2012.

8. IEC (Ed.) *International Standard IEC 60601-2-51:2003 Medical Electrical Equipment. Particular Requirements for Safety, Including Essential Performance, of Recording and Analysing Single Channel and Multichannel Electrocardiographs*; International Electrotechnical Commission (IEC): Geneva, Switzerland, 2003; p. 86.
9. Young, B.; Schmid, J.J. Updates to IEC/AAMI ECG standards, a new hybrid standard. *J. Electrocardiol.* **2018**, *51*, S103–S105. [CrossRef] [PubMed]
10. ANSI/AAMI. *EC11:1991/(R)2001 Diagnostic Electrocardiographic Devices*; ANSI/AAMI: Arlington, VA, USA, 1992; p. 39.
11. ANSI/AAMI. *EC13:2002 Cardiac Monitors, Heart Rate Meters, and Alarms*; ANSI/AAMI: Arlington, VA, USA, 2002; p. 79.
12. ANSI/AAMI. *EC38:1998 Ambulatory Electrocardiographs*; ANSI/AAMI: Arlington, VA, USA, 1999; p. 58.
13. IEC (Ed.) *International Standard 60601-2-25:2011 Medical Electrical Equipment. Particular Requirements for Safety, Including Essential Performance, of Recording and Analysing Single Channel and Multichannel Electrocardiographs*; International Electrotechnical Commission (IEC): Geneva, Switzerland, 2011; p. 190.
14. Dower, G.E.; Yakush, A.; Nazzal, S.B.; Jutzy, R.V.; Ruiz, C.E. Deriving the 12-lead electrocardiogram from four (EASI) electrodes. *J. Electrocardiol.* **1988**, *21*, S182–S187. [CrossRef]
15. Drew, B.J.; Pelter, M.M.; Brodnick, D.E.; Yadav, A.V.; Dempel, D.; Adams, M.G. Comparison of a new reduced lead set ECG with the standard ECG for diagnosing cardiac arrhythmias and myocardial ischemia. *J. Electrocardiol.* **2002**, *35*, 13–21. [CrossRef] [PubMed]
16. Hongo, R.H.; Goldschlager, N. Status of Computerized Electrocardiography. *Cardiol. Clin.* **2006**, *24*, 491–504. [CrossRef] [PubMed]
17. Smulyan, H. The Computerized ECG: Friend and Foe. *Am. J. Med.* **2019**, *132*, 153–160. [CrossRef]
18. Willems, J.L. Common standards for quantitative electrocardiography. Second progress report. In Proceedings of the CSE Conference, Leuven, Belgium, 20 November 1982; pp. 1–246.
19. Willems, J.L.; Arnaud, P.; van Bemmel, J.H.; Bourdillon, P.J.; Brohet, C.; Dalla Volta, S.; Andersen, J.D.; Degani, R.; Denis, B.; Demeester, M.; et al. Assessment of the performance of electrocardiographic computer programs with the use of a reference data base. *Circulation* **1985**, *71*, 523–534. [CrossRef]
20. Zywietz, C.; Mertins, V.; Willems, J.L. Quality assurance of interpretative electrocardiographs: Development of testing services in the European communities-first results. In Proceedings of the [1990] Proceedings Computers in Cardiology, Chicago, IL, USA, 23–26 September 1990; pp. 181–184.
21. Kligfield, P.; Gettes, L.S.; Bailey, J.J.; Childers, R.; Deal, B.J.; Hancock, E.W.; van Herpen, G.; Kors, J.A.; Macfarlane, P.; Mirvis, D.M.; et al. Recommendations for the Standardization and Interpretation of the Electrocardiogram. Part I: The Electrocardiogram and Its Technology. A Scientific Statement From the American Heart Association Electrocardiography and Arrhythmias Committee, Council on Clinical Cardiology; the American College of Cardiology Foundation; and the Heart Rhythm Society. Endorsed by the International Society for Computerized Electrocardiology. *Circulation* **2007**, *115*, 1325–1332.
22. Schläpfer, J.; Wellens Hein, J. Computer-Interpreted Electrocardiograms. *J. Am. Coll. Cardiol.* **2017**, *70*, 1183–1192. [CrossRef] [PubMed]
23. Salerno, S.M.; Alguire, P.C.; Waxman, H.S. Competency in interpretation of 12-lead electrocardiograms: A summary and appraisal of published evidence. *Ann. Intern. Med.* **2003**, *138*, 751–760. [CrossRef] [PubMed]
24. Laks, M.M.; Selvester, R.H.S. Computerized Electrocardiography—An Adjunct to the Physician. *N. Engl. J. Med.* **1991**, *325*, 1803–1804. [CrossRef] [PubMed]
25. Tsai, T.L.; Fridsma, D.B.; Gatti, G. Computer decision support as a source of interpretation error: The case of electrocardiograms. *J. Am. Med. Inform. Assoc.* **2003**, *10*, 478–483. [CrossRef] [PubMed]
26. Shah, A.P.; Rubin, S.A. Errors in the computerized electrocardiogram interpretation of cardiac rhythm. *J. Electrocardiol.* **2007**, *40*, 385–390. [CrossRef] [PubMed]
27. DeCamilla, J.; Xia, X.; Wang, M.; Wade, J.; Mykins, B.; Zareba, W.; Couderc, J.P. The multiple arrhythmia dataset evaluation database (MADAE). *J. Electrocardiolog* **2018**, *51*, S106–S112. [CrossRef] [PubMed]
28. FDA (Ed.) *Artificial Intelligence/Machine Learning (AI/ML)-Based Software as a Medical Device (SAMD) Action Plan*; FDA: Silver Spring, MD, USA, 2021.
29. Corabian, P. *Accuracy and Reliability of Using Computerized Interpretation of Electrocardiograms for Routine Examinations*; HTA 25: Series A; Alberta Heritage Foundation for Medical Research: Edmonton, AB, Canada, 2002.
30. Farrell, R.M.; Rowlandson, G.I. The effects of noise on computerized electrocardiogram measurements. *J. Electrocardiol.* **2006**, *39*, S165–S173. [CrossRef] [PubMed]
31. Zywietz, C.; Willems, J.L.; Arnaud, P.; van Bemmel, J.H.; Degani, R.; Macfarlane, P.W. Stability of computer ECG amplitude measurements in the presence of noise. *Comput. Biomed. Res.* **1990**, *23*, 10–31. [CrossRef]

 *hearts*

Review

Computer Modeling of the Heart for ECG Interpretation—A Review

Olaf Dössel *, Giorgio Luongo, Claudia Nagel and Axel Loewe

Institute of Biomedical Engineering, Karlsruhe Institute of Technology (KIT), 76131 Karlsruhe, Germany; Giorgio.Luongo@kit.edu (G.L.); claudia.nagel@kit.edu (C.N.); axel.loewe@kit.edu (A.L.)
* Correspondence: Olaf.doessel@ibt.kit.edu

Abstract: Computer modeling of the electrophysiology of the heart has undergone significant progress. A healthy heart can be modeled starting from the ion channels via the spread of a depolarization wave on a realistic geometry of the human heart up to the potentials on the body surface and the ECG. Research is advancing regarding modeling diseases of the heart. This article reviews progress in calculating and analyzing the corresponding electrocardiogram (ECG) from simulated depolarization and repolarization waves. First, we describe modeling of the P-wave, the QRS complex and the T-wave of a healthy heart. Then, both the modeling and the corresponding ECGs of several important diseases and arrhythmias are delineated: ischemia and infarction, ectopic beats and extrasystoles, ventricular tachycardia, bundle branch blocks, atrial tachycardia, flutter and fibrillation, genetic diseases and channelopathies, imbalance of electrolytes and drug-induced changes. Finally, we outline the potential impact of computer modeling on ECG interpretation. Computer modeling can contribute to a better comprehension of the relation between features in the ECG and the underlying cardiac condition and disease. It can pave the way for a quantitative analysis of the ECG and can support the cardiologist in identifying events or non-invasively localizing diseased areas. Finally, it can deliver very large databases of reliably labeled ECGs as training data for machine learning.

Keywords: in silico; electrophysiology; electrocardiogram; ECG; cardiac disease; arrhythmia; ischemia

Citation: Dössel, O.; Luongo, G.; Nagel, C.; Loewe, A. Computer Modeling of the Heart for ECG Interpretation—A Review. *Hearts* **2021**, *2*, 350–368. https://doi.org/10.3390/hearts2030028

Academic Editor: Peter Macfarlane

Received: 27 May 2021
Accepted: 14 July 2021
Published: 26 July 2021

1. Introduction

This article reviews research aimed at building a bridge between computerized modeling of the electrophysiology of the human heart and the analysis of the electrocardiogram (ECG). Potential applications of computer modeling for better interpretation of the ECG are demonstrated and an outlook for further research is given.

The research field of computerized modeling of the electrophysiology of the heart has reached a mature state. The healthy heart can be replicated in a computer model with various degrees of detail, starting with the ion channels and ending with the spread of a depolarization wave through the atria and the ventricles. Several diseases have been the focus of this research but many open questions remain: modeling can only be as good as our basic understanding of the pathologies of the heart.

On the other hand, after more than 100 years of ECG interpretation, the clinical knowledge about ECG and what it can tell us about cardiac diseases has reached an expert level. Most often, this knowledge is based on personal experience or empirical studies and only coarse attempts are made to relate a decisive feature in the ECG to its pathological origin inside the heart. The classical heart vector is a valuable tool for understanding the general shape of the ECG, but it is not good enough to follow details of the spatial spread of de- and repolarization.

It is astounding that the number of articles where modeling of the heart is extended to the calculation of the ECG and where this is used for better ECG interpretation is limited. Table 1 shows the result of a literature survey.

Table 1. Literature survey of research about modeling of the heart together with the corresponding ECG.

Topic	Modeling Challenge	References
healthy heart—QRS	modeling the Purkinje tree	[1–9]
healthy heart—T-wave	modeling heterogeneity of repolarization	[10–13]
healthy heart—P-wave	modeling sinus node excitation and pathways from right to left atrium, anatomical variability	[14–22]
ischemia and infarction	modeling the effect of hyperkalemia, acidosis, hypoxia and cell-to-cell uncoupling	[23–30]
ventricular ectopic beats	localization with 12-lead ECG	[31–34]
ventricular tachycardia	localization of exit points with 12-lead ECG	[29,35]
cardiomyopathy	modeling typical changes of QRS- and T-wave	[36]
bundle branch blocks LBBB and RBBB	modeling asynchrony	[37–40]
atrial ectopic beats	localization with 12-lead ECG	[32–34]
atrial tachycardia, flutter	modeling all types of flutter	[20,41–46]
atrial fibrillation and fibrosis	modeling fibrosis and its distribution	[47–51]
genetic diseases	modeling LQT, SQT, Brugada	[52–54]
imbalance of electrolytes	hyper- and hypokalemia and hyper- and hypocalcemia	[13,55–60]
drug-induced changes	effect of various channel blockers	[61–63]

Methods for calculation of the ECG (and the body surface potential map, BSPM) from the source distribution in the heart have been described in several articles. The main differences between the approaches are (a) the cellular models and their parameters, (b) the method to calculate the spread of depolarization (bidomain, monodomain, eikonal, reaction eikonal) and (c) the method of forward calculation (finite difference, finite element, boundary element methods; homogeneous torso versus different organs considered). The forward problem is obviously related to the inverse problem of ECG as used in ECG imaging (ECGi). Thus, all articles dealing with fast and realistic methods of calculating the lead field matrix that maps the sources in the heart to the body surface ECG are related to the topic of this article but are not discussed in detail.

Published in 2004, the software package ECGSim allows for a fast and easy relation of source patterns on the heart to the corresponding 12-lead ECG. The user can modify local activation times, repolarization times and the slope of the transmembrane voltage [64]. Thus, these source distributions can be realistic or not—no model of excitation spread is running in the background. Meanwhile, advanced software packages to simulate the electrophysiology of the heart are available: openCARP [65,66], acCELLerate [67], FEniCS [68], Chaste [69], propag-5 [70], and LifeV [71]. They all have been verified in an N-version benchmark activity initiated by Niederer [72].

The literature survey yielded several articles that do not focus on a specific disease but rather deal with the general concept of calculating the ECG from computer models of the heart. Lyon et al. gave an outline of a computational pipeline, listed examples of modeling diseases together with the ECG and showed up several applications of modeling in ECG interpretation [73]. Potse suggested a fast method for realistic ECG simulation without oversimplifying the torso model by using a lead-field approach [74]. Building upon this approach, Pezzuto et al. found an even faster method that allows for implementation on a general-purpose graphic processing unit (GPGPU) [75]. Keller et al. investigated the influence of tissue conductivities on the resulting ECG [76]. Schuler et al. [77] found a way to downsample the fine grid necessary for calculating the spread of depolarization for the forward calculation of the ECG—further reducing calculation time. Neic et al. developed a reaction eikonal algorithm that simulates the spread of depolarization very fast and still delivers realistic ECGs [78].

Calculation times for computing the spread of depolarization and repolarization, the lead field matrix and the body surface potentials including the ECG strongly depend on the methods employed: highly detailed cell models versus simplified phenomenological models, high versus low spatial resolutions, etc. They can range from one day down to one second. As an example, the calculation times of the P-waves shown in Figure 1 were 27 h for the full bidomain model and the Courtemanche cell model, 1 h and 24 min for a pseudo-bidomain model and 40 min for a monodomain simulation (heart mesh with 4.7 million elements and 920 k nodes, desktop computer with 12 cores at 1.4 GHz). Fast calculation times are important for the researcher aiming at the identification of new features in the ECG, for creating a training dataset for machine learning and for personalization of a heart model. They are not relevant any more if, for example, a machine learning algorithm is finally used in clinics.

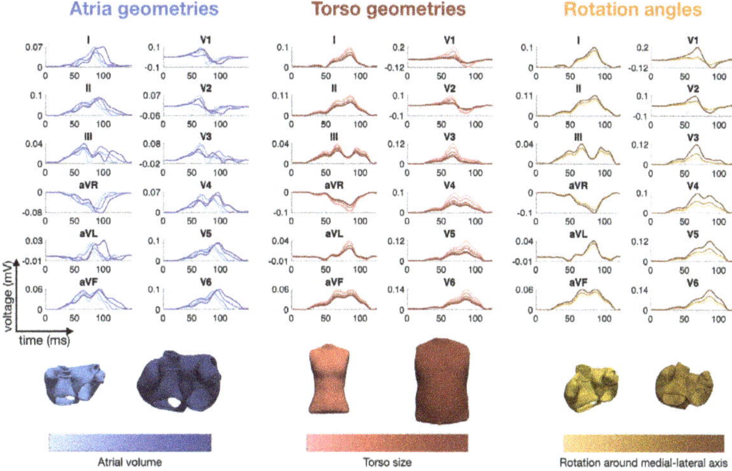

Figure 1. Simulated P-waves of the 12-lead ECG with various atrial shapes, several orientations of the atria inside the torso and a variety of body shapes. The colors represent the total atrial volume in blue, the torso size in red and the orientation angle around the medial-lateral axis in orange [50].

2. Modeling the ECG of a Healthy Heart

2.1. The QRS Complex and the Purkinje Tree

The morphology of the QRS complex is strongly determined by the topology of the His-Purkinje system in the ventricular subendocardial layer [2]. An earlier approach to fit a Punkinje network model to a measured ECG was published by Keller et al. in 2009 [1]. In total, 744 Purkinje muscle junctions were distributed across the ventricular endocardial surface following specific rules. Other publications followed this scheme and implemented "root points" coupled to thin endocardial layers with very fast conduction [3].

Automatic and reproducible manipulation of root node locations is facilitated by chamber-specific coordinate systems [79–81]. Mincholé et al. investigated the impact of anatomical variability on simulated QRS complexes [6]. They found that QRS duration is mainly determined by myocardial volume and not affected by the position of the heart in the torso. The latter influences QRS morphology in the precordial leads, whereas ventricular anatomy dominates in the limb leads. Cranford et al. carried out a sensitivity study: they implemented 1 to 4 "seed stimuli" and up to 384 "regional stimuli" and observed the changes in the QRS complex while changing the number and topology of excitation sites [5]. The topology of four seed stimuli at adapted positions was more relevant than a large number of regional stimuli. Pezzuto et al. were able to reproduce the QRS complex of 11 patients using up to 11 of the earliest activation sites on the endocardium and adapted the conduction velocity in the ventricles and in a fast endocardial layer [7]. Gillette et al. proposed a comprehensive workflow to optimize the positioning of five root disks, timing, and endocardial conduction velocities (10 parameters) to reproduce the QRS complex with a personalized model [8].

The healthy QRS complex can be reproduced faithfully, meaning that adapted heart models show an ECG with high correlation to measured ECGs. This does not prove that the modeled spread of depolarization is the one present in the patient, but it is good to see that there are no inconsistencies. Some relevant questions are: Which parameters in the model are responsible for the natural variability of QRS complexes—both inter- and intra-patient wise [4,6] (see also Section 2.5)? Is the heart axis that is visible in the ECG mainly determined by the geometrical axis of the heart or by the properties of the His-Purkinje system? Which simplifications of the thorax model are acceptable and where do we need detailed models?

2.2. The T-Wave and the Repolarization

Modeling the T-wave is a challenging task. If all myocytes in the ventricles would follow the same action potential, the T-wave would have the opposite sign as the R-peak. Heterogeneity is a necessary condition for concordant T-waves. Keller et al. investigated various schemes of heterogeneity of the I_{Ks} repolarization current (transmural, apico-basal, left–right) and found that both transmural and apico-basal gradients can lead to realistic T-waves, whereas a pure left–right heterogeneity creates a notch in the T-wave [11]. Even though the focus of an article of Bukhari et al. was on the changes in the T-wave during dialysis (see Section 3.9), this article also reported on the heterogeneity that is needed to obtain a realistic T-wave in healthy hearts. They assume a solely transmural dispersion of ion channel conductivities [13]. Xue et al. analyzed how transmural and apico-basal heterogeneities change the morphology of the T-wave. They included heterogeneities of the following ion channels: I_{Ks}, I_{Kr}, I_{to}. They concluded that mainly apico-basal gradients contribute to a positive T-Wave [10]. The modeled heterogeneity scenarios are informed by experimental cellular data [82]. However, the available data do not allow us to draw definite conclusions and different heterogeneity patterns can lead to the similar T-wave morphologies.

While contraction of the heart happens only after the P-wave and the early QRS complex, it can influence the source distribution during the repolarization. How the contraction of the heart affects the morphology of the simulated T-wave was investigated by Moss et al. They observed an 8% increase in amplitude and a shift of the T-wave peak by 7 ms [12].

2.3. The P-Wave

A review about computerized modeling of the atria including the corresponding ECG was given by Doessel et al. in 2012 [15]. Krueger et al. were the first to set up an atrial model that included realistic fiber orientation [14,83]. They also investigated the influence of atrial heterogeneities on the morphology of the P-wave, created personalized models and compared the ECGs of several patients [16].

The contribution of the left and right atria to the P-wave was analyzed by Loewe et al. [17] and Jacquemet et al. [18]. Even in the last third of the P-wave, one-third of the signal stems from the right atrium [17]. Potse et al. discovered that a jigging morphology of the P-wave, which was observed in computer simulations, was not an artefact but could be observed in a similar way in healthy volunteers when carefully preventing smoothing through filtering or averaging [19]. Loewe et al. investigated the influence of the earliest site of activation in the right atrium (i.e., the sinus node exit site) on the morphology of the P-wave [20] and could demonstrate that small shifts in the earliest excitation site and its proximity to the inter-atrial connections can significantly change the terminal phase of the P-wave. Andlauer et al. dissected the differential effects of atrial dilation and hypertrophy on the morphology of the P-wave [21] and showed that left atrial dilation did not influence P-wave duration significantly, but instead had a strong effect on P-wave amplitude and thus P-wave Terminal Force in lead V1 (PTF-V1).

A literature survey of simulations of the P-wave and in particular of the P-wave in patients suffering from paroxysmal atrial fibrillation (AFib) was published by Filos et al. [43]. All the effects described in the literature that have an influence on the morphology of the P-wave of AFib patients are outlined. Despite the very large number of articles, we conclude that there is still a way to go before these results can be routinely used in clinical practice.

Nagel et al. analyzed the inter- and intra-patient variability of the P-wave in the Physionet ECG database, aiming at the optimization of a simulated database of P-waves [84] (see also Section 2.5). Figure 1 shows several examples of P-waves with various atrial shapes, several orientations of the atria inside the torso and a variety of body shapes.

2.4. Modeling Rhythmic Features and Heart Rate Variability

Modeling of a heart beat most often starts with a stimulation either from an area around the sinus node (atria) or from the model of the Purkinje tree (ventricles, see Section 2.1). Modeling of the sinus node is an interesting research topic that goes beyond the scope of this article.

The ECG fluctuates from beat to beat even in the healthy state. Both the RR interval and also the morphology of the P-, QRS- and T-wave are not completely periodic. The fluctuations of the RR interval are well known and analyzed by means of heart rate variability (HRV), as reviewed by Rajendra et al. [85]. HRV is high in normal hearts and low when there is a cardiac problem. The variation in the beat-to-beat RR interval is usually studied in both the time domain and frequency domain. Not many clinicians make use of this measure in daily clinical practice. On the modeling side, only few articles describe simulations of beta-adrenergic and vagal tones on the sinus node [86]. It seems as if there is still a "missing link" between computerized modeling of the heart and HRV interpretation [87].

2.5. Modeling Inter- and Intra-Patient Variability

The variety of ECG morphologies observed in a cohort of healthy humans is large. This can be explained by different geometries of the heart [88], different rotation inside the thorax [6], and different shapes of the torso. Moreover, differences in electrophysiology also contribute to the variability (see, for example, the discussion about the QRS morphology and the Purkinje tree in Section 2.1).

As already mentioned in Section 2.3, Nagel et al. investigated the inter- and intra-patient variability of the P-wave [84]. The beat-to-beat variability of the P-wave in case of atrial fibrillation was investigated by Pezzuto et al. [42]. Already small variations (1 to 5 mm) in the location of the earliest activation site lead to changes in the morphology of the P-wave. This effect was significantly enhanced if slow conducting regions were near the earliest activation site.

3. Modeling Diseases and the Corresponding ECG

3.1. Ischemia and Infarction

Loewe at al. gave an outline of how computer modeling can support comprehension of cardiac ischemia and discussed the link to the corresponding ECG [28]. Figure 2 shows several examples of ischemic regions together with the corresponding ECG. The parameters of the ten Tusscher–Panfilov cell model which reflect the degree and temporal stage of the occlusion were summarized by Wilhelms et al. [25]. They considered the cellular effects due to hyperkalemia, acidosis and hypoxia as well as due to cellular uncoupling. After clarifying the origin of ST-segment elevation (and depression), they also demonstrated how several ischemic scenarios will not show any ST-segment change [26]. Thus, they were able to explain the large group of non-ST-segment elevation myocardial infarctions (NSTEMI). Potyagaylo et al. showed that these scenarios are not only electrically but also magnetically "silent" [89]. Loewe at al.—using computer modeling—investigated whether additional electrodes, optimized electrode placement or improved analysis of the ST segment could lead to better diagnosis of patients with acute ischemia. They suggest the deviation from baseline at the K-point as being superior to J-point analysis [27].

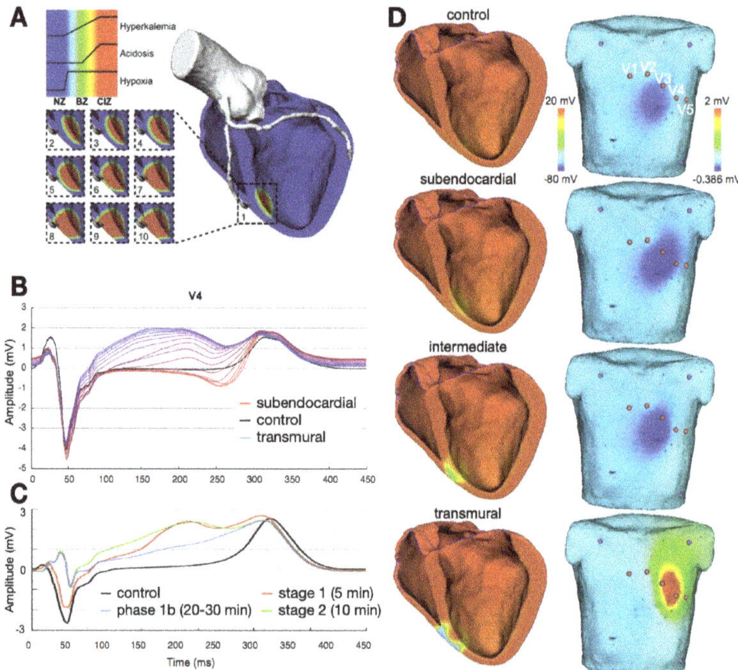

Figure 2. Examples of ischemic regions with varying transmural extent due to occlusion of the left anterior descending coronary artery and the related levels of hyperkalemia, acidosis, and hypoxia (**A**). ECG lead V4 for ischemia of varying transmural extent in temporal stage 2 (**B**) and varying duration of a transmural ischemia (**C**). Ventricular transmembrane voltage and body surface potential distribution during the action potential plateau (t = 200 ms) for ischemia of varying transmural extent in stage 2 (**D**). The QRS complex was not optimized in this study. (Images reproduced with permission from [28].)

Ledezma et al. created populations of control and ischemic cell strands and observed the corresponding pseudo-ECGs (which is the voltage between two virtual electrodes at or near to the ends of a tissue strand immersed in an infinite homogeneous volume conductor).

Based on these data, they trained an artificial neural network that was able to determine severity (mild or severe) and size of the ischemic region from the pseudo-ECG [30].

All these articles deal with ischemia and "fresh" infarctions (not older than a couple of hours). The modeling of "old" infarction scars seems to be straightforward: the scar areas cannot depolarize, they should be "switched off" during modeling. Basically. the QRS complex will change. In particular, small bridges of viable tissue within a scar area are of interest since they are likely to lead to ventricular tachycardia (VT). Lopez-Perez et al. were able to set up a personalized model of a patient with an old infarction with strong emphasis on modeling border zones. They were able to reproduce the 12-lead ECG of a patient with a history of infarction both in sinus rhythm and during VT (see Section 3.3) [29].

Electrocardiographic imaging of myocardial infarction was the subject of the Challenge of the Computing in Cardiology conference in 2007. BSPMs of patients were provided for the participants. Ghasemi et al. were very successful in finding the location and extent of the infarction using only the heart vector and a very simplified model of the distribution of depolarization during systole [90]. Farina et al. employed a model-based approach to solve the task [23] and also reached very good results; however, based on the full BSPM. Jiang et al. investigated the best electrode arrangements to localize an infarcted area in the heart [24]. A dense set of electrodes including and extending the precordial leads was essential. Optimal results were obtained when using at least 64 electrodes.

3.2. Ventricular Ectopic Beats and Extrasystoles

The localization of ventricular ectopic beats (premature ventricular contractions, PVCs) is a major topic of the inverse problem of ECG. Any knowledge about the site of origin can guide the cardiologist during an ablation procedure and thus shorten the duration of the invasive procedure. Many publications contain chapters on calculating the body surface potential map of ventricular ectopic beats using simulations of the spread of depolarization (see, for example, Potyagaylo et al. [31]). Most of them assume that the individual body shape and cardiac geometry is known, which is, however, not the setting of traditional ECG analysis.

Figure 3 shows, as an example, a fast simulation of the spread of depolarization (transmembrane voltage and epicardial potentials) and the corresponding 12-lead ECG for three different ventricular extrasystoles.

3.3. Ventricular Tachycardia

Sapp et al. were able to localize the exit point of a ventricular tachycardia (VT) from a 12-lead ECG based on an empirical study including 38 patients [91]. No modeling was employed; nevertheless, an accuracy of 10 mm was achieved. It might be that the number of patients was too small for drawing general conclusions. In 2007, Segal et al. suggested features in the ECG that point to specific areas of exit points of VT (e.g., antero-lateral, antero-apical, mid-septum, etc.) without any computer model [92]. Kania et al. localized the exit point of a ventricular tachycardia using a personalized geometry for modeling and only three ECG leads (Frank VCG) with an accuracy of 11.7 mm [35]. Additionally, in this section, the article of Lopez-Perez et al. has to be mentioned: they reproduced the ECG during VT of a personalized model of a patient with an old infarction [29].

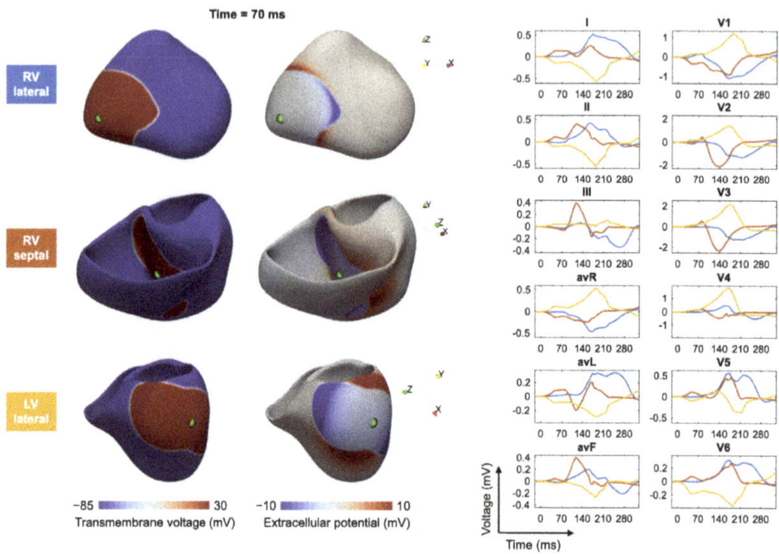

Figure 3. Modeling of ectopic beats and the corresponding ECG: for three different trigger locations in the right ventricle (RV) and left ventricle (LV), the transmembrane voltage (**left column**), the extracellular potentials (**middle column**) and corresponding ECGs (**right column**) are shown. Excitation propagation was computed by solving the anisotropic Eikonal equation.

3.4. Cardiomyopathy

Lyon et al. investigated ECG phenotypes resulting from hypertrophic cardiomyopathy using computer modeling [36]. The objective was to better identify patients at high risk of sudden cardiac death. An inverted T-wave with normal QRS was obtained with increased apico-basal repolarization gradients in the septum and the apex. Lateral QRS abnormalities were only obtained with abnormal Purkinje-myocardium coupling.

3.5. Bundle Branch Blocks

The most important feature visible in the ECG that points towards a bundle branch block is a prolonged QRS complex. Modeling in this respect is mainly aiming at optimization of cardiac resynchronization therapy. The bundle branch block is most often simulated by only stimulating the Purkinje system in the right ventricle (left bundle branch block, LBBB) or left ventricle (RBBB). Potse et al. investigated different hypotheses about activation times for the LV endocardium, morphology of electrograms and ECG features in patients with heart failure and LBBB using personalized computer modeling [37]. Simulated and measured ECGs matched quite well (correlation coefficient between measured and simulated activation times was r = 0.91 in one case and 0.87 in another case). Giffard-Roisin et al. were able to match the 12-lead ECG of a patient with LBBB in a personalized geometry by adjusting only 3 conduction velocities (CVs): in the myocardium, in the left ventricular Purkinje system and in the right ventricular Purkinje system [39]. Personalized models of five patients with heart failure and LBBB were created by Nguyen et al. [38]. They reached a very good agreement between simulated and measured 12-lead ECG and demonstrated that the correct heart position and orientation has a strong impact on several features of the ECG. A combined clinical (21 patients) and in-silico (3 simulated cases) study of LBBB was presented by Nguyen et al. in 2018 [40]: the computer models were personalized using endocardial mapping and the 12-lead ECG.

3.6. Atrial Ectopic Beats

The localization of atrial ectopic beats is of interest since the knowledge of their origin could guide an ablation procedure. Localization of ectopic foci on the left atrium was enabled by model-based computation of the spread of depolarization, forward calculation to the body surface and comparison with "measured" BSPM by Potyagaylo et al. [93]. A "full search" method was applied, which cannot only deliver the location of the focus but also a confidence region on the atria. Using only the first half of the P-wave improved results. Feng et al. optimized a method to classify atrial ectopic beats, stable re-entries and sinus rhythm in the 12-lead ECG by using simulated data [32]. Yang et al. trained a convolutional neural network with simulated ectopic beats and applied it to nine patients to localize ectopic beats from the 12-lead ECG by identifying the most likely origin from 25 cardiac segments. They achieved a localization error of 11 mm when using a personalized model of the thorax of the patient [33]. Additionally, Ferrer-Albero et al. were able to localize atrial ectopic beats using simulated body surface P-wave integral maps and a machine learning approach (support vector machine). The computed ECGs compare well with clinical ECGs. Dividing the atria in up to six regions (clusters), the origin of an ectopic beat could be localized reliably, with the best accuracy of 96% achieved with four clusters [34].

3.7. Atrial Tachycardia, Flutter and Fibrillation

Atrial flutter (AFlut) can be clearly diagnosed from the ECG. Typical and atypical AFlut can be reliably recognised from the ECG by a trained cardiologist. However, the question of which type of atypical atrial flutter the patient suffers from cannot be answered directly from the ECG up to date. Again, this knowledge would speed up ablation procedures and could inform clinical decision making. It also could inform the cardiologist about two alternating classes of AFlut that might be present. It could give evidence that the type of flutter that is observed in the electrophysiology lab is identical with the type that the patient suffers from out of hospital. Zhou et al. employed ECGi to classify several types of AFlut in seven patients based on full BSPMs and personalized torso models [41]. On the other hand, Medi et al. were able to separate a few classes of atypical AFlut just from the ECG without any modeling [94] such as, for example, focal versus macro-re-entry and right atrial versus left atrial macro-re-entry. An approach with similar objectives—classification of macro-re-entrant atrial tachycardia—but without any modeling should be mentioned: Ruiperez-Campillo et al. analyzed loops of the atrial vectorcardiogram, created "archetypes" of the loops of the four most frequent types of AFlut and determined the similarity of the vectorcardiogram loop of a patient with the archetypes [95].

Luongo et al. were able to classify 20 different types of AFlut using the 12-lead ECG based on a recurrence quantification analysis [45] in a computational study with a clinical proof of concept. Figure 4 shows two examples. They also investigated the influence of different atrial and torso geometries on the outcome of their algorithm [46]. Figure 4 shows two examples of simulated AFlut scenarios and the corresponding BSPMs and 12-lead ECGs.

Pezzuto et al. presented simulations of beat-to-beat variations of the P-waves in patients with paroxysmal atrial fibrillation (AFib). They suggest that variations of the exit location of the sinus node are responsible for P-wave fluctuations observed in patients [42] (see also [20]). Filos et al. conducted a literature survey of articles about simulations of the P-wave, the beat-to-beat variations with a particular focus on P-waves of patients suffering from paroxysmal AFib. They showed how advances in computer models and high performance computing could lead to the use of electrophysiological models of the heart to improve quality of life of patients with AFib and optimize AFib treatment [43] (see also Section 2.3).

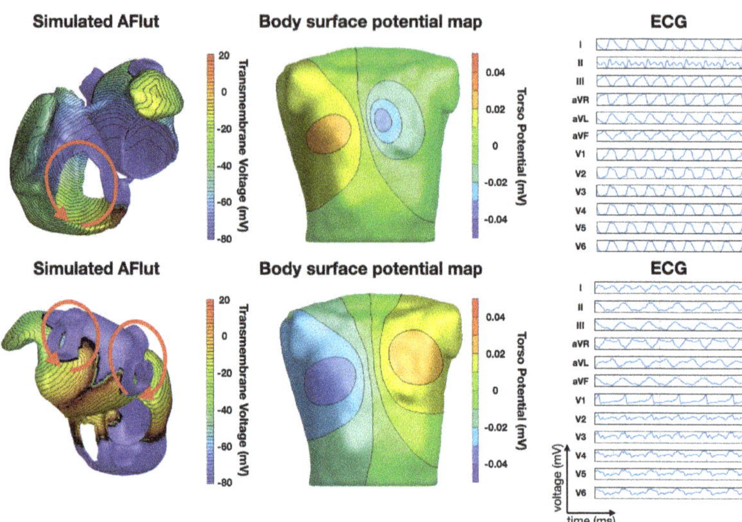

Figure 4. Examples of simulated AFlut transmembrane voltage distributions and the corresponding BSPMs and 12-lead ECGs. Top row: AFlut around the tricuspid valve in the counter-clockwise direction. Bottom row: figure-8 macro-re-entry around the left and right pulmonary veins in the anterior direction of rotation [45].

Atrial fibrillation remains a challenge regarding the relation of ECG features with patterns of depolarization. In case several (more than three) re-entry mechanisms are active simultaneously, it will probably be impossible to discriminate them from the ECG only. Rodrigo et al. used BSPMs computed in personalized torso models and found that areas of high dominant frequency and areas of large rotors can be localized [48]. However, as mentioned before, this is not the setting of classical ECG interpretation. Nevertheless, if only one ectopic center or one rotor is the driver of AFib, then important characteristics can be deduced from the ECG. Luongo et al. were able to classify whether an AFib driver is located in the vicinity of the pulmonary veins or not, and if not, whether it is in the right or left atrium using a machine learning approach [51]. A decision tree was trained purely on simulated data and performed favorably when subsequently being applied to a clinical dataset comprising 46 patients.

Using computerized modeling, a method was proposed by Saha et al. that can monitor an ablation procedure [47]. In total, 20 ablation patterns were investigated. A simulated 16-lead ECG (including V3R, V4R, V8 and V9) was analyzed in respect to P-wave duration (PWD) and P-wave area. Several reconnections could be detected by comparing pre- and post-ablation ECGs, but the settings of detection thresholds were quite demanding (5 ms for PWD and 0.1 µVs for P-wave area). Regarding a real-time assessment of a typical ablation procedure in case of perimitral flutter, the anterior mitral line, was demonstrated by Lehrmann et al. in a clinical study that was enriched and mechanistically underpinned using computer simulations [44]. The time from the stimulus in the left atrial appendage to the P-wave peak in ECG lead V1 ("V1 delay") can be used as a reliable measure for blocking of the anterior mitral line. Even more predictive is the sudden jump of the P-wave ("V1 jump") upon completion of the line.

3.8. Genetic Diseases and Channelopathies

Modeling of genetic diseases leading to cardiac arrhythmias can be straightforward. If the affected ion channel and its modification are known, the modified model of the ion channel can be built into the computer model replacing the normal ion channel. If the dynamics of the ion channel are not affected, just the maximum channel conductance is replaced by a modified value. Sometimes, it is not as easy as that: depending on how the mutation affects the channel, a complete reparametrization or even reformulation of the model can be required. Most frequent cases are the long QT syndromes (LQT1 to LQT16), the short QT syndrome (SQT) and the Brugada syndrome. A review about the modifications of ion channels for theses channelopathies was given by Schwartz et al. [96].

In a very early article, Shimizu and Antzelevitch described the link between transmural heterogeneity in LQT1 syndrome and features of the ECG, in particular concerning the T-wave [97]. Numerous articles followed dealing with empirical clinical studies of ECG morphology in LQT, SQT and Brugada patients but with no link to computer modeling. On the other hand, many articles deal with computational studies about LQT (LQT1 to LQT16), SQT and Brugada syndrome without calculating the corresponding ECG. A computed pseudo-transmural ECG is shown sometimes but it only allows one to identify gross morphological changes but no details in the ECG (for Long QT: Seemann et al., [52], for Short QT: Weiss et al. [53]).

Seemann et al. carried out simulations of LQT1, LQT2 and LQT3 and calculated the corresponding ECGs using a realistic torso model. The T-waves of a healthy control group can clearly be distinguished from LQT patients and the simulated T-waves show good correspondence to clinical data. However, the objective to classify the type of LQT from the ECG was not achieved [54].

3.9. Imbalance of Electrolytes

Imbalance of electrolytes is dangerous in many respects. In particular, it can lead to cardiac arrhythmias. Large deviations from homeostasis can be observed in patients suffering from chronic kidney disease. A continuous monitoring of electrolyte plasma concentration via ECG could be an important aid for patients. Several changes in the ECG due to electrolyte imbalance are well known (see, for example, Wald [98]). Hyper- and hypocalcemia and hyper- and hypokalemia show typical alterations in ECGs.

Bukhari et al. investigated the changes in morphology of the T-wave during hemodialysis, both with modeling and with a clinical study [13]. They identified morphological T-wave features (e.g., amplitude, upslope, and morphological variability) to reconstruct $[Ca^{2+}]$ and $[K^+]$ concentrations. They also pointed out that the heterogeneity of action potentials in the ventricles leading to the T-wave shows strong inter-individual variability [99]. This complicates quantitative analysis and it will probably require a personalized model to determine the electrolyte concentration from the ECG.

A review article about the classification and quantification of potassium and calcium disorders using an ECG was published by Pilia et al. [59]. It contains a chapter on modeling electrolyte disorders and the characteristic features in the ECG that can be derived from modeling [55]. It also points out that classical ventricular cell models are not prepared to show reasonable results for electrolyte concentrations that are far away from homeostasis—they have to be adapted [100]. This group also presented a method to estimate blood calcium concentration [58] and they suggested an optimized selection of features of the T-wave and a polynomial regression method to reconstruct the potassium concentration from the ECG [60]. Further on, they investigated various lead reduction techniques to extract the most important information from a 12-lead ECG using computer simulations [56].

Loewe at al. contributed to the comprehension of the electrolytes on the sinus node and thus on the heart rate [57]. They could demonstrate how the heart rate can be slowed down during hypocalcemia.

3.10. Drug-Induced Changes in the ECG

Several antiarrhythmic drugs used today aim at specific modifications of the ion channels. Class I agents are blockers of the fast sodium channel. Class II antiarrhythmics are known to block β-receptors and thus attenuate the sympathetic nervous system. Class III drugs block one or several outward potassium currents and class IV drugs block calcium channels. These modifications can be introduced into the electrophysiological cell models of the heart and the specific changes in the action potential can be observed. Often times, integrative modeling is a means to predict the net effect of multi-channel blockers when experimental data are only available on the single channel level [101]. Predicting the effect of specific ion channel modifications on the vulnerability of the heart for arrhythmias is a very active field of research [102–104]. The effect on the ECG is not the major focus of most of these studies.

Since drugs that prolong the QT interval are thought to be responsible for life threatening arrhythmias (e.g., torsades de pointes), the effect of a drug on the QT interval is investigated intensively even beyond antiarrhythmic drugs [105]. For that purpose, a strongly simplified model of a strand of cells (e.g., $1 \times 1 \times 20$ mm) representing a transmural wedge can be sufficient under specific conditions. The electric voltage measured from one end to the other is called a "transmural pseudo-ECG". The morphology of this signal cannot be directly compared to a real ECG, but the changes in QT interval due to a drug can be represented. A prerequisite is to model the transmural heterogeneity properly, as already discussed in Section 2.2.

One of the first studies which closed the gap between modeling drug effects on the cellular and tissue level and the corresponding ECG was presented by Wilhelms et al. [61]. The effects of amiodarone and cisapride on human ventricular electrophysiology and the corresponding ECG were analyzed. For cisapride, only a block of I_{Kr} is reported. In contrast, amiodarone affects currents through several ion channels (I_{Kr}, I_{Na}, I_{NaK}, I_{CaL}, I_{NaCa}, I_{Ks}). While cisapride only prolongs the QT interval, amiodarone in addition modifies the QRS complex. Similar observations were confirmed by Zemzemi et al. [62]. They investigated the influence of a 50% hERG channel block (I_{Kr}) and of a 50% block of the fast sodium channel (I_{Na}). Both led to a prolongation of the QT interval but the hERG blocker did not affect the QT interval, whereas the 50% sodium channel block prolonged the QRS complex by 12%. Zemzemi et al. also investigated the effect of an I_{CaL} blocker and the combined effect with an I_{Kr} blocker at various combinations of doses [63]. The dose–response dependency of the degree of block on concentration is modeled using the classical Hill curve in these studies causing a pore block (by reducing g_{max}) but not representing any drug-induced changes in the channel dynamics.

A review about various approaches to additionally include the channel dynamics using Markov-type models was published by Yuan et al. [106]. The action of amiodarone and cisapride at the organ level is shown and pseudo-ECGs are presented. Several articles followed with a large variety of objectives (e.g., unmasking LQT syndrome [107], designing antiarrhythmic drugs for SQT patients [108], uncovering cardiac drug toxicity beyond QT prolongation [109], understanding the antiarrhythmic effect of dofetelide [110], analyzing the interaction of drugs with the cardiac conduction system [111], assessment of drug safety regarding late sodium current (I_{NaL}) blockers [112]) but they all end at the pseudo-ECG level. They are important building blocks for in silico drug assessment and design, but they are not in the focus of this article. The reader is also referred to the "Comprehensive In Vitro Proarrhythmia Assay"(CiPA) initiative, promoting cardiac drug safety assessments based on four approaches: ion channel data, myocyte data, human studies, and in silico investigations [113].

In summary, modeling drug effects so far most often stopped at the pseudo-ECG level and did not consider the realistic ECG derived by computing the field distribution in the thorax induced by the spatio-temporal source distribution in the full heart. The potential to validate in silico modeling with measured ECGs and the potential to enable the individual optimization of the dose via quantification of features in the ECG is not fully realized yet.

4. Options of Modeling for Better Interpretation of the ECG

Computer modeling can contribute to a better understanding of the relation between features in the ECG and the underlying cardiac condition or disease. This was demonstrated in cases of genetic diseases, effect of drugs and imbalance of electrolytes. It can pave the way for quantitative analysis of the ECG. As an example, the quantitative level of extracellular $[K^+]$ can be estimated from the shape of the T-wave.

Model assisted ECG interpretation can guide the cardiologist to better localize ectopic events or diseased areas. This was demonstrated for the localization of regions of infarction, location of the origin of ectopic beats (atrial and ventricular) and the exit points of a VT.

Moreover, modeling can be a means to generate a very large database of perfectly annotated ECGs as a training dataset for machine learning. Millions of ventricular extrasystoles with exact knowledge about the site of origin can be created for the development of machine learning-based localization. Millions of atria with increasing levels of fibrosis can be created to estimate the degree of fibrosis in the atria from the P-wave. Even though it has to be stated that measured patient data would always be the ideal option for a database, we have to accept that measured data are scarce, often contain incorrect annotations and nearly always have a bias because the variety of diseases is rarely equally distributed among study populations.

The objectives of linking computer modeling with the ECG can also be classified into (a) personalization of a heart model ("digital twin"), (b) the investigation of fundamental mechanisms using generic models and (c) the investigation of cohorts of patients [50,114]. All three play their role in better ECG interpretation but considering variability will become more and more important to derive universally valid conclusions.

5. Summary and Outlook

Establishing a stronger link between computer modeling of the heart and the ECG holds great potential. To consequently add at least the calculation of endocardial electrograms and compare with clinical data from the electrophysiology lab would add more evidence to computerized modeling of the heart. Moreover, the forward calculation of the ECG on the body surface is possible and allows for a comparison with the clinical ECG that is most often available. It is a valuable test of the consistency of the modeling approach and can lead to new insights about the relation between electrophysiological phenomena in the heart and the corresponding ECG.

Likewise, if new (and most often computerized) methods of ECG analysis are proposed, it would be important to make the results explainable by mechanistically underpinning the results, e.g., by backing up the hypotheses with state-of-the-art computer simulations. In many cases, a "rule of thumb" using the classical heart vector for an explanation can be misleading. If a feature in the ECG can be clearly linked to a source pattern on the heart, the diagnostic value of ECG can be increased.

It might be possible to construct personalized models of the heart from the 12–lead ECG [7,8]. However, often there will be ambiguities and spatially higher resolved BSPMs or intracardiac electrograms will be needed. There are also other options for personalization: e.g., measuring the ECG of a patient for one or two electrolytes or drug concentrations and using computer modeling to predict (interpolate) intermediate values can enable a quantitative interpretation of the ECG.

For making general conclusions about features in the ECG that point to specific diseases, the analysis of computer simulations with just one geometry of heart and torso will not be sufficient in the long run. A variety of heart shapes and body shapes can be created using published shape models (for the atria see [22,115,116], for the ventricles see [117,118], for the torso see [119]). Features in the ECG that are discovered with computer simulations using only one heart and body geometry might easily lose their applicability due to the large inter-patient variability of ECGs. Only features that can be discriminated from the natural variety are useful.

In summary, bridging the gap between computerized modeling of the heart and ECG analysis (as well as intracardiac electrograms) holds great potential to lead to better comprehension of cardiac diseases, better diagnosis and optimized therapy planning.

Author Contributions: Conceptualization: O.D.; writing—original draft: O.D.; writing—review and editing: A.L., C.N. and G.L.; visualization: A.L., C.N. and G.L. All authors have read and agreed to the final version of the manuscript.

Funding: This work was supported by the EMPIR programme co-financed by the participating states and from the European Union's Horizon 2020 research and innovation programme under grant MedalCare 18HLT07 and by the European Union's Horizon 2020 programme under grant agreement No. 766082, MY-ATRIA project. We gratefully acknowledge support by Deutsche Forschungsgemeinschaft (DFG) (project ID 391128822, LO 2093/1-1). We acknowledge support by the KIT-Publication Fund of the Karlsruhe Institute of Technology.

Institutional Review Board Statement: Not applicable.

Informed Consent Statement: Not applicable.

Data Availability Statement: Not applicable.

Acknowledgments: The authors thank Steffen Schuler for preparing Figure 3.

Conflicts of Interest: The authors declare no conflict of interest. The funders had no role in the design of the study; in the collection, analyses, or interpretation of data; in the writing of the manuscript, or in the decision to publish the results.

References

1. Keller, D.U.J.; Kalayciyan, R.; Dössel, O.; Seemann, G. Fast creation of endocardial stimulation profiles for the realistic simulation of body surface ECGs. In Proceedings of the IFMBE World Congress on Medical Physics and Biomedical Engineering, Munich, Germany, 7–12 September 2009; Volume 25/4, pp. 145–148.
2. Vigmond, E.J.; Stuyvers, B.D. Modeling our understanding of the His-Purkinje system. *Prog. Biophys. Mol. Biol.* **2016**, *120*, 179–188. [CrossRef]
3. Cardone-Noott, L.; Bueno-Orovio, A.; Mincholé, A.; Zemzemi, N.; Rodriguez, B. Human ventricular activation sequence and the simulation of the electrocardiographic QRS complex and its variability in healthy and intraventricular block conditions. *EP Eur.* **2016**, *18*, iv4–iv15. [CrossRef]
4. Kahlmann, W.; Poremba, E.; Potyagaylo, D.; Dössel, O.; Loewe, A. Modelling of patient-specific Purkinje activation based on measured ECGs. *Curr. Dir. Biomed. Eng.* **2017**, *3*, 171–174. [CrossRef]
5. Cranford, J.P.; O'Hara, T.J.; Villongco, C.T.; Hafez, O.M.; Blake, R.C.; Loscalzo, J.; Fattebert, J.L.; Richards, D.F.; Zhang, X.; Glosli, J.N.; et al. Efficient Computational Modeling of Human Ventricular Activation and Its Electrocardiographic Representation: A Sensitivity Study. *Cardiovasc. Eng. Technol.* **2018**, *9*, 447–467. [CrossRef]
6. Mincholé, A.; Zacur, E.; Ariga, R.; Grau, V.; Rodriguez, B. MRI-Based Computational Torso/Biventricular Multiscale Models to Investigate the Impact of Anatomical Variability on the ECG QRS Complex. *Front. Physiol.* **2019**, *10*, 1103. [CrossRef]
7. Pezzuto, S.; Prinzen, F.W.; Potse, M.; Maffessanti, F.; Regoli, F.; Caputo, M.L.; Conte, G.; Krause, R.; Auricchio, A. Reconstruction of three-dimensional biventricular activation based on the 12-lead electrocardiogram via patient-specific modelling. *EP Eur.* **2021**, *23*, 640–647. [CrossRef]
8. Gillette, K.; Gsell, M.A.; Prassl, A.J.; Karabelas, E.; Reiter, U.; Reiter, G.; Grandits, T.; Peyer, C.; Štern, D.; Urschler, M.; et al. A Framework for the Generation of Digital Twins of Cardiac Electrophysiology from Clinical 12-leads ECGs. *Med. Image Anal.* **2021**, *71*, 102080. [CrossRef] [PubMed]
9. Grandits, T.; Effland, A.; Pock, T.; Krause, R.; Plank, G.; Pezzuto, S. GEASI: Geodesic-based Earliest Activation Sites Identification in cardiac models. *arXiv* **2021**, arXiv:2102.09962v1.
10. Xue, J.; Chen, Y.; Han, X.; Gao, W. Electrocardiographic morphology changes with different type of repolarization dispersions. *J. Electrocardiol.* **2010**, *43*, 553–559. [CrossRef]
11. Keller, D.U.J.; Weiss, D.L.; Dossel, O.; Seemann, G. Influence of I(Ks) heterogeneities on the genesis of the T-wave: A computational evaluation. *IEEE Trans. Biomed. Eng.* **2012**, *59*, 311–322. [CrossRef]
12. Moss, R.; Moritz Wülfers, E.; Seemann, G. T-Wave Changes Due to Cardiac Deformation Are Dependent on the Temporal Relationship Between Repolarization and Diastolic Phase. *Comput. Cardiol.* **2018**, *45*, 1–4. [CrossRef]
13. Bukhari, H.A.; Palmieri, F.; Ramirez, J.; Laguna, P.; Ruiz, J.E.; Ferreira, D.; Potse, M.; Sanchez, C.; Pueyo, E. Characterization of T Wave Amplitude, Duration and Morphology Changes During Hemodialysis: Relationship with Serum Electrolyte Levels and Heart Rate. *IEEE Trans. Biomed. Eng.* **2020**, *1*, ahead of print. [CrossRef]
14. Krueger, M.W.; Schmidt, V.; Tobón, C.; Weber, F.M.; Lorenz, C.; Keller, D.U.J.; Barschdorf, H.; Burdumy, M.; Neher, P.; Plank, G.; et al. Modeling atrial fiber orientation in patient-specific geometries: A semi-automatic rule-based approach. In *Functional Imaging and Modeling of the Heart 2011*; Lecture Notes in Computer Science; Axel, L., Metaxas, D., Eds.; Springer: Berlin/Heidelberg, Germany, 2011; Volume 6666, pp. 223–232. [CrossRef]
15. Dössel, O.; Krueger, M.W.; Weber, F.M.; Wilhelms, M.; Seemann, G. Computational modeling of the human atrial anatomy and electrophysiology. *Med. Biol. Eng. Comput.* **2012**, *50*, 773–799. [CrossRef]

16. Krueger, M.W.; Dorn, A.; Keller, D.U.J.; Holmqvist, F.; Carlson, J.; Platonov, P.G.; Rhode, K.S.; Razavi, R.; Seemann, G.; Dössel, O. In-silico modeling of atrial repolarization in normal and atrial fibrillation remodeled state. *Med. Biol. Eng. Comput.* **2013**, *51*, 1105–1119. [CrossRef] [PubMed]

17. Loewe, A.; Krueger, M.W.; Platonov, P.G.; Holmqvist, F.; Dössel, O.; Seemann, G. Left and Right Atrial Contribution to the P-wave in Realistic Computational Models; In *Lecture Notes in Computer Science*; Berlin/Heidelberg, Germany, 2015; Volume 9126, pp. 439–447. [CrossRef]

18. Jacquemet, V. Modeling left and right atrial contributions to the ECG: A dipole-current source approach. *Comput. Biol. Med.* **2015**, *65*, 192–199. [CrossRef]

19. Potse, M.; Lankveld, T.A.R.; Zeemering, S.; Dagnelie, P.C.; Stehouwer, C.D.A.; Henry, R.M.; Linnenbank, A.C.; Kuijpers, N.H.L.; Schotten, U. P-wave complexity in normal subjects and computer models. *J. Electrocardiol.* **2016**, *49*, 545–553. [CrossRef]

20. Loewe, A.; Krueger, M.W.; Holmqvist, F.; Dössel, O.; Seemann, G.; Platonov, P.G. Influence of the earliest right atrial activation site and its proximity to interatrial connections on P-wave morphology. *EP Eur.* **2016**, *18*, iv35–iv43. [CrossRef]

21. Andlauer, R.; Seemann, G.; Baron, L.; Dössel, O.; Kohl, P.; Platonov, P.; Loewe, A. Influence of left atrial size on P-wave morphology: Differential effects of dilation and hypertrophy. *EP Eur.* **2018**, *20*, iii36–iii44. [CrossRef]

22. Nagel, C.; Schuler, S.; Dössel, O.; Loewe, A. A bi-atrial statistical shape model for large-scale in silico studies of human atria: Model development and application to ECG simulations. *arXiv* **2021**, arXiv:2102.10838.

23. Farina, D.; Dössel, O. Model-based approach to the localization of infarction. *Comput. Cardiol.* **2007**, *34*, 173–176. [CrossRef]

24. Jiang, Y.; Qian, C.; Hanna, R.; Farina, D.; Dössel, O. Optimization of the electrode positions of multichannel ECG for the reconstruction of ischemic areas by solving the inverse electrocardiographic problem. *Int. J. Bioelectromagn.* **2009**, *11*, 27–37

25. Wilhelms, M.; Dössel, O.; Seemann, G. Comparing Simulated Electrocardiograms of Different Stages of Acute Cardiac Ischemia. In Proceedings of the International Conference on Functional Imaging and Modeling of the Hear, LNCS, New York, NY, USA, 25–27 May 2011; Volume 6666, pp. 11–19. [CrossRef]

26. Wilhelms, M.; Dössel, O.; Seemann, G. In silico investigation of electrically silent acute cardiac ischemia in the human ventricles. *IEEE Trans. Biomed. Eng.* **2011**, *58*, 2961–2964. [CrossRef]

27. Loewe, A.; Schulze, W.H.W.; Jiang, Y.; Wilhelms, M.; Luik, A.; Dössel, O.; Seemann, G. ECG-Based Detection of Early Myocardial Ischemia in a Computational Model: Impact of Additional Electrodes, Optimal Placement, and a New Feature for ST Deviation. *BioMed Res. Int. Artic.* **2014**, *2015*, 530352. [CrossRef]

28. Loewe, A.; Wülfers, E.M.; Seemann, G. Cardiac ischemia-insights from computational models. *Herzschrittmachertherapie Elektrophysiologie* **2018**, *29*, 48–56. [CrossRef]

29. Lopez-Perez, A.; Sebastian, R.; Izquierdo, M.; Ruiz, R.; Bishop, M.; Ferrero, J.M. Personalized Cardiac Computational Models: From Clinical Data to Simulation of Infarct-Related Ventricular Tachycardia. *Front. Physiol.* **2019**, *10*, 580. [CrossRef]

30. Ledezma, C.A.; Zhou, X.; Rodríguez, B.; Tan, P.J.; Díaz-Zuccarini, V. A modeling and machine learning approach to ECG feature engineering for the detection of ischemia using pseudo-ECG. *PLoS ONE* **2019**, *14*, e0220294. [CrossRef] [PubMed]

31. Potyagaylo, D.; Chmelevsky, M.; van Dam, P.; Budanova, M.; Zubarev, S.; Treshkur, T.; Lebedev, D. ECG Adapted Fastest Route Algorithm to Localize the Ectopic Excitation Origin in CRT Patients. *Front. Physiol.* **2019**, *10*, 183. [CrossRef]

32. Feng, Y.; Roney, C.; Hocini, M.; Niederer, S.; Vigmond, E. Robust Atrial Ectopic Beat Classification From Surface ECG Using Second-Order Blind Source Separation. *Comput. Cardiol.* **2020**, *47*. [CrossRef]

33. Yang, T.; Yu, L.; Jin, Q.; Wu, L.; He, B. Localization of Origins of Premature Ventricular Contraction by Means of Convolutional Neural Network From 12-Lead ECG. *IEEE Trans. Biomed. Eng.* **2018**, *65*, 1662–1671. [CrossRef]

34. Ferrer-Albero, A.; Godoy, E.J.; Lozano, M.; Martínez-Mateu, L.; Atienza, F.; Saiz, J.; Sebastian, R. Non-invasive localization of atrial ectopic beats by using simulated body surface P-wave integral maps. *PLoS ONE* **2017**, *12*, e0181263. [CrossRef] [PubMed]

35. Kania, M.; Coudière, Y.; Cochet, H.; Haissaguerre, M.; Jais, P.; Potse, M. Prediction of the Exit Site of Ventricular Tachycardia Based on Different ECG Lead Systems. *Comput. Cardiol.* **2017**, *44*. [CrossRef]

36. Lyon, A.; Bueno-Orovio, A.; Zacur, E.; Ariga, R.; Grau, V.; Neubauer, S.; Watkins, H.; Rodriguez, B.; Mincholé, A. Electrocardiogram phenotypes in hypertrophic cardiomyopathy caused by distinct mechanisms: Apico-basal repolarization gradients vs. Purkinje-myocardial coupling abnormalities. *EP Eur.* **2018**, *20*, iii102–iii112. [CrossRef]

37. Potse, M.; Krause, D.; Kroon, W.; Murzilli, R.; Muzzarelli, S.; Regoli, F.; Caiani, E.; Prinzen, F.W.; Krause, R.; Auricchio, A. Patient-specific modelling of cardiac electrophysiology in heart-failure patients. *EP Eur.* **2014**, *16* (Suppl. 4), iv56–iv61. [CrossRef]

38. Nguyen, U.C.; Potse, M.; Regoli, F.; Caputo, M.L.; Conte, G.; Murzilli, R.; Muzzarelli, S.; Moccetti, T.; Caiani, E.G.; Prinzen, F.W.; et al. An in-silico analysis of the effect of heart position and orientation on the ECG morphology and vectorcardiogram parameters in patients with heart failure and intraventricular conduction defects. *J. Electrocardiol.* **2015**, *48*, 617–625. [CrossRef] [PubMed]

39. Giffard-Roisin, S.; Fovargue, L.; Webb, J.; Molléro, R.; Lee, J.; Delingette, H.; Ayache, N.; Razavi, R.; Sermesant, M. Estimation of Purkinje Activation from ECG: An Intermittent Left Bundle Branch Block Study. *Lect. Notes Comput. Sci.* **2017**, *10124*, 135–142. [CrossRef]

40. Nguyên, U.C.; Potse, M.; Vernooy, K.; Mafi-Rad, M.; Heijman, J.; Caputo, M.L.; Conte, G.; Regoli, F.; Krause, R.; Moccetti, T.; et al. A left bundle branch block activation sequence and ventricular pacing influence voltage amplitudes: An in vivo and in silico study. *EP Eur.* **2018**, *20*, iii77–iii86. [CrossRef]

41. Zhou, Z.; Jin, Q.; Yu, L.; Wu, L.; He, B. Noninvasive Imaging of Human Atrial Activation during Atrial Flutter and Normal Rhythm from Body Surface Potential Maps. *PLoS ONE* **2016**, *11*, e0163445. [CrossRef] [PubMed]
42. Pezzuto, S.; Gharaviri, A.; Schotten, U.; Potse, M.; Conte, G.; Caputo, M.L.; Regoli, F.; Krause, R.; Auricchio, A. Beat-to-beat P-wave morphological variability in patients with paroxysmal atrial fibrillation: An in silico study. *EP Eur.* **2018**, *20*, iii26–iii35. [CrossRef]
43. Filos, D.; Tachmatzidis, D.; Maglaveras, N.; Vassilikos, V.; Chouvarda, I. Understanding the Beat-to-Beat Variations of P-Waves Morphologies in AF Patients During Sinus Rhythm: A Scoping Review of the Atrial Simulation Studies. *Front. Physiol.* **2019**, *10*, 742. [CrossRef]
44. Lehrmann, H.; Jadidi, A.S.; Minners, J.; Chen, J.; Müller-Edenborn, B.; Weber, R.; Dössel, O.; Arentz, T.; Loewe, A. Novel Electrocardiographic Criteria for Real-Time Assessment of Anterior Mitral Line Block. *JACC Clin. Electrophysiol.* **2018**, *4*, 920–932. [CrossRef]
45. Luongo, G.; Schuler, S.; Luik, A.; Almeida, T.P.; Soriano, D.C.; Dossel, O.; Loewe, A. Non-Invasive Characterization of Atrial Flutter Mechanisms Using Recurrence Quantification Analysis on the ECG: A Computational Study. *IEEE Trans. Biomed. Eng.* **2021**, *68*, 914–925. [CrossRef]
46. Luongo, G.; Schuler, S.; Rivolta, M.W.; Dössel, O.; Sassi, R.; Loewe, A. Automatic ECG-based Discrimination of 20 Atrial Flutter Mechanisms: Influence of Atrial and Torso Geometries. *Comput. Cardiol.* **2020**. [CrossRef]
47. Saha, M.; Conte, G.; Caputo, M.L.; Regoli, F.; Krause, R.; Auricchio, A.; Jacquemet, V. Changes in P-wave morphology after pulmonary vein isolation: Insights from computer simulations. *Europace* **2016**, *18*, iv23–iv34. [CrossRef] [PubMed]
48. Rodrigo, M.; Climent, A.M.; Liberos, A.; Fernández-Avilés, F.; Berenfeld, O.; Atienza, F.; Guillem, M.S. Highest dominant frequency and rotor positions are robust markers of driver location during noninvasive mapping of atrial fibrillation: A computational study. *Heart Rhythm* **2017**, *14*, 1224–1233. [CrossRef]
49. Irakoze, E.; Jacquemet, V. Simulated P wave morphology in the presence of endo-epicardial activation delay. *EP Eur.* **2018**, *20*, iii16–iii25. [CrossRef] [PubMed]
50. Nagel, C.; Luongo, G.; Azzolin, L.; Schuler, S.; Dössel, O.; Loewe, A. Non-Invasive and Quantitative Estimation of Left Atrial Fibrosis Based on P Waves of the 12-Lead ECG—A Large-Scale Computational Study Covering Anatomical Variability. *J. Clin. Med.* **2021**, *10*, 1797. [CrossRef]
51. Luongo, G.; Azzolin, L.; Schuler, S.; Rivolta, M.W.; Almeida, T.P.; Martínez, J.P.; Soriano, D.C.; Luik, A.; Müller-Edenborn, B.; Jadidi, A.; et al. Machine learning enables noninvasive prediction of atrial fibrillation driver location and acute pulmonary vein ablation success using the 12-lead ECG. *Cardiovasc. Digit. Health J.* **2021**, *2*, 126–136. [CrossRef]
52. Seemann, G.; Weiß, D.L.; Sachse, F.B.; Dössel, O. Simulation of the Long-QT Syndrome in a Model of Human Myocardium. *Comput. Cardiol.* **2003**, *30*, 287–290.
53. Weiss, D.L.; Seemann, G.; Sachse, F.B.; Dössel, O. Modelling of the short QT syndrome in a heterogeneous model of the human ventricular wall. *EP Eur.* **2005**, *7s2*, 105–117. [CrossRef]
54. Seemann, G.; Alvarez de Eulate, M.; Konrad, N.; Maier, J.; Wilhelms, M.; Keller, D.; Dössel, O.; Scholz, E. Evaluating body surface ECG differences of simulated long-QT syndromes. *Comput. Cardiol.* **2013**, *40*, 345–348.
55. Hernández Mesa, M.; Pilia, N.; Dössel, O.; Severi, S.; Loewe, A. Effects of Serum Calcium Changes on the Cardiac Action Potential and the ECG in a Computational Model. *Curr. Dir. Biomed. Eng.* **2018**, *4*, 251–254. [CrossRef]
56. Hernández Mesa, M.; Pilia, N.; Dössel, O.; Loewe, A. Influence of ECG Lead Reduction Techniques for Extracellular Potassium and Calcium Concentration Estimation. *Curr. Dir. Biomed. Eng.* **2019**, *5*, 69–72. [CrossRef]
57. Loewe, A.; Lutz, Y.; Nairn, D.; Fabbri, A.; Nagy, N.; Toth, N.; Ye, X.; Fuertinger, D.H.; Genovesi, S.; Kotanko, P.; et al. Hypocalcemia-Induced Slowing of Human Sinus Node Pacemaking. *Biophys. J.* **2019**, *117*, 2244–2254. [CrossRef] [PubMed]
58. Pilia, N.; Hernandez Mesa, M.; Dössel, O.; Loewe, A. ECG-based Estimation of Potassium and Calcium Concentrations: Proof of Concept with Simulated Data. In Proceedings of the 2019 41st Annual International Conference of the IEEE Engineering in Medicine and Biology Society (EMBC), Berlin, Germany, 23–27 July 2019; pp. 2610–2613. [CrossRef]
59. Pilia, N.; Severi, S.; Raimann, J.G.; Genovesi, S.; Dössel, O.; Kotanko, P.; Corsi, C.; Loewe, A. Quantification and classification of potassium and calcium disorders with the electrocardiogram: What do clinical studies, modeling, and reconstruction tell us? *APL Bioeng.* **2020**, *4*, 041501. [CrossRef]
60. Pilia, N.; Corsi, C.; Severi, S.; Dössel, O.; Loewe, A. Reconstruction of Potassium Concentrations with the ECG on Imbalanced Datasets. *arXiv* **2020**, arXiv:2006.05212.
61. Wilhelms, M.; Rombach, C.; Scholz, E.P.; Doessel, O.; Seemann, G. Impact of amiodarone and cisapride on simulated human ventricular electrophysiology and electrocardiograms. *EP Eur.* **2012**, *14*, v90–v96. [CrossRef] [PubMed]
62. Zemzemi, N.; Bernabeu, M.O.; Saiz, J.; Cooper, J.; Pathmanathan, P.; Mirams, G.R.; Pitt-Francis, J.; Rodriguez, B. Computational assessment of drug-induced effects on the electrocardiogram: from ion channel to body surface potentials. *Br. J. Pharmacol.* **2013**, *168*, 718–733. [CrossRef]
63. Zemzemi, N.; Rodriguez, B. Effects of L-type calcium channel and human ether-a-go-go related gene blockers on the electrical activity of the human heart: A simulation study. *EP Eur.* **2015**, *17*, 326–333. [CrossRef] [PubMed]
64. Oosterom, A.V.; Oostendorp, T.F. ECGSIM: An interactive tool for studying the genesis of QRST forms. *Heart* **2004**, *90*, 165–168. [CrossRef]

65. Plank, G.; Loewe, A.; Neic, A.; Augustin, C.; Huang, Y.L.; Gsell, M.A.; Karabelas, E.; Nothstein, M.; Prassl, A.J.; Sánchez, J.; et al. The openCARP simulation environment for cardiac electrophysiology. *Comput. Methods Programs Biomed.* **2021**, *208*, 106223. [CrossRef]
66. Sánchez, J.; Nothstein, M.; Neic, A.; Huang, Y.L.; Prassl, A.J.; Klar, J.; Ulrich, R.; Bach, F.; Zschumme, P.; Selzer, M.; et al. openCARP: An Open Sustainable Framework for In-Silico Cardiac Electrophysiology Research. *Comput. Cardiol.* **2020**, *47*. [CrossRef]
67. Seemann, G.; Sachse, F.B.; Karl, M.; Weiss, D.L.; Heuveline, V.; Dössel, O. Framework for modular, flexible and efficient solving the cardiac bidomain equation using PETSc. *Math. Ind.* **2010**, *15*, 363–369. [CrossRef]
68. Sundnes, J.; Nielsen, B.F.; Mardal, K.A.; Cai, X.; Lines, G.T.; Tveito, A. On the computational complexity of the bidomain and the monodomain models of electrophysiology. *Ann. Biomed. Eng.* **2006**, *34*, 1088–1097. [CrossRef] [PubMed]
69. Cooper, F.; Baker, R.; Bernabeu, M.; Bordas, R.; Bowler, L.; Bueno-Orovio, A.; Byrne, H.; Carapella, V.; Cardone-Noott, L.; Cooper, J.; et al. Chaste: Cancer, Heart and Soft Tissue Environment. *J. Open Source Softw.* **2020**, *5*, 1848. [CrossRef]
70. Krause, D.; Potse, M.; Dickopf, T.; Krause, R.; Auricchio, A.; Prinzen, F. Hybrid Parallelization of a Large-Scale Heart Model. *Lect. Notes Comput. Sci.* **2012**, *7174*, 120–132. [CrossRef]
71. Quarteroni, A.; Manzoni, A.; Vergara, C. The cardiovascular system: Mathematical modelling, numerical algorithms and clinical applications. *Acta Numer.* **2017**, *26*, 365–590. [CrossRef]
72. Niederer, S.A.; Kerfoot, E.; Benson, A.P.; Bernabeu, M.O.; Bernus, O.; Bradley, C.; Cherry, E.M.; Clayton, R.; Fenton, F.H.; Garny, A.; et al. Verification of cardiac tissue electrophysiology simulators using an N-version benchmark. *Philos. Trans. R. Soc. A Math. Phys. Eng. Sci.* **2011**, *369*, 4331–4351. [CrossRef]
73. Lyon, A.; Mincholé, A.; Martínez, J.P.; Laguna, P.; Rodriguez, B. Computational techniques for ECG analysis and interpretation in light of their contribution to medical advances. *J. R. Soc. Interface* **2018**, *15*, 20170821. [CrossRef] [PubMed]
74. Potse, M. Scalable and Accurate ECG Simulation for Reaction-Diffusion Models of the Human Heart. *Front. Physiol.* **2018**, *9*, 370. [CrossRef] [PubMed]
75. Pezzuto, S.; Kal'avský, P.; Potse, M.; Prinzen, F.W.; Auricchio, A.; Krause, R. Evaluation of a Rapid Anisotropic Model for ECG Simulation. *Front. Physiol.* **2017**, *8*, 265. [CrossRef] [PubMed]
76. Keller, D.U.J.; Weber, F.M.; Seemann, G.; Dössel, O. Ranking the Influence of Tissue Conductivities on Forward-Calculated ECGs. *IEEE Trans. Biomed. Eng.* **2010**, *57*, 1568–1576. [CrossRef]
77. Schuler, S.; Tate, J.D.; Oostendorp, T.F.; MacLeod, R.S.; Dössel, O. Spatial Downsampling of Surface Sources in the Forward Problem of Electrocardiography. In *Functional Imaging and Modeling of the Heart*; Lecture Notes in Computer Science; Coudière, Y., Ozenne, V., Vigmond, E., Zemzemi, N., Eds.; Springer International Publishing: Berlin, Germany, 2019; Volume 11504, pp. 29–36. [CrossRef]
78. Neic, A.; Campos, F.O.; Prassl, A.J.; Niederer, S.A.; Bishop, M.J.; Vigmond, E.J.; Plank, G. Efficient computation of electrograms and ECGs in human whole heart simulations using a reaction-eikonal model. *J. Comput. Phys.* **2017**, *346*, 191–211. [CrossRef] [PubMed]
79. Schuler, S.; Pilia, N.; Potyagaylo, D.; Loewe, A. Cobiveco: Consistent biventricular coordinates for precise and intuitive description of position in the heart—With MATLAB implementation. *arXiv* **2021**, arXiv:2102.02898.
80. Bayer, J.; Prassl, A.J.; Pashaei, A.; Gomez, J.F.; Frontera, A.; Neic, A.; Plank, G.; Vigmond, E.J. Universal ventricular coordinates: A generic framework for describing position within the heart and transferring data. *Med. Image Anal.* **2018**, *45*, 83–93. [CrossRef] [PubMed]
81. Roney, C.H.; Pashaei, A.; Meo, M.; Dubois, R.; Boyle, P.M.; Trayanova, N.A.; Cochet, H.; Niederer, S.A.; Vigmond, E.J. Universal atrial coordinates applied to visualisation, registration and construction of patient specific meshes. *Med. Image Anal.* **2019**, *55*, 65–75. [CrossRef] [PubMed]
82. Opthof, T.; Remme, C.A.; Jorge, E.; Noriega, F.; Wiegerinck, R.F.; Tasiam, A.; Beekman, L.; Alvarez-Garcia, J.; Munoz-Guijosa, C.; Coronel, R.; et al. Cardiac activation-repolarization patterns and ion channel expression mapping in intact isolated normal human hearts. *Heart Rhythm* **2017**, *14*, 265–272. [CrossRef]
83. Wachter, A.; Loewe, A.; Krueger, M.W.; Dössel, O.; Seemann, G. Mesh structure-independent modeling of patient-specific atrial fiber orientation. *Curr. Dir. Biomed. Eng.* **2015**, *1*, 409–412. [CrossRef]
84. Nagel, C.; Pilia, N.; Loewe, A.; Dössel, O. Quantification of Interpatient 12-lead ECG Variabilities within a Healthy Cohort. *Curr. Dir. Biomed. Eng.* **2020**, *6*, 493–496. [CrossRef]
85. Rajendra Acharya, U.; Paul Joseph, K.; Kannathal, N.; Lim, C.M.; Suri, J.S. Heart rate variability: A review. *Med. Biol. Eng. Comput.* **2006**, *44*, 1031–1051. [CrossRef]
86. Wilders, R.; Hoekstra, M.; van Ginneken, A.C.G.; Verkerk, A.O. Beta-adrenergic modulation of heart rate: Contribution of the slow delayed rectifier K+ current (IKs). *Comput. Cardiol.* **2010**, *37*, 629–631.
87. Dössel, O.; Reumann, M.; Seemann, G.; Weiss, D. The missing link between cardiovascular rhythm control and myocardial cell modeling. *Biomed. Tech.* **2006**, *51*, 205–209. [CrossRef]
88. Rodero, C.; Strocchi, M.; Marciniak, M.; Longobardi, S.; Whitaker, J.; O'Neill, M.D.; Gillette, K.; Augustin, C.; Plank, G.; Vigmond, E.J.; et al. Linking statistical shape models and simulated function in the healthy adult human heart. *PLoS Comput. Biol.* **2021**, *17*, e1008851. [CrossRef]

89. Potyagaylo, D.; Seemann, G.; Schulze, W.; Dössel, O. Magnetocardiography did not uncover electrically silent ischemia in an in-silico study case. *Comput. Cardiol.* **2015**, *42*, 1145–1148.
90. Ghasemi, M. Electrocardiographic imaging of myocardial infarction using heart vector analysis. *Comput. Cardiol.* **2007**. [CrossRef]
91. Sapp, J.L.; Bar-Tal, M.; Howes, A.J.; Toma, J.E.; El-Damaty, A.; Warren, J.W.; MacInnis, P.J.; Zhou, S.; Horáček, B.M. Real-Time Localization of Ventricular Tachycardia Origin From the 12-Lead Electrocardiogram. *JACC Clin. Electrophysiol.* **2017**, *3*, 687–699. [CrossRef] [PubMed]
92. Segal, O.R.; Chow, A.W.C.; Wong, T.; Trevisi, N.; Lowe, M.D.; Davies, D.W.; Della Bella, P.; Packer, D.L.; Peters, N.S. A novel algorithm for determining endocardial VT exit site from 12-lead surface ECG characteristics in human, infarct-related ventricular tachycardia. *J. Cardiovasc. Electrophysiol.* **2007**, *18*, 161–168. [CrossRef]
93. Potyagaylo, D.; Loewe, A.; van Dam, P.; Dössel, O. ECG imaging of focal atrial excitation: Evaluation in a realistic simulation setup. *Comput. Cardiol.* **2016**, *43*, 113–116. [CrossRef]
94. Medi, C.; Kalman, J.M. Prediction of the atrial flutter circuit location from the surface electrocardiogram. *EP Eur.* **2008**, *10*, 786–796. [CrossRef]
95. Ruipérez-Campillo, S.; Castrejón, S.; Martínez, M.; Cervigón, R.; Meste, O.; Merino, J.L.; Millet, J.; Castells, F. Non-invasive characterisation of macroreentrant atrial tachycardia types from a vectorcardiographic approach with the slow conduction region as a cornerstone. *Comput. Methods Programs Biomed.* **2021**, *200*, 105932. [CrossRef]
96. Schwartz, P.J.; Ackerman, M.J.; Antzelevitch, C.; Bezzina, C.R.; Borggrefe, M.; Cuneo, B.F.; Wilde, A.A.M. Inherited cardiac arrhythmias. *Nat. Rev. Dis. Prim.* **2020**, *6*, 767–779. [CrossRef]
97. Shimizu, W.; Antzelevitch, C. Cellular basis for the ECG features of the LQT1 form of the long-QT syndrome: Effects of [beta]-adrenergic agonists and antagonists and sodium channel blockers on transmural dispersion of repolarization and torsade de pointes. *Circulation* **1998**, *98*, 2314–2322. [CrossRef]
98. Wald, D.A. ECG manifestations of selected metabolic and endocrine disorders. *Emerg. Med. Clin. N. Am.* **2006**, *24*, 145–157. [CrossRef]
99. Bukhari, H.A.; Palmieri, F.; Ferreira, D.; Potse, M.; Ramírez, J.; Laguna, P.; Sánchez, C.; Pueyo, E. Transmural Ventricular Heterogeneities Play a Major Role in Determining T-Wave Morphology at Different Extracellular Potassium Levels. *Comput. Cardiol.* **2019**, 1–4. [CrossRef]
100. Loewe, A.; Hernandez Mesa, M.; Pilia, N.; Severi, S.; Dössel, O. A heterogeneous formulation of the Himeno et al. human ventricular myocyte model for simulation of Body Surface ECGs. *Comput. Cardiol.* **2018**, *45*. [CrossRef]
101. Cavero, I.; Holzgrefe, H. CiPA: Ongoing testing, future qualification procedures, and pending issues. *J. Pharmacol. Toxicol. Methods* **2015**, *76*, 27–37. [CrossRef] [PubMed]
102. Corrias, A.; Jie, X.; Romero, L.; Bishop, M.J.; Bernabeu, M.; Pueyo, E.; Rodriguez, B. Arrhythmic risk biomarkers for the assessment of drug cardiotoxicity: From experiments to computer simulations. *Philos. Trans. Ser. A Math. Phys. Eng. Sci.* **2010**, *368*, 3001–3025. [CrossRef] [PubMed]
103. Passini, E.; Britton, O.J.; Lu, H.R.; Rohrbacher, J.; Hermans, A.N.; Gallacher, D.J.; Greig, R.J.H.; Bueno-Orovio, A.; Rodriguez, B. Human Drug Trials Demonstrate Higher Accuracy than Animal Models in Predicting Clinical Pro-Arrhythmic Cardiotoxicity. *Front. Physiol.* **2017**, *8*, 668. [CrossRef] [PubMed]
104. Loewe, A.; Lutz, Y.; Wilhelms, M.; Sinnecker, D.; Barthel, P.; Scholz, E.P.; Dössel, O.; Schmidt, G.; Seemann, G. In-silico assessment of the dynamic effects of amiodarone and dronedarone on human atrial patho-electrophysiology. *EP Eur.* **2014**, *16*, iv30–iv38. [CrossRef]
105. Mamoshina, P.; Rodriguez, B.; Bueno-Orovio, A. Toward a broader view of mechanisms of drug cardiotoxicity. *Cell Rep. Med.* **2021**, *2*, 100216. [CrossRef]
106. Yuan, Y.; Bai, X.; Luo, C.; Wang, K.; Zhang, H. The virtual heart as a platform for screening drug cardiotoxicity. *Br. J. Pharmacol.* **2015**, *172*, 5531–5547. [CrossRef]
107. Romero, L.; Trenor, B.; Yang, P.C.; Saiz, J.; Clancy, C.E. In silico screening of the impact of hERG channel kinetic abnormalities on channel block and susceptibility to acquired long QT syndrome. *J. Mol. Cell. Cardiol.* **2014**, *72*, 126–137. [CrossRef] [PubMed]
108. Luo, C.; Wang, K.; Zhang, H. Modelling the effects of chloroquine on KCNJ2-linked short QT syndrome. *Oncotarget* **2017**, *8*, 106511–106526. [CrossRef] [PubMed]
109. Jie, X.; Rodriguez, B.; Pueyo, E. A new ECG biomarker for drug toxicity: A combined signal processing and computational modeling study. In Proceedings of the 2010 Annual International Conference of the IEEE Engineering in Medicine and Biology, Buenos Aires, Argentina, 31 August–4 September 2010; pp. 2565–2568. [CrossRef]
110. Saiz, J.; Gomis-Tena, J.; Monserrat, M.; Ferrero, J.M.J.; Cardona, K.; Chorro, J. Effects of the antiarrhythmic drug dofetilide on transmural dispersion of repolarization in ventriculum. A computer modeling study. *IEEE Trans. Biomed. Eng.* **2011**, *58*, 43–53. [CrossRef] [PubMed]
111. Dux-Santoy, L.; Sebastian, R.; Felix-Rodriguez, J.; Ferrero, J.M.; Saiz, J. Interaction of specialized cardiac conduction system with antiarrhythmic drugs: A simulation study. *IEEE Trans. Biomed. Eng.* **2011**, *58*, 3475–3478. [CrossRef]
112. Trenor, B.; Gomis-Tena, J.; Cardona, K.; Romero, L.; Rajamani, S.; Belardinelli, L.; Giles, W.R.; Saiz, J. In silico assessment of drug safety in human heart applied to late sodium current blockers. *Channels* **2013**, *7*, 249–262. [CrossRef]
113. Park, J.S. Introduction to in silico model for proarrhythmic risk assessment under the CiPA initiative. *Transl. Clin. Pharmacol.* **2019**, *27*, 12–18. [CrossRef]

114. Niederer, S.A.; Aboelkassem, Y.; Cantwell, C.D.; Corrado, C.; Coveney, S.; Cherry, E.M.; Delhaas, T.; Fenton, F.H.; Panfilov, A.V.; Pathmanathan, P.; et al. Creation and application of virtual patient cohorts of heart models. *Philos. Trans. R. Soc. A Math. Phys. Eng. Sci.* **2020**, *378*, 20190558. [CrossRef]
115. Nagel, C.; Schuler, S.; Dössel, O.; Loewe, A. A Bi-atrial Statistical Shape Model and 100 Volumetric Anatomical Models of the Atria. *Zenodo* **2020**. [CrossRef]
116. Corrado, C.; Razeghi, O.; Roney, C.; Coveney, S.; Williams, S.; Sim, I.; O'Neill, M.; Wilkinson, R.; Oakley, J.; Clayton, R.H.; et al. Quantifying atrial anatomy uncertainty from clinical data and its impact on electro-physiology simulation predictions. *Med. Image Anal.* **2020**, *61*, 101626. [CrossRef]
117. Schuler, S.; Loewe, A. Biventricular statistical shape model of the human heart adapted for computer simulations. *Zenodo* **2021**. [CrossRef]
118. Bai, W.; Shi, W.; de Marvao, A.; Dawes, T.J.; O'Regan, D.P.; Cook, S.A.; Rueckert, D. A bi-ventricular cardiac atlas built from 1000+ high resolution MR images of healthy subjects and an analysis of shape and motion. *Med. Image Anal.* **2015**, *26*, 133–145. [CrossRef] [PubMed]
119. Pishchulin, L.; Wuhrer, S.; Helten, T.; Theobalt, C.; Schiele, B. Building statistical shape spaces for 3D human modeling. *Pattern Recognit.* **2017**, *67*, 276–286. [CrossRef]

 *hearts*

Article

Electrocardiographic Predictors of Mortality: Data from a Primary Care Tele-Electrocardiography Cohort of Brazilian Patients

Gabriela M. M. Paixão [1], Emilly M. Lima [1], Paulo R. Gomes [1], Derick M. Oliveira [2], Manoel H. Ribeiro [3] and Jamil S. Nascimento [1], Antonio H. Ribeiro [4], Peter W. Macfarlane [5] and Antonio L. P. Ribeiro [1,*,†]

[1] Telehealth Network of Minas Gerais, Hospital das Clínicas and Faculdade de Medicina, Universidade Federal de Minas Gerais, Avenida Professor Alfredo Balena 110, Belo Horizonte 30130-100, Brazil; gabimiana@gmail.com (G.M.M.P.); emillymalveira@gmail.com (E.M.L.); prxgomes@gmail.com (P.R.G.); manoelhortaribeiro@gmail.com (M.H.R.); jamilhcufmg@gmail.com (J.S.N.)

[2] Computer Science Department, Universidade Federal de Minas Gerais, Belo Horizonte 31270-901, Brazil; derickmath@dcc.ufmg.br

[3] School of Computer and Communication Sciences, École Polytechnique Fédérale de Lausanne, 1015 Lausanne, Switzerland; manoel.hortaribeiro@epfl.ch

[4] Department of Information Technology, Uppsala University, 752 37 Uppsala, Sweden; antonio.horta.ribeiro@it.uu.se

[5] Division of Cardiovascular and Medical Sciences, University of Glasgow, Scotland G12 8QQ, UK; Peter.Macfarlane@glasgow.ac.uk

* Correspondence: alpr@ufmg.br; Tel.: +55-31-3307-9201

† Courrent Address: Telehealth Center, Hospital das Clínicas da UFMG Av. Alfredo Balena, 110, Sala 106 Sul, Belo Horizonte 30130-100, Brazil.

Citation: Paixão, G.M.M.; Lima, E.M.; Gomes, P.R.; Oliveira, D.M.; Ribeiro, M.H.; Nascimento, J.S.; Ribeiro, A.H.; Macfarlane, P.W.; Ribeiro, A.L.P. Electrocardiographic Predictors of Mortality: Data from a Primary Care Tele-Electrocardiography Cohort of Brazilian Patients. *Hearts* **2021**, *2*, 449–458. https://doi.org/10.3390/hearts2040035

Academic Editor: Gaetano Santulli

Received: 27 August 2021
Accepted: 27 September 2021
Published: 29 September 2021

Publisher's Note: MDPI stays neutral with regard to jurisdictional claims in published maps and institutional affiliations.

Abstract: Computerized electrocardiography (ECG) has been widely used and allows linkage to electronic medical records. The present study describes the development and clinical applications of an electronic cohort derived from a digital ECG database obtained by the Telehealth Network of Minas Gerais, Brazil, for the period 2010–2017, linked to the mortality data from the national information system, the Clinical Outcomes in Digital Electrocardiography (CODE) dataset. From 2,470,424 ECGs, 1,773,689 patients were identified. A total of 1,666,778 (94%) underwent a valid ECG recording for the period 2010 to 2017, with 1,558,421 patients over 16 years old; 40.2% were men, with a mean age of 51.7 [SD 17.6] years. During a mean follow-up of 3.7 years, the mortality rate was 3.3%. ECG abnormalities assessed were: atrial fibrillation (AF), right bundle branch block (RBBB), left bundle branch block (LBBB), atrioventricular block (AVB), and ventricular pre-excitation. Most ECG abnormalities (AF: Hazard ratio [HR] 2.10; 95% CI 2.03–2.17; RBBB: HR 1.32; 95%CI 1.27–1.36; LBBB: HR 1.69; 95% CI 1.62–1.76; first degree AVB: Relative survival [RS]: 0.76; 95% CI0.71–0.81; 2:1 AVB: RS 0.21 95% CI0.09–0.52; and RS 0.36; third degree AVB: 95% CI 0.26–0.49) were predictors of overall mortality, except for ventricular pre-excitation (HR 1.41; 95% CI 0.56–3.57) and Mobitz I AVB (RS 0.65; 95% CI 0.34–1.24). In conclusion, a large ECG database established by a telehealth network can be a useful tool for facilitating new advances in the fields of digital electrocardiography, clinical cardiology and cardiovascular epidemiology.

Keywords: electronic cohort; electrocardiogram; mortality; big data; telehealth

1. Introduction

Cardiovascular diseases are the main cause of mortality both worldwide and in Brazil, and are responsible for 31.2% of total deaths and a mortality rate standardized by age of 256.0 per 100,000 inhabitants [1]. The electrocardiogram (ECG) is a low-cost, easy-access and non-invasive exam used for cardiovascular assessment, and possesses both diagnostic and prognostic value.

Epidemiological studies using the ECG began in the 1940s with the first cardiovascular cohorts [2]. However, ECG reports were very heterogeneous due to the lack of an established coding system appropriate to epidemiological and population-based studies [3]. The Minnesota Code [4] was created in 1960 to standardize ECG classification and enable comparison between different populations. In the following decades, many papers were published on the use of the ECG in population-based studies, showing the prognostic value of different electrocardiographic abnormalities [5–11].

Simultaneously, the evolution of computerized ECG and automated interpretation had a great impact on cardiovascular epidemiological studies [12,13]. Systems that are capable of transmitting electrocardiographic tracings over the Internet and software packages that enable automatic analysis and coding of tracings have revolutionized the electrocardiography of population-based studies, enhancing its applications and facilitating the study of large populations [14–17].

The identification of new electrocardiographic variables as predictors for cardiovascular events is an important objective of research among electronic cohorts, especially when performing ECG for population screening remains controversial [18,19] and the benefit of the traditional ECG markers with cardiovascular risk scores for discrimination and reclassification is questionable [18,20]. The use of new technologies such as artificial intelligence (AI) is a promising tool in this field for the recognition of potential non-traditional electrocardiographic risk factors.

Despite many studies on ECG abnormalities and their prognostic value having been published, their data are usually from cohorts that typically include hundreds or thousands of patients, or even from secondary care or an inpatient setting, resulting in very specific populations. Big data sets with over one million patients are relatively new, especially in the outpatient setting, and can provide more precise estimates of the risk related to each ECG abnormality in the community setting. This information should be useful for physicians in the primary care setting, and may help to support clinical decisions. Thus, the present study aims to describe the development and clinical applications of an electronic cohort, entitled the Clinical Outcomes in Digital Electrocardiography (CODE) study [21]. This cohort is derived from a digital ECG database obtained by the Telehealth Network of Minas Gerais (TNMG), Brazil [22], from 2010 to 2017, and linked to the mortality data from the national information system with more than 1.5 million patients.

2. Methods

2.1. Study Design

This study is based on a retrospective cohort of primary care patients from Minas Gerais, Brazil, whose ECGs were analyzed by the Telehealth Network of Minas Gerais (TNMG) cardiologists between 2010 and 2017. TNMG currently covers 817 of the 853 counties in Minas Gerais and nearly 400 in other Brazilian states. It has already acquired more than five million ECGs since its implementation [23].

2.2. Inclusion Criteria

Patients older than 16 years with 12-lead ECGs performed at TNMG between 2010 and 2017 were included in the study. For the specific analysis of ventricular pre-excitation, all age groups were included.

2.3. Exclusion Criteria

Isoelectric recordings and those with interference, reversal or poor positioning of electrodes, which compromised the analysis, were excluded (6.03%). For the analysis of electrocardiographic changes, patients who underwent more than one ECG had only the first exam analyzed; subsequent recordings were excluded (28.20%).

2.4. Data Collection

ECGs were performed by the local primary care professional using digital electrocardiographs manufactured by Tecnologia Eletrônica Brasileira model ECGPC (São Paulo, Brazil) or Micromed Biotecnologia model ErgoPC 13 (Brasilia, Brazil).

Clinical data (age, sex and comorbidities) were collected using a standardized questionnaire. Clinical conditions included self-reported smoking, hypertension, diabetes, dyslipidemia, Chagas disease, previous myocardial infarction and chronic obstructive pulmonary disease.

Specific software, developed in-house, was able to capture an ECG tracing, upload the ECG and the patient's clinical history, and then transmit the data to the TNMG analysis center via the internet. The clinical information, ECG tracings and reports were stored in a specific database. All data managed and transferred followed the national law for security and protection of the database. For the purpose of the present study, the Glasgow 12-lead ECG analysis program (license 28.4.1, approved for use on 16 June 2009) was used to automatically interpret all ECGs available in the database, exporting the diagnosis as interpreted by both Glasgow and Minnesota codes.

2.5. Data Analysis

Major Electrocardiographic Abnormalities

The major electrocardiographic abnormalities included were atrial fibrillation (AF), right bundle branch block (RBBB), left bundle branch block (LBBB), first, second and third degree atrioventricular blocks (AVB) and ventricular pre-excitation [24].

ECGs were analyzed by a team of fourteen trained cardiologists using standardized criteria [24]. Each ECG was interpreted by only one cardiologist.

The ECG report was recorded as an unstructured free text. To recognize ECG abnormalities among these million reports, a computational linguistics program was used. First, the cardiologist's text was preprocessed by removing "stop-words" (such as: the, is, at, which and on) and generating n-grams, defined as a contiguous sequence of n items from a given sample of text or speech. Then, we used a self-supervised learning classification model based on artificial intelligence, using a recurrent neural network as a classifier [25,26], which was built with a 2800-sample dictionary manually created by specialists based on text from real diagnoses. The final report with the ECG abnormalities was obtained by imputing the classifier results for recognition of each ECG abnormality. The classification model was tested on 4557 medical reports manually labeled by two cardiologists with 80.7% positive predictive value, 94.3% sensitivity and 87.0% F1 score for AF; 86.1% positive predictive value, 95.4% sensitivity and 90.9% F1 score for RBBB; 91.4% positive predictive value, 86.0% sensitivity and 88.6% F1 score for LBBB; 75.6% positive predictive value, 93.5% sensitivity and 83.6% F1 score for AVB, and 96.7% positive predictive value, 96.7% sensitivity and 96.7% F1 score for ventricular pre-excitation [27]. F1 score is a measure of the model's accuracy and it is calculated from the positive predictive value and the sensitivity of the test.

The diagnosis of electrocardiographic abnormalities was accepted, without manual review, when there was agreement in the cardiologist's report with one of the automatic systems (Minnesota or Glasgow). The ECGs in which the abnormality was reported by the cardiologist only or by the two automatic systems were manually reviewed by trained staff (Figure 1). For LBBB and RBBB, 17,903 ECGs were revised, while for AVB 9038, AF 4343 and ventricular pre-excitation 1090 tracings were amended. This represents 1.3% of the total number processed, or 2.4 million ECGs.

2.6. Outcomes

The primary end point was all-cause mortality. All International Classification of Diseases (ICD) codes were considered for all-cause mortality. The secondary end point was cardiovascular mortality, defined by nine groups of cardiovascular disease through ICD coding: rheumatic heart disease (I01–I01.9, I02.0, I05–I09.9), ischemic heart disease (I20–I25.9), cerebrovascular disease (G45–G46.8, I60–I61.9, I62.0, I63–I63.9, I65–I66.9, I67.0–I67.3, I67.5–I67.6, I68.1–I68.2, I69.0–I69.3), hypertensive heart disease (I11), myocarditis (A39.52, B33.2–B33.24, D86.85, I40–I43.9, I51.4–I51.5), atrial fibrillation or flutter (I48), aortic aneurysm (I71), peripheral artery disease (I70.2–I70.7, I73–I73.9) and endocarditis (A39.51, I33–I33.9, I38–I39.9). The secondary outcome was evaluated only for AF, LBBB and RBBB diagnoses.

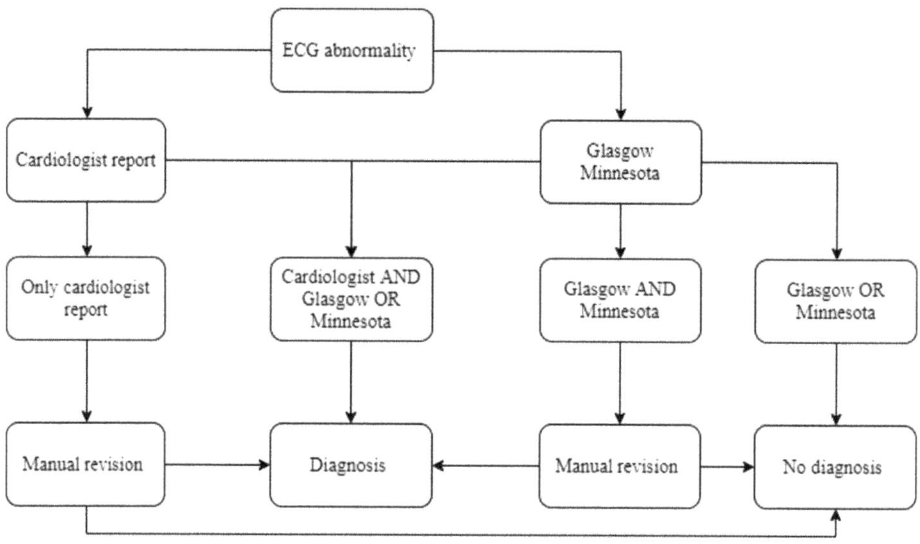

Figure 1. Diagram for ECG abnormality diagnosis. Concordance between the cardiologist's report and one of the automatic systems (Glasgow or Minnesota) was required for a diagnosis to be accepted without manual revision.

2.7. Probabilistic Linkage

The electronic cohort was obtained linking data from the ECG exams (name, sex, date of birth, city of residence) and those from the national mortality information system, using standard probabilistic linkage methods (FRIL: Fine-grained record linkage software, v.2.1.5, Atlanta, GA, USA) [21,28].

2.8. Statistical Analysis

Qualitative variables were described by frequency distribution. Data obtained from continuous quantitative variables were expressed as mean and standard deviation or median with interquartile range.

For the analysis of the electrocardiographic abnormalities, the time elapsed between the date of the electrocardiogram (index event) and the event of interest (date of death) was considered a dependent variable. The presence of the electrocardiographic abnormality was an independent variable, along with the clinical characteristics of the population. The comparison group was patients without major electrocardiographic changes, which included both those with a normal ECG and those with all other abnormalities. Patients who did not present with an event of interest by the end of follow-up were censored, but were included in our analysis with follow-up time until the study's end date (September 2017).

The non-parametric Kaplan–Meier method was used to calculate survival. The level of statistical significance was defined for p values less than 0.05, calculated by the Log rank test. The Cox proportional regression multivariate model was used for all analyses, except for AVB, in which we used the Log-normal model, since the assumptions of the

Cox model could not be achieved. Hazard ratio (HR) with 95% confidence interval was used for the ECG abnormalities analysis, except for the AVB survival analysis, in which relative survival risk (RS) was used. RS under 1 means lower survival rate, while RS over 1 means higher survival rate. Analyses were adjusted for age, sex and comorbidities. The R statistical program (version 3.4.3, Vienna, Austria) was used for all analyses.

3. Results

3.1. CODE Cohort

From 2,470,424 ECGs, 1,773,689 patients were identified. A total of 1,666,778 (94%) underwent a valid ECG recording from 2010 to 2017, with 1,558,421 patients over 16 years old. Most patients were women (60.8%), and mean age was 51.6 (SD ±17.6) years. The overall mortality rate was 3.31% in a mean follow-up of 3.7 years. The clinical conditions of all adult patients and the prevalence of the studied abnormalities are described in Table 1.

Table 1. Prevalence of comorbidities and ECG abnormalities from a total of 1,558,421 patients.

	N	%
Comorbidities		
Hypertension	492,637	31.611
Diabetes	101,470	6.511
Current smoking	108,814	6.982
Dyslipidemia	60,590	3.888
Chagas disease	34,590	2.220
Myocardial infarction	11,604	0.745
COPD	11,266	0.723
ECG abnormalities		
Atrial fibrillation	20,782	1.334
Left bundle branch block	20,226	1.298
Right bundle branch block	37,031	2.376
Ventricular pre-excitation *	1090	0.065
First degree AVB	20,644	1.325
Second degree AVB Mobitz I	212	0.014
Second degree AVB 2:1	61	0.004
Third degree AVB	621	0.040

COPD: chronic obstructive pulmonary disease; AVB: atrioventricular block; * 1,666,778 patients were included in the ventricular pre-excitation analysis.

3.2. Survival Analysis: ECG Abnormalities

All ECG abnormalities, with the exception of ventricular pre-excitation and second degree AVB Mobitz I, were associated with higher mortality for all causes. Patients with AF and LBBB were also at higher risk of cardiovascular mortality (Table 2, Figure 2).

Table 2. Prognostic value of ECG abnormalities, adjusted for age, sex and clinical conditions *.

ECG Abnormalities	Survival Analysis Estimate	Overall Mortality 95% CI	*p*-Value	Cardiovascular Mortality [HR; 95% CI]	*p*-Value
Atrial fibrillation	HR	2.10 [2.03–2.17]	<0.001	2.06 [1.86–2.29]	<0.001
Left bundle branch block	HR	1.69 [1.62–1.76]	<0.001	1.76 [1.55–2.01]	<0.001
Right bundle branch block	HR	1.32 [1.27–1.36]	<0.001	1.12 [0.99–1.28]	0.06
Ventricular pre-excitation	HR	1.41 [0.56–3.57]	0.470	NA	
First degree AVB	RS	0.76 [0.71–0.81]	<0.001	NA	
Second degree AVB Mobitz I	RS	0.65 [0.34–1.24]	0.269	NA	
Second degree AVB 2:1	RS	0.21 [0.09–0.52]	0.005	NA	
Third degree AVB	RS	0.36 [0.26–0.49]	<0.001	NA	

HR: Hazard ratio; RS: relative survival; AVB: atrioventricular block; NA: not available; * age, sex, hypertension, diabetes, current smoking, dyslipidemia, Chagas disease, myocardial infarction and chronic obstructive pulmonary disease.

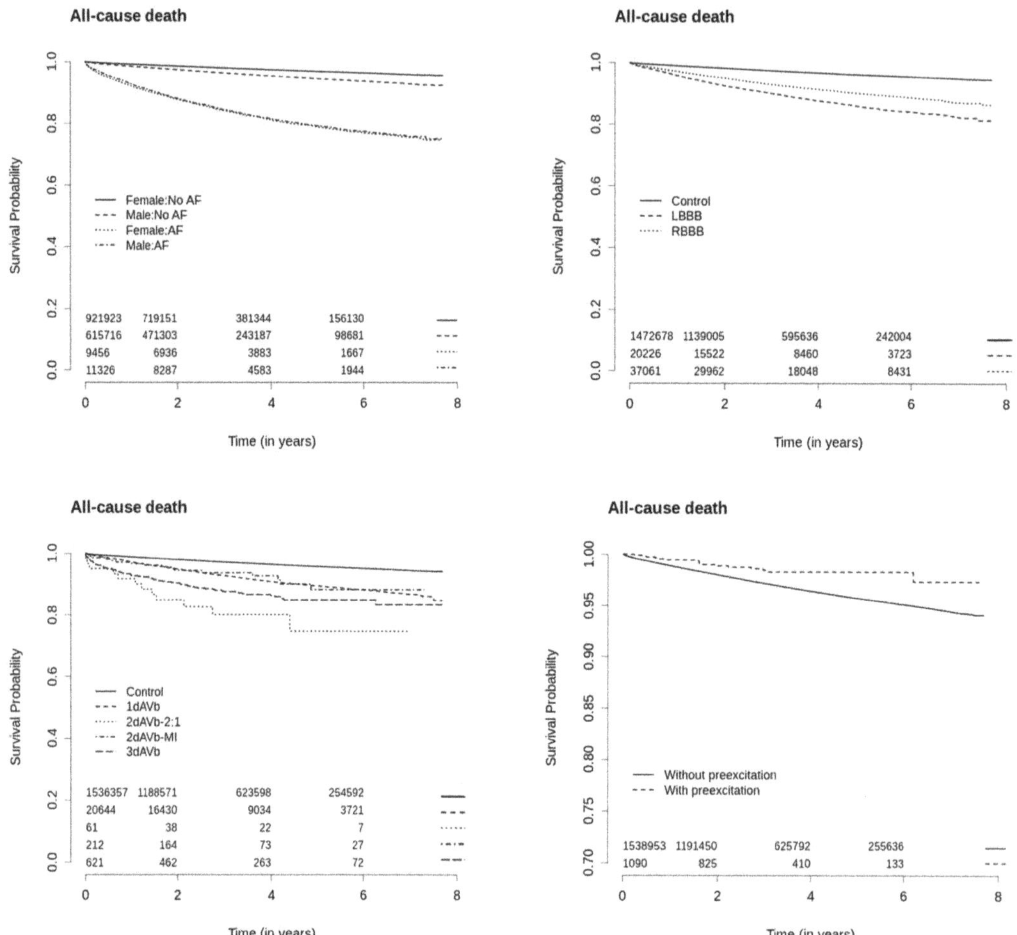

Figure 2. Kaplan–Meier curves for overall mortality.

4. Discussion

The resulting dataset has several potential applications, both for technical and clinical-epidemiological studies. Previous studies from our group showed that ECG abnormalities that are considered important, such as pre-excitation syndrome, have no prognostic impact in a community setting. On the other hand, the risk of dying for a person with RBBB is almost as high as with LBBB, the latter being considered a much stronger marker of risk in general cardiology practice [28–30]. Patients with AF were at a higher risk of mortality compared to the other abnormalities. First degree AVB was a more severe ECG abnormality than Mobitz I, which had a benign prognosis in this population. A 2:1 AVB in the 12-lead ECG was associated with 79% reduction of relative survival, probably indicating an infranodal block.

According to the World Health Organization, primary health care is an integral part of a country's health system, with a main focus on the social and economic development of the community [31]. Its essence is to treat people, not specific diseases and conditions. Actions related to health promotion and both primary and secondary prevention of cardiovascular diseases are necessary to improve collective health. In this context, the search for new features that are capable of predicting individual cardiovascular risk and, therefore, stimulating development of cost-effective preventive actions, is a matter of great importance.

Several tests, such as coronary calcium score, carotid and vertebral echodoppler, and serum measurement of ultrasensitive C-reactive protein have already been recommended for re-stratification of cardiovascular risk [32], although their cost-effectiveness is questionable [32], especially in the context of public health. On the other hand, an inexpensive and widely available exam, such as the ECG, could diagnose abnormalities such as AF, RBBB, LBBB and AVB that imply a higher risk for mortality regardless of age, sex or previous comorbidities.

Stratification of cardiovascular risk by ECG could be a potentially useful tool for clinical practice, especially in primary health care. Identifying the patient who will benefit most from tighter control of blood pressure, diabetes, and cholesterol levels may prevent cardiac events in the future. Electrocardiographic abnormalities draw attention to the potential severity of the patient's condition and the importance of more intensive treatment. In addition, they may help to rationalize and prioritize referrals to secondary or tertiary referral centers.

Electronic cohorts with a large amount of data are powerful sources for the development of population based studies, and, therefore, provide more strong evidence to be used in healthcare. Information on ECG parameters or abnormalities from big data sets [33] may have a major impact by distinguishing between benign and potentially life-threatening cardiac conditions. Each population has specific features, such as social, racial and lifestyle characteristics, that have an impact in their health [34]. Chagas disease, for example, is prevalent in Brazil and is associated with major ECG abnormalities [35], while it is very rare in United States and Europe.

AI in healthcare is the future pathway to managing big data from electronic cohorts. The development of machine learning (ML) models for disease prediction and diagnosis is in a state of exponential growth. In electrocardiography, AI algorithms have been extensively studied for both the automatic diagnosis of electrocardiographic alterations [36] as well as for the prediction of cardiovascular events and identification of new cardiovascular risk factors [37]. Estimation of age and sex by electrocardiographic tracing alone has also been demonstrated [38]. Furthermore, the isolated analysis of the 12-lead ECG can predict mortality within one year with good accuracy, even in tracings reported as normal [39]. AI can extract information from the electrocardiogram that is undervalued and/or unrecognized by conventional methods of analysis, adding diagnostic and prognostic value.

The CODE study is now also working with ML techniques. We found good performance of a deep neural network in the recognition of six ECG abnormalities [36]. In the field of prognosis and health promotion, the concept of an electrocardiographic age via AI, compared with the patient's biological age, is promising [40]. This new promising cardiac biomarker can summarize the individual electrocardiographic characteristics simply and intuitively. It has the potential to provide patients with accessible and understandable information about their cardiovascular risk. More of our results will soon be available and should highlight the importance of ECG epidemiological studies with both traditional and AI methods.

Our study has limitations. Data on comorbidities were self-reported, and thus might have been under-reported. The clinical data came from a predetermined questionnaire not tailored for this study. Therefore, some important variables with impacts on the cardiovascular prognosis, such as heart failure, were unavailable and not considered as comorbidities in the multivariate analysis. The AI classifier used for ECG report classification had good accuracy, sensitivity and positive predictive value, but can make errors. In order to minimize this problem, we included the automatic classification of Glasgow and Minnesota in the diagnostic algorithm. Furthermore, manual revision was done in more than 30,000 ECGs to confirm the presence of the ECG abnormality. The quality of the data from the national mortality information system varies according to region within the state of Minas Gerais; therefore, the information from the national mortality system is heterogeneous among the regions of Minas Gerais such that misclassification of the basic cause of death can occur. The probabilistic linkage also has some issues, such as less than perfect sensitivity and the possibility of false pairs. We defined a high cut off point (94 of 100) for true pairs and made manual revisions in doubtful cases.

5. Conclusions

Electrocardiographic markers are predictors of mortality in the TNMG population. AF, LBBB, RBBB and AVB are associated with a higher risk of death from all causes, regardless of age, sex and associated comorbidities. AF and LBBB are independent predictors of higher cardiovascular mortality. Ventricular pre-excitation and Mobitz I second-degree AVB are not associated with higher overall mortality. An electronic cohort with a large amount of ECG data can be a useful prognostic tool and provide a stimulus for future developments in the fields of digital electrocardiography, clinical cardiology and cardiovascular epidemiology.

Author Contributions: Conceptualization, G.M.M.P. and A.L.P.R.; methodology, E.M.L., G.M.M.P. and A.L.P.R.; software, A.H.R., D.M.O., M.H.R., P.R.G.; validation, G.M.M.P., P.R.G., J.S.N.; formal analysis, E.M.L.; resources, A.L.P.R.; data curation, P.R.G.; writing—original draft preparation, G.M.M.P.; writing—review and editing, G.M.M.P., P.W.M. and A.L.P.R.; visualization, P.W.M. and A.L.P.R.; supervision, A.L.P.R.; project administration, A.L.P.R.; funding acquisition, A.L.P.R. All authors have read and agreed to the published version of the manuscript.

Funding: This research received no external funding.

Institutional Review Board Statement: This study complied with all relevant ethical regulations. The CODE Study was approved by the Research Ethics Committee of the Universidade Federal de Minas Gerais, protocol 49368496317.7.0000.5149. Since this is a secondary analysis of anonymized data stored in the TNMG, informed consent was not required by the Research Ethics Committee for the present study. All researchers who deal with datasets signed terms of confidentiality and data utilization.

Informed Consent Statement: Not applicable.

Data Availability Statement: Researchers affiliated with educational or research institutions can make requests to access the datasets. Requests should be made to the corresponding author of this paper. They will be forwarded and considered on an individual basis by the Telehealth Network of Minas Gerais. The estimated time needed for data access requests to be evaluated is three months. If approved, any data use will be restricted to non-commercial research purposes. The data will only be made available on the execution of appropriate data use agreements.

Acknowledgments: This research was partly supported by the Brazilian Agencies CNPq, CAPES, and FAPEMIG, and also by project IATS. A.H.R. and A.L.P.R. are recipients of unrestricted research scholarships from CNPq; E.M.L. and A.H.R. received scholarships from CAPES and CNPq; and D.M.O., M.H.R. and A.L.P.R. received a Google Latin America Research Award scholarship.

Conflicts of Interest: None of the authors has financial or non-financial competing interests.

References

1. Brant, L.C.C.; Nascimento, B.R.; Passos, V.M.A.; Duncan, B.B.; Bensenõr, I.J.M.; Malta, D.C.; Souza, M.F.M.; Ishitani, L.H.; França, E.; Oliveira, M.S.; et al. Variações e diferenciais da mortalidade por doença cardiovascular no Brasil e em seus estados, em 1990 e 2015: Estimativas do Estudo Carga Global de Doença. *Rev. Bras. Epidemiol.* **2017**, *20* (Suppl. 1), 116–128. [CrossRef]
2. Dawber, T.R.; Kannel, W.B.; Love, D.E.; Streeper, R.B. The electrocardiogram in heart disease detection; a comparison of the multiple and single lead procedures. *Circulation* **1952**, *5*, 559–566. [CrossRef]
3. Blackburn, H.; Keys, A.; Simonson, E.; Rautaharju, P.; Punsar, S. The electrocardiogram in population studies. A classification system. *Circulation* **1960**, *5*, 559–566.
4. Marty, A.T. Minnesota Code Manual of Electrocardiographic Findings. *Crit. Care Med.* **1983**, *11*, 583. [CrossRef]
5. Blackburn, H.; Taylor, H.L.; Keys, A. Coronary heart disease in seven countries. XVI. The electrocardiogram in prediction of five-year coronary heart disease incidence among men aged forty through fifty-nine. *Circulation* **1970**, *41* (Suppl. 4), 154–161. [CrossRef] [PubMed]

6. Kannel, W.B.; Gordon, T.; Castelli, W.P.; Margolis, J.R. Electrocardiographic left ventricular hypertrophy and risk of coronary heart disease. The Framingham study. *Ann. Intern. Med.* **1970**, *72*, 813–822. [CrossRef] [PubMed]
7. Pooling Project ResearchGroup. Relationship of blood pressure, serum cholesterol, smoking habit, relative weight and ECG abnormalities to incidence of major coronary events: Final report of the pooling project. *J. Chronic Dis.* **1978**, *31*, 201–306. [CrossRef]
8. Zhang, Z.M.; Prineas, R.J.; Soliman, E.Z.; Baggett, C.; Heiss, G. Prognostic significance of serial Q/ST-T changes by the Minnesota Code and Novacode in the Atherosclerosis Risk in Communities [ARIC] study. *Eur. J. Prev. Cardiol.* **2012**, *19*, 1430–1436. [CrossRef]
9. Zhang, Z.; Prineas, R.J.; Eaton, C.B. Evaluation and Comparison of the Minnesota Code and Novacode for Electrocardiographic Q-ST Wave Abnormalities for the Independent Prediction of Incident Coronary Heart Disease and Total Mortality [from the Women's Health Initiative]. *Am. J. Cardiol.* **2010**, *106*, 18–25. [CrossRef]
10. Machado, D.B.; Crow, R.S.; Boland, L.L.; Hannan, P.J.; Taylor, H.A.; Folsom, A.R. Electrocardiographic Findings and Incident Coronary Heart Disease Among Participants in the Atherosclerosis Risk in Communities [ARIC] Study. *Am. J. Cardiol.* **2006**, *97*, 1176–1181. [CrossRef]
11. Greenland, P.; Xie, X.; Liu, K.; Colangelo, L.; Liao, Y.; Daviglus, M.L.; Agulnek, A.N.; Stamler, J. Impact of minor electrocardiographic ST-segment and/or T-wave abnormalities on cardiovascular mortality during long-term follow-up. *Am. J. Cardiol.* **2003**, *91*, 1068–1074. [CrossRef]
12. Savage, D.D.; Rautaharju, P.M.; Bailey, J.J.; Horton, M.R.; Hadden, W.; Lacroix, A.Z.; Haynes, S.G.; Wolf, H.K.; Prineas, R.J. The emerging prominence of computer electrocardiography in large population-based surveys. *J. Electrocardiol.* **1987**, *20*, 48–52.
13. Ribeiro, A.L.; da Cunha Pereira, S.V.; Bergmann, K.; Ladeira, R.M.; Oliveira, R.A.M.; Lotufo, P.A.; Mill, J.G.; Barreto, S.M. Challenges to implementation of the ECG reading center in ELSA-Brazil. *Rev. Saude Publica* **2013**, *47* (Suppl. 2), 87–94. [CrossRef] [PubMed]
14. Van der Ende, M.Y.; Siland, J.E.; Snieder, H.; Van der Harst, P.; Rienstra, M. Population-based values and abnormalities of the electrocardiogram in the general Dutch population: The LifeLines Cohort Study. *Clin. Cardiol.* **2017**, *40*, 865–872. [CrossRef] [PubMed]
15. Palhares, D.M.F.; Marcolino, M.S.; Santos, T.M.M.; da Silva, J.L.P.; Gomes, P.R.; Ribeiro, L.B.; Macfarlane, P.W.; Ribeiro, A.L.P. Normal limits of the electrocardiogram derived from a large database of Brazilian primary care patients. *BMC Cardiovasc. Disord.* **2017**, *17*, 152. [CrossRef] [PubMed]
16. Goldman, A.; Hod, H.; Chetrit, A.; Dankner, R. Incidental abnormal ECG findings and long-term cardiovascular morbidity and all-cause mortality: A population based prospective study. *Int. J. Cardiol.* **2019**, *295*, 36–41. [CrossRef]
17. Strauss, D.G.; Mewton, N.; Verrier, R.L.; Nearing, B.D.; Marchlinski, F.E.; Killian, T.; Moxley, J.; Tereshchenko, L.G.; Wu, K.C.; Winslow, R.; et al. Screening entire health system ecg databases to identify patients at increased risk of death. *Circ. Arrhythmia Electrophysiol.* **2013**, *6*, 1156–1162. [CrossRef]
18. Curry, S.J.; Krist, A.H.; Owens, D.K.; Barry, M.J.; Caughey, A.B.; Davidson, K.W.; Doubeni, C.A.; Epling, J.W., Jr.; Kemper, A.R.; Kubik, M.; et al. Screening for cardiovascular disease risk with electrocardiography us preventive services task force recommendation statement. *JAMA J. Am. Med. Assoc.* **2018**, *319*, 2308–2314. [CrossRef] [PubMed]
19. Goff, D.C., Jr.; Lloyd-Jones, D.M.; Bennett, G.; Coady, S.; D'Agostino, R.B.; Gibbons, R.; Greenland, P.; Lackland, D.T.; Levy, D.; O'Donnell, C.J.; et al. American College of Cardiology/American Heart Association Task Force on Practice Guidelines. 2013 ACC/AHA guideline on the assessment of cardiovascular risk: A report of the American College of Cardiology/American Heart Association Task Force on Practice Guidelines. *Circulation* **2014**, *129* (Suppl. 2), 49–73. [CrossRef]
20. Shah, A.J.; Vaccarino, V.; Janssens, A.C.J.W.; Flanders, W.D.; Kundu, S.; Veledar, E.; Wilson, P.W.F.; Soliman, E.Z. An electrocardiogram-based risk equation for incident cardiovascular disease from the National Health and Nutrition Examination Survey. *JAMA Cardiol.* **2016**, *1*, 779–786. [CrossRef]
21. Ribeiro, A.L.P.; Paixão, G.M.M.; Gomes, P.R.; Ribeiro, M.H.; Ribeiro, A.H.; Canazart, J.A.; Oliveira, D.M.; Ferreira, M.P.; Lima, E.M.; Moraes, J.L.; et al. Tele-electrocardiography and bigdata: The CODE (Clinical Outcomes in Digital Electrocardiography) study. *J. Electrocardiol.* **2019**, *57*, 75–78. [CrossRef] [PubMed]
22. Alkmim, M.B.; Figueira, R.M.; Marcolino, M.S.; Cardoso, C.D.; Abreu, M.P.; Cunha, L.; Cunha, D.F.; Antunes, A.P.; Resende, A.G.A.; Resende, E.S.; et al. Improving patient access to specialized health care: The Telehealth Network of Minas Gerais, Brazil. *Bull. World Health Organ.* **2012**, *90*, 373–378. [CrossRef] [PubMed]
23. Alkmim, M.B.; Silva, C.B.G.; Figueira, R.M.; Santos, D.V.V.; Ribeiro, L.B.; Da Paixão, M.C.; Marcolino, M.S.; Paiva, J.C.; Ribeiro, A.L.P. Brazilian national service of telediagnosis in electrocardiography. *Stud. Health Technol. Inform.* **2019**, *264*, 1635–1636. [CrossRef] [PubMed]
24. Kligfield, P.; Gettes, L.S.; Bailey, J.J.; Childers, R.; Deal, B.J.; Hancock, E.W.; van Herpen, G.; Kors, J.A.; Macfarlane, P.W.; Mirvis, D.M.; et al. Recommendations for the Standardization and Interpretation of the Electrocardiogram. *Circulation* **2007**, *115*, 1306–1324. [CrossRef]
25. Hochreiter, S.; Schmidhuber, J. Long Short-term Memory. *Neural Comput.* **1997**, *9*, 1735–1780. [CrossRef]
26. Mikolov, T.; Karafiát, M.; Burget, L.; Cernocký, J.; Khudanpur, S. Recurrent neural network based language model. In Proceedings of the 11th Annual Conference of the International Speech Communication Association, Interspeech, Chiba, Japan, 26–30 September 2010; Volume 2, pp. 1045–1048.

27. Pedrosa, J.A.O.; Oliveira, D.; Meira, W., Jr.; Ribeiro, A. Automated classification of cardiology diagnoses based on textual medical reports. In Proceedings of the Symposium On Knowledge Discovery, Mining And Learning (Kdmile), Porto Alegre, Brazil, 20 October 2020; Volume 8, pp. 185–192. [CrossRef]
28. Paixão, G.M.M.; Silva, L.G.S.; Gomes, P.R.; Lima, E.M.; Ferreira, M.P.F.; Oliveira, D.M.; Ribeiro, M.H.; Ribeiro, A.H.; Nascimento, J.N.; Canazart, J.A.; et al. Evaluation of Mortality in Atrial Fibrillation: Clinical Outcomes in Digital Electrocardiography [CODE] Study. *Glob. Heart* **2020**, *15*, 48. [CrossRef]
29. Paixão, G.M.M.; Lima, E.M.; Gomes, P.R.; Ferreira, M.P.F.; Oliveira, D.M.; Ribeiro, M.H.; Ribeiro, A.H.; Nascimento, J.N.; Canazart, J.A.; Ribeiro, L.B.; et al. Evaluation of mortality in bundle branch block patients from an electronic cohort: Clinical Outcomes in Digital Electrocardiography [CODE] study. *J. Electrocardiol.* **2019**, *57*, 56–60. [CrossRef]
30. Paixão, G.M.M.; Lima, E.M.; Batista, L.M.; Santos, L.F.; Araujo, S.L.O.; Araujo, R.M.; Oliveira, D.M.; Nascimento, J.N.; Gomes, P.R.; Ribeiro, A.L.P. Ventricular pre-excitation in primary care patients: Evaluation of the risk of mortality. *J. Cardiovasc. Electrophysiol.* **2021**, *32*, 1290–1295. [CrossRef]
31. Faquim, J.P.S.; Guerra, L.D.S.; Carnut, L.; Zilbovicius, C. Atenção Primária à Saúde. *J. Manag. Prim. Health Care* **2021**, *12*, 1–7. [CrossRef]
32. Lin, J.S.; Evans, C.V.; Johnson, E.; Redmond, N.; Coppola, E.L.; Smith, N. Nontraditional risk factors in cardiovascular disease risk assessment: Updated evidence report and systematic review for the US preventive services task force. *JAMA J. Am. Med. Assoc.* **2018**, *320*, 281–297. [CrossRef]
33. Chen, M.; Mao, S.; Liu, Y. Big data: A survey. *Mob. Netw. Appl.* **2014**, *19*, 171–209. [CrossRef]
34. Ferreira, J.P.; Rossignol, P.; Dewan, P.; Lamiral, Z.; White, W.B.; Pitt, B.; Mc Murray, J.J.V.; Zannad, F. Income level and inequality as complement to geographical differences in cardiovascular trials. *Am. Heart J.* **2019**, *218*, 66–74. [CrossRef] [PubMed]
35. Marcolino, M.S.; Palhares, D.M.; Ferreira, L.R.; Ribeiro, A.L. Electrocardiogram and Chagas Disease A Large Population Database of Primary Care Patients. *Glob. Heart* **2015**, *10*, 167–172. [CrossRef] [PubMed]
36. Ribeiro, A.H.; Ribeiro, M.H.; Paixão, G.M.M.; Oliveira, D.M.; Gomes, P.R.; Canazart, J.A.; Ferreira, M.P.S.; Andersson, C.R.; Macfarlane, P.W.; Meira, W., Jr. Automatic diagnosis of the 12-lead ECG using a deep neural network. *Nat. Commun.* **2020**, *11*, 1760. [CrossRef] [PubMed]
37. Attia, Z.I.; Noseworthy, P.A.; Lopez-Jimenez, F.; Asirvatham, S.J.; Deshmukh, A.J.; Gersh, B.J.; Carter, R.E.; Yao, X.; Rabinstein, A.A.; Erickson, B.J.; et al. An artificial intelligence-enabled ECG algorithm for the identification of patients with atrial fibrillation during sinus rhythm: A retrospective analysis of outcome prediction. *Lancet* **2019**, *394*, 861–867. [CrossRef]
38. Attia, Z.I.; Friedman, P.A.; Noseworthy, P.A.; Lopez-Jimenez, F.; Ladewig, D.J.; Satam, G.; Pellikka, P.A.; Munger, T.M.; Asirvatham, S.J.; Scott, C.G.; et al. Age and Sex Estimation Using Artificial Intelligence From Standard 12-Lead ECGs. *Circ. Arrhythm Electrophysiol.* **2019**, *12*, e007284. [CrossRef]
39. Raghunath, S.; Ulloa Cerna, A.E.; Jing, L.; Van Maanen, D.P.; Stough, J.; Hartzel, D.N.; Nemani, A.; Carbonati, T.; Johnson, K.W.; Young, K.; et al. Prediction of mortality from 12-lead electrocardiogram voltage data using a deep neural network. *Nat. Med.* **2020**, *26*, 886–891. [CrossRef]
40. Lima, E.M.; Ribeiro, A.H.; Paixão, G.M.M.; Ribeiro, M.H.; Pinto-Filho, M.M.; Gomes, P.R.; Oliveira, D.M.; Sabino, E.C.; Duncan, B.B.; Giatti, L.; et al. Deep neural network-estimated electrocardiographic age as a mortality predictor. *Nat. Commun.* **2021**, *12*, 5117. [CrossRef] [PubMed]

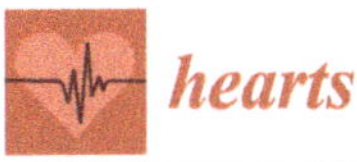

Review

ECG Interpretation: Clinical Relevance, Challenges, and Advances

Nikita Rafie [1], Anthony H. Kashou [2,*] and Peter A. Noseworthy [2]

1   Department of Medicine, Mayo Clinic, 200 First Street SW, Rochester, MN 55905, USA; rafie.nikita@mayo.edu
2   Division of Cardiovascular Diseases, Mayo Clinic, 200 First Street SW, Rochester, MN 55905, USA; noseworthy.peter@mayo.edu
*   Correspondence: kashou.anthony@mayo.edu

Abstract: Since its inception, the electrocardiogram (ECG) has been an essential tool in medicine. The ECG is more than a mere tracing of cardiac electrical activity; it can detect and diagnose various pathologies including arrhythmias, pericardial and myocardial disease, electrolyte disturbances, and pulmonary disease. The ECG is a simple, non-invasive, rapid, and cost-effective diagnostic tool in medicine; however, its clinical utility relies on the accuracy of its interpretation. Computer ECG analysis has become so widespread and relied upon that ECG literacy among clinicians is waning. With recent technological advances, the application of artificial intelligence-augmented ECG (AI-ECG) algorithms has demonstrated the potential to risk stratify, diagnose, and even interpret ECGs—all of which can have a tremendous impact on patient care and clinical workflow. In this review, we examine (i) the utility and importance of the ECG in clinical practice, (ii) the accuracy and limitations of current ECG interpretation methods, (iii) existing challenges in ECG education, and (iv) the potential use of AI-ECG algorithms for comprehensive ECG interpretation.

Keywords: electrocardiogram; ECG interpretation; artificial intelligence; machine learning

Citation: Rafie, N.; Kashou, A.N.; Noseworthy, P.A. ECG Interpretation: Clinical Relevance, Challenges, and Advances. *Hearts* **2021**, *2*, 505–513. https://doi.org/10.3390/hearts2040039

Academic Editor: Peter Macfarlane

Received: 21 September 2021
Accepted: 20 October 2021
Published: 2 November 2021

Publisher's Note: MDPI stays neutral with regard to jurisdictional claims in published maps and institutional affiliations.

1. Introduction

Since its development over a century ago, the ECG remains the cornerstone of cardiovascular screening, evaluation, and diagnosis. The ECG is one of the most widely used tools in medicine today—nearly 200 million ECGs are recorded annually worldwide [1]. While the ECG tracing itself has remained relatively unchanged since its inception, our ability to leverage the "humble" ECG to detect and diagnose various pathologies continues to evolve [2]. Its use is imperative in the evaluation and management of an array of cardiovascular diseases, including arrhythmias, pericardial and myocardial disease, as well as many non-cardiac conditions including electrolyte disturbances and pulmonary disease among many others. Advances in cardiology have been accelerated by technology, and the ECG is at the center of this dependency [3].

The ECG has enabled countless advances in the field of cardiology, bearing witness to the genesis of entire subspecialties. For instant, prior to the development of the ECG, arrhythmias were poorly understood but the ECG is now central to the burgeoning field of cardiac electrophysiology [2]. Similarly, the ECG facilitated monumental strides in the recognition, management, and treatment of acute myocardial infarction (MI) [2]. Our reliance on the ECG to identify life-threatening arrhythmias and acute MIs has become so widespread and critically important that it has become part of the core curriculum for most medical trainees.

Over the last few decades, computing power and digitized data availability have made exponential gains. This has led to the application of artificial intelligence (AI) to the ECG. In recent years, the use of AI-enabled ECG (AI-ECG) algorithms for various risk stratification, diagnostic evaluation, and clinical interpretation have emerged. Researchers have shown some algorithms to be capable of rhythm identification [4] and even perform comprehensive 12-lead ECG interpretation [5]. In fact, the accuracy appears to be better than that of the currently implemented ECG software [5].

In this review, we examine (i) the utility and importance of the ECG in clinical practice, (ii) the accuracy and limitations of current ECG interpretation methods, (iii) existing challenges in ECG education, and (iv) the potential use of AI-ECG algorithms for comprehensive ECG interpretation.

2. ECG and Its Clinical Utility

The expansive application of the ECG has been well established since its invention and continues to evolve over a century later. Its clinical utility includes, but is not limited to, the detection of acute and chronic myocardial injury, cardiac arrhythmias, structural heart disease, and inflammatory processes (e.g., pericarditis) [3,6–8]. In addition, its use spans throughout nearly all specialties (e.g., cardiology, emergency medicine, internal medicine) and clinical settings (e.g., emergency department, primary care clinic, intensive care unit) in medicine. The ECG is also used to assess whether a medication is safe to prescribe, monitor the effects of a drug on the heart, predict the risk of life-threatening arrhythmias, evaluate for coronary artery disease, and even determine if implantation of a device would improve cardiac function and quality of life.

One of the first clinical applications of the ECG that directly impacted patient care was its ability to identify acute myocardial injury [2]. This helped clinicians to also differentiate cardiac chest pain from non-cardiac chest pain mimickers. Studies comparing various ECG patterns in humans and animals following myocardial infarction led to recognition of ST-segment elevation patterns in patients with myocardial infarction [2]. The reproducible myocardial injury patterns have enabled computer program algorithms to identify ST-segment elevation myocardial infarctions (STEMIs). Furthermore, the concept of waveform changes associated with MI and myocardial injury have advanced the utility of the ECG to detect demand ischemia (e.g., exercise-induced ischemia before an episode of acute myocardial infarction) [3]. With this advancement, coronary artery disease and its impact can be detected before a potentially fatal event, leading to improvement in cardiac morbidity and overall mortality.

Arrhythmia detection is one of the most common clinical uses of the ECG. The identification of atrial fibrillation on the ECG remains a key diagnostic component that alters patient management [2]. The ECG also aids in differentiating supraventricular and ventricular arrhythmias [9]. In addition to rhythm detection, the ECG can localize the origin of a rhythm in many instances [3]. This information can aid in patient care (e.g., accessory pathway ablation of an accessory pathway in Wolff–Parkinson–White syndrome).

The ECG can detect structural changes within the heart (e.g., atrial enlargement and ventricular hypertrophy). The identification of left atrial enlargement may help identify patients at risk for developing atrial fibrillation, whereas the presence of left ventricular hypertrophy may identify patients at greater risk of cardiovascular disease from systemic hypertension. In fact, the changes in electrical conduction in a hypertrophied myocardium can help prognosticate cardiovascular disease in patients with hypertension [6]. The diagnostic ECG criteria for ventricular hypertrophy are rather straightforward, which has made it a programmable feature for computer interpretation algorithms to identify. While many ECG criteria for structural abnormalities are rather nonspecific, such findings may direct a clinician towards more aggressive treatment of chronic comorbidities (e.g., hypertension treatment if left ventricular hypertrophy is evident) or further diagnostic evaluation (e.g., echocardiogram).

The ECG can also aid in the diagnosis, recurrence, and resolution of inflammatory conditions, namely, pericarditis. Early and accurate diagnosis and initiation of treatment of pericarditis is imperative in decreasing recurrence and demand of resources from complications [8]. Following the diagnosis and initiation of treatment, the dynamic ECG changes can aid in monitoring for resolution [8].

Despite advances in imaging modalities, the ECG remains one of the most rapid and non-invasive tools to aid in the diagnosis of pericardial effusions and cardiac tamponade. The sensitivity of ECG in detecting cardiac tamponade remains quite low despite advances in interpretation so it cannot be used to rule out cardiac tamponade clinically. However, the high specificity of these ECG changes makes the ECG, in conjunction with clinical correlation, a key tool in ruling in cardiac tamponade [7]. As with all other cardiovascular diseases, the ECG can be used to augment clinical suspicion without requiring advanced imaging or invasive procedures, again emphasizing the importance of diagnosis for rapid and accurate diagnosis and treatment.

The simple, non-invasive, rapid, and cost-effective nature of obtaining an ECG allows for its widespread use and powerful diagnostic abilities in medicine. However, the story behind its interpretation is not the same as its simple and routine acquisition.

3. The Advent and Consequences of Computer-Aided ECG Interpretation

Major technological, imaging, and procedural advances have flourished across multiple disciplines in medicine over the last century, and the interpretation of the ECG is no exception. Over 50 years ago, computer analysis of the ECG was introduced to extract, analyze, and interpret ECGs [4]. The overarching goals were to improve interpretation accuracy, efficiency, and clinical workflow. What once required knowledge and detailed analysis was now being replaced by computer software programs. The computer processing and analyzing of ECGs has become a mainstay in many clinical settings even to this day. However, this is certainly not without consequences.

Over the years, medical providers have become more reliant on the computerized ECG interpretation for aiding in clinical decision making. Unfortunately, this is an issue given the notorious inaccuracies associated with such ECG interpretation software. Perhaps an inadvertent byproduct of such reliance is less focus around ECG education across medical training programs. This has merely reduced ECG competency to a small minority of skilled providers such that the skill of ECG interpretation has become a lost art.

Countless studies have demonstrated significant inaccuracies and limitations of computer ECG interpretation software [10]. This is a major problem given the reliance of medical providers on the software and the subsequent direct impact it can have on patient care. Despite attempts to improve the accuracy of computerized ECG interpretation algorithms, the final ECG interpretation continues to rely on a physician's over-read [10–13]. In addition, even with efforts to standardize ECG interpretation in the United States and Europe, there has yet to be an all-encompassing internationally accepted standard for computerized ECG interpretation [14].

4. ECG Interpretation Accuracy and Limitations

Although the ECG is central to practice, some have argued that expert ECG interpretation may be a dying art. Stewart Hart and Calvin Smith warned of the detail and precision that is required to analyze and interpret ECGs, and that this invaluable skill must be "practised regularly, systematically, and faithfully" [2].

Medical trainees are expected to learn a vast array of subjects and acquire countless clinical skills in the few short years that are dedicated to formal medical education. Only a fraction of this time is dedicated to formal ECG education and, unfortunately, opportunities for formal ECG training are lacking once these students enter the professional workforce [15]. In addition to the lack of formal ECG education available, there is an everlasting challenge for medical educators. Educators must develop a curriculum that is not only broad enough to capture the many complexities of ECG analysis, but also specific enough to provide learners with the details needed to be deemed competent in their evaluation [15]. A formal curriculum on ECG interpretation once trainees have entered the professional workforce also faces the constant challenge of accommodating the demanding resident schedule.

These obstacles, in addition to countless others, have hindered formal ECG education and have led to an ever-growing population of healthcare professionals who do not feel confident or are not competent in their ECG analysis abilities. For these reasons, it does not come as a surprise that many have come to rely on the computerized ECG interpretation rather than their own skills for ECG analysis. The solution to reviving the lost art of ECG interpretation must begin at the most basic level of education in early pre-clinical years of medical school and gradually build on this foundation until competency is achieved.

Although the computerized ECG interpretation algorithms are regularly improved, their diagnostic accuracy is limited [14]. Computer ECG analysis has been criticized for many years. In 1991, a systemic assessment of computer-based ECG interpretation showed that the computer program's accuracy was 6.6% lower than the comparative cardiologists [14]. Additionally, the variability in diagnostic performance varied more greatly between computer models than between cardiologists [14].

Computerized ECG interpretation can contribute to diagnostic inaccuracies. The consequence of these inaccuracies is not limited only to simply a delay in diagnosis but may also have detrimental consequence from treatment delay. One of these detrimental delays can be seen in delay in diagnosis of STEMI, resulting in delaying the door to balloon time of coronary artery reperfusion. One study found a high rate of false negative results in computer diagnosis of STEMI advising against computerized ECG interpretation for such use [16]. Given the detrimental consequence of false negative ECGs for STEMI, the need for clinician interpretation of final ECG diagnosis has become common practice in patients presenting with acute coronary syndrome symptom [17].

The necessity of clinician over-read of ECG in the diagnosis is essential and widely accepted among most healthcare institutions. One of the most relied upon features of the computerized ECG analysis is the measurement of intervals in the rhythm strip. The automated measurement of a QT interval is generally longer than the QT interval manually measured [14]. The over estimation of the QT interval may have significant implications on various cardiac and non-cardiac disease states and management of patients. Over estimation of the QT interval may lead to unnecessary dose adjustments or treatment changes due to the concern of QT interval prolongation. This can lead to suboptimal patient care and leave patients without the best treatment options due to the inaccurate interval estimation. Additionally, in families with long QT syndrome, an inaccurately recorded QT interval may lead to misdiagnosis.

The inaccuracies of ECG interpretation are not consistent across all disease states; the computer-based algorithm interpretation varies significantly in accuracy based on the underlying electrocardiographic rhythm. The positive predictive value of computer interpretation of sinus rhythm was 95%, but only 53.5% in non-sinus rhythms and un-interpretable in 2% of ECG tracings in one study of 2112 randomly selected ECGs [14]. Furthermore, the computer algorithm misinterpreted 75% of pacemaker rhythms in older studies, which has led to improvement in pacemaker algorithms [14]. Despite the well-established inaccuracies of computer interpretation of ECGs, this analysis remains trusted by many clinicians and relied on for diagnostic accuracy.

5. AI-ECG Interpretation

Advances in machine learning have enabled a potential avenue to improve comprehensive ECG interpretation. Multiple studies have demonstrated the ability of AI-ECG algorithms to not only perform detection of various arrhythmias from single-lead ECGs [17–19] but also provide comprehensive, human-like 12-lead ECG interpretation [4]. This is exemplified in Figure 1, where the AI-ECG interpretation provides a more thorough ECG interpretation when compared to the traditional computer-generated interpretation. Moreover, the AI-ECG interpretation of the ECG in Figure 1 accurately identifies premature atrial complexes, while the traditional computer-generated interpretation labels the premature atrial complex as sinus arrhythmia and overestimates the QT. Figure 2 also demonstrates the AI-ECG model identifying first-degree AV block that was not labeled by the traditional computer ECG interpretation. When the performance of an AI-ECG algorithm was evaluated by cardiac electrophysiologists and compared to an existing standard automated computer program used in clinical practice, it was noted to outperform it and better approximate expert over-reads than an existing, widely used computerized ECG interpretation software [5].

The AI-ECG algorithm was applied to ECGs with 66 diagnoses codes by Kashou et al. [4]. In this study, the AI-ECG algorithm was able to generate diagnoses codes consistent with the interpretations of cardiologists. The model performed well for a wide range of ECG diagnoses codes including rhythm, conduction, ischemia, and waveform morphology [4]. Kashou et al. went on to emphasize that the abilities of the AI-ECG algorithm will continue to improve its interpretation skills as more high-quality raw data is incorporated into its algorithm [4].

In a similar study, the accuracy of AI-ECG was compared to traditional computer interpretation and final clinician interpretation of 500 ECGs [5]. Expert over-reading cardiologist in this study rated the interpretation of the traditional computer interpretation, AI-ECG interpretation, and clinician interpretation as unacceptable, acceptable, or ideal based on the accuracy of the interpretation. The results of this study showed that 202 (13.5%) of the traditional computer interpreted, 123 (8.2%) of the AI-ECG interpreted, and 90 (6.0%) of the clinician interpreted ECGs were deemed as unacceptable and required edits [5]. Conversely, 958 (63.9%) of the traditional computer interpreted, 1058 (70.5%) of the AI-ECG interpreted, and 1118 (74.5%) of the clinician interpreted ECGs were considered ideal and did not require edits [5]. Analysis on this data demonstrated that AI-ECG algorithms outperformed traditional computerized interpretation. In addition, AI-ECG algorithms were a better approximation of expert cardiologist over-read [5]. This further exemplifies the suggestion that AI-ECG interpretation may serve as an alternative, more accurate ECG interpretation compared to traditional computer algorithms.

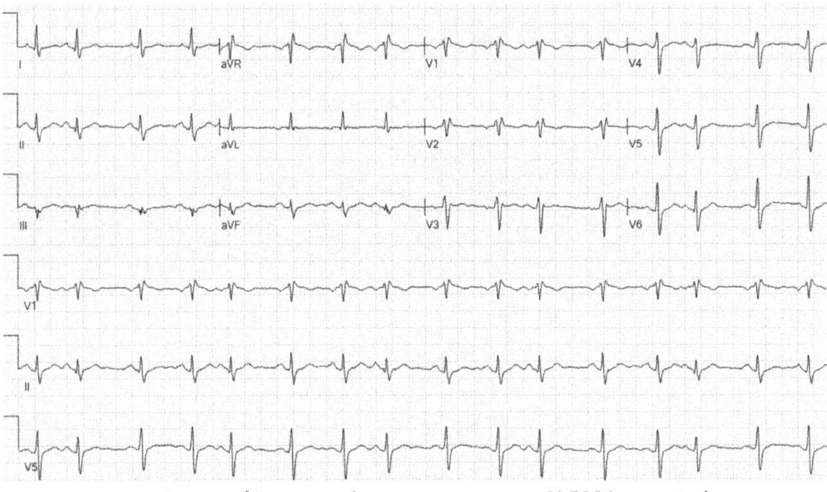

Figure 1. ECG showing traditional computer-generated interpretation versus AI-ECG interpretation. The AI-ECG interpretation provides a more accurate interpretation of the ECG, while the traditional computer-generated interpretation mislabels sinus arrhythmia for premature atrial complexes and overestimates the QT interval.

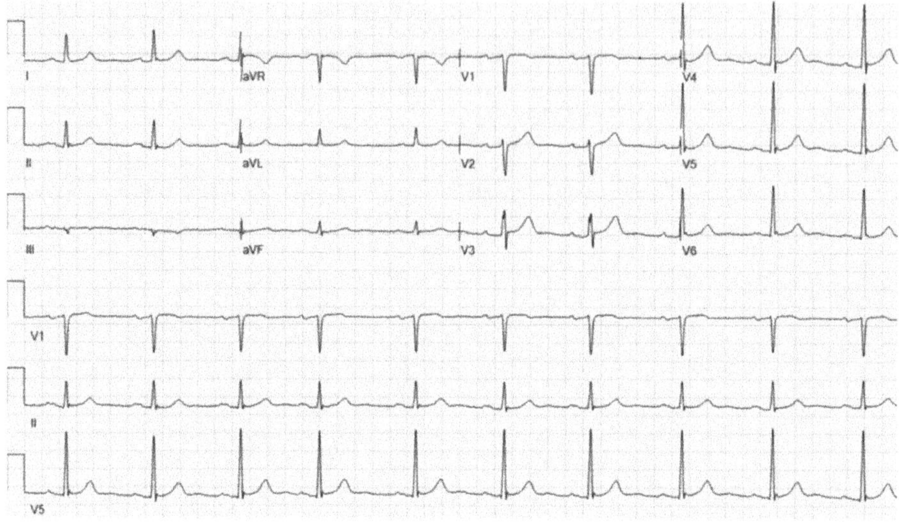

Computer-Generated Interpretation
- Sinus bradycardia
- Otherwise normal ECG

AI-ECG Interpretation
- Sinus bradycardia
- 1st degree AV block

Figure 2. ECG showing traditional computer-generated interpretation versus AI-ECG interpretation. The AI-ECG interpretation provides a more specific and accurate interpretation of the ECG, while the computer-generated interpretation does not identify the first-degree AV block. While the AI-ECG algorithm is able to identify the AV conduction defect, it does not report a PR interval duration like traditional computer-generated algorithms.

It is still unclear how such an AI-ECG algorithm capable of comprehensive 12-lead ECG interpretation would perform against other conventional computer-devised models, if its performance would be preserved in various populations, and how it would be used in the clinical setting. Figure 3 illustrates a potential clinical workflow incorporation of AI-ECG into the clinical workflow including its ability to enhance AI-ECG interpretation as well as incorporate a novel AI-ECG prediction variable to improve the delivery of patient care. Additional research is required to improve AI-ECG model interpretation accuracy, better understand how to seamlessly incorporate it into clinical practice, and to expand access in resource-scarce regions.

The importance of advancing and improving AI-ECG algorithms can be seen and highlighted in institutions where there is a high volume of ECGs recorded daily. This is exemplified at our institution where over 100 trained ECG technologists are always present and reading ECGs with physician oversight. The vast amount of resources that such an ECG lab requires can be mitigated with the incorporation of advancing AI-ECG interpretations. AI-ECG interpretation will allow for the potential to increase overall accuracy and decrease the demand on ECG technologists and physicians.

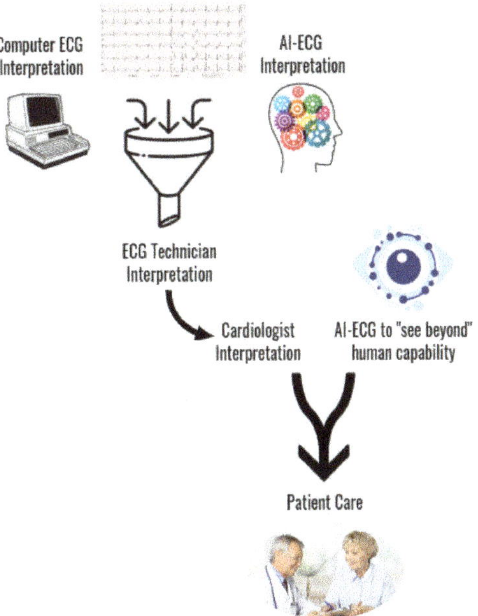

Figure 3. Proposed incorporation of AI-ECG into the clinical workflow.

6. A Look Ahead

With the emergence of mobile cardiac monitoring modalities into the marketplace, the ability to obtain electrocardiographic signals at any moment is now feasible. This instantaneous capture of high-quality cardiac signals can help personalize, expedite, and optimize patient care. Patients experiencing intermittent palpitations can record their heart's rhythm at the onset of symptoms on their own personal device without having to consult a medical provider to obtain a remote monitoring device. These signals could then be displayed at an in-person appointment or transmitted to a medical provider to help better understand if an underlying arrhythmia is contributing to the patient's symptoms and whether further investigation and/or a change in management is warranted. This will not only save the patient and healthcare institution time and finances but also be used to capture rare and intermittent symptoms that may not necessarily be captured over the standard 24 to 48 h of remote monitoring.

Despite these advances in data collection, there remain barriers as to how to ideally transmit, interpret, and use this ECG data clinically. With the ever-growing list of devices, a user-friendly means for patients to transmit the data to a provider is needed. Such a platform that ingests the ECG data must be able to accept various devices and be able to scale as new innovative technologies come to market. Additionally, quality controls must be in place to ensure the data are acceptable for clinical use. Lastly, and perhaps a future daunting some clinicians, is the foreseeable burden of data and increasing patient expectations. Actions must be in place to help alleviate the burden, compensate their work, and allow innovation to benefit all parties.

7. Conclusions

The performance and interpretation of the ECG is vital to the practice of medicine. In fact, this century-old diagnostic tool is experiencing a renaissance as novel technologies and potential clinical utility come to light. With the ever-growing utility of the various cardiac devices becoming available, it is more evident than ever that the ECG literacy is an essential skill. The commonly accepted conventional computerized ECG interpretation algorithms have many pitfalls and require physician oversight. As AI-ECG algorithms continue to improve their diagnostic accuracy, there is the possibility to improve ECG interpretation accuracy, improve clinical workflow, and better serve under-resourced areas. While advances and refinement in AI-ECG algorithms may help minimize ECG interpretation inaccuracies, ongoing efforts to improve ECG education for all medical providers will remain essential to provide high-quality patient care.

Author Contributions: Conceptualization, A.H.K. and P.A.N.; writing—original draft preparation, N.R. and A.H.K.; writing—review and editing, A.H.K. and P.A.N.; visualization, A.H.K.; supervision, A.H.K. and P.A.N. All authors have read and agreed to the published version of the manuscript.

Funding: This work was supported by the Department of Cardiovascular Medicine at Mayo Clinic in Rochester, MN. The authors also acknowledge support by NIH T32 HL007111.

Institutional Review Board Statement: Not applicable.

Informed Consent Statement: Not applicable.

Conflicts of Interest: A.H.K. and P.A.N. are potential beneficiaries of intellectual property discussed in this article. The remaining author declares no conflicts of interest.

References

1. Reichlin, T.; Abächerli, R.; Twerenbold, R.; Kühne, M.; Schaer, B.; Müller, C.; Sticherling, C.; Osswald, S. Advanced ECG in 2016: Is there more than just a tracing? *Swiss. Med. Wkly.* **2016**, *146*, w14303. [CrossRef] [PubMed]
2. Lüderitz, B.; de Luna, A.B. The history of electrocardiography. *J. Electrocardiol.* **2017**, *50*, 539. [CrossRef] [PubMed]
3. Fye, W.B. A history of the origin, evolution, and impact of electrocardiography. *Am. J. Cardiol.* **1994**, *73*, 937–949. [CrossRef]
4. Kashou, A.H.; Ko, W.-Y.; Attia, Z.I.; Cohen, M.S.; Friedman, P.A.; Noseworthy, P.A. A comprehensive artificial intelligence–enabled electrocardiogram interpretation program. *Cardiovasc. Digit. Health J.* **2020**, *1*, 62–70. [CrossRef]
5. Kashou, A.H.; Mulpuru, S.K.; Deshmukh, A.J.; Ko, W.-Y.; Attia, Z.I.; Carter, R.E.; Friedman, P.A.; Noseworthy, P.A. An artificial intelligence–enabled ECG algorithm for comprehensive ECG interpretation: Can it pass the 'Turing test'? *Cardiovasc. Digit. Health J.* **2021**, *2*, 164–170. [CrossRef]
6. Gosse, P.; Jan, E.; Coulon, P.; Cremer, A.; Papaioannou, G.; Yeim, S. ECG detection of left ventricular hypertrophy: The simpler, the better? *J. Hypertens.* **2012**, *30*, 990–996. [CrossRef] [PubMed]
7. Ang, K.P.; Nordin, R.B.; Lee, S.C.Y.; Lee, C.Y.; Lu, H.T. Diagnostic value of electrocardiogram in cardiac tamponade. *Med. J. Malaysia* **2019**, *74*, 51–56. [PubMed]
8. McNamara, N.; Ibrahim, A.; Satti, Z.; Ibrahim, M.; Kiernan, T.J. Acute pericarditis: A review of current diagnostic and management guidelines. *Future Cardiol.* **2019**, *15*, 119–126. [CrossRef] [PubMed]
9. Clementy, N.; Fourquet, A.; Andre, C.; Bisson, A.; Pierre, B.; Fauchier, L.; Babuty, D.; Angoulvant, D. Benefits of an early management of palpitations. *Medicine* **2018**, *97*, e11466. [CrossRef] [PubMed]
10. Martínez-Losas, P.; Higueras, J.; Gómez-Polo, J.C.; Brabyn, P.; Ferrer, J.M.; Cañadas, V.; Villacastín, J.P. The influence of computerized interpretation of an electrocardiogram reading. *Am. J. Emerg. Med.* **2016**, *34*, 2031–2032. [CrossRef] [PubMed]
11. Novotny, T.; Bond, R.; Andrsova, I.; Koc, L.; Sisakova, M.; Finlay, D.; Guldenring, D.; Spinar, J.; Malik, M. The role of computerized diagnostic proposals in the interpretation of the 12-lead electrocardiogram by cardiology and non-cardiology fellows. *Int. J. Med. Inform.* **2017**, *101*, 85–92. [CrossRef] [PubMed]
12. Bogun, F.; Anh, D.; Kalahasty, G.; Wissner, E.; Bou Serhal, C.; Bazzi, R.; Weaver, W.D.; Schuger, C. Misdiagnosis of atrial fibrillation and its clinical consequences. *Am. J. Med.* **2004**, *117*, 636–642. [CrossRef]
13. Anh, D.; Krishnan, S.; Bogun, F. Accuracy of electrocardiogram interpretation by cardiologists in the setting of incorrect computer analysis. *J. Electrocardiol.* **2006**, *39*, 343–345. [CrossRef] [PubMed]
14. Schläpfer, J.; Wellens, H.J. Computer-Interpreted Electrocardiograms: Benefits and Limitations. *J. Am. Coll. Cardiol.* **2017**, *70*, 1183–1192. [CrossRef]
15. Kashou, A.; May, A.; DeSimone, C.; Noseworthy, P. The essential skill of ECG interpretation: How do we define and improve competency? *Postgrad. Med. J.* **2020**, *96*, 125–127. [CrossRef] [PubMed]
16. Garvey, J.L.; Zegre-Hemsey, J.; Gregg, R.; Studnek, J.R. Electrocardiographic diagnosis of ST segment elevation myocardial infarction: An evaluation of three automated interpretation algorithms. *J. Electrocardiol.* **2016**, *49*, 728–732. [CrossRef] [PubMed]
17. Acharya, U.R.; Fujita, H.; Lih, O.S.; Hagiwara, Y.; Tan, J.H.; Adam, M. Automated detection of arrhythmias using different intervals of tachycardia ECG segments with convolutional neural network. *Inf. Sci.* **2017**, *405*, 81–90. [CrossRef]

18. Xiong, Z.; Stiles, M.K.; Zhao, J. Robust ECG signal classification for detection of atrial fibrillation using a novel neural network. In Proceedings of the 2017 Computing in Cardiology (CinC), Rennes, France, 24–27 September 2017; pp. 1–4.
19. Hannun, A.Y.; Rajpurkar, P.; Haghpanahi, M.; Tison, G.H.; Bourn, C.; Turakhia, M.P.; Ng, A.Y. Cardiologist-level arrhythmia detection and classification in ambulatory electrocardiograms using a deep neural network. *Nat. Med.* **2019**, *25*, 65–69. [CrossRef] [PubMed]

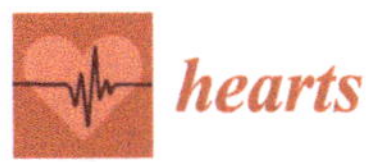

MDPI

Review

Applications of Machine Learning in Ambulatory ECG

Joel Xue [1,*] and **Long Yu** [2]

1 Department of Biomedical Informatics, Emory University, Atlanta, GA 30322, USA
2 Diagnostic Cardiology, GE Healthcare, Wauwatosa, WI 53226, USA; long.yu@ge.com
* Correspondence: joel.q.xue@emory.edu

Simple Summary: The ambulatory ECG (AECG) is an important diagnostic tool for many heart electrophysiology-related cases. This review covers some key Ambulatory ECG applications of Machine Learning algorithms, which include both statistical learning and neural network-based deep learning algorithms.

Abstract: The ambulatory ECG (AECG) is an important diagnostic tool for many heart electrophysiology-related cases. AECG covers a wide spectrum of devices and applications. At the core of these devices and applications are the algorithms responsible for signal conditioning, ECG beat detection and classification, and event detections. Over the years, there has been huge progress for algorithm development and implementation thanks to great efforts by researchers, engineers, and physicians, alongside the rapid development of electronics and signal processing, especially machine learning (ML). The current efforts and progress in machine learning fields are unprecedented, and many of these ML algorithms have also been successfully applied to AECG applications. This review covers some key AECG applications of ML algorithms. However, instead of doing a general review of ML algorithms, we are focusing on the central tasks of AECG and discussing what ML can bring to solve the key challenges AECG is facing. The center tasks of AECG signal processing listed in the review include signal preprocessing, beat detection and classification, event detection, and event prediction. Each AECG device/system might have different portions and forms of those signal components depending on its application and the target, but these are the topics most relevant and of greatest concern to the people working in this area.

Keywords: ambulatory ECG; machine learning; deep learning; pattern recognition; noise reduction; Holter ECG

Citation: Xue, J.; Yu, L. Applications of Machine Learning in Ambulatory ECG. *Hearts* **2021**, *2*, 472–494. https://doi.org/10.3390/hearts2040037

Academic Editor: Peter Macfarlane

Received: 2 August 2021
Accepted: 8 October 2021
Published: 13 October 2021

Publisher's Note: MDPI stays neutral with regard to jurisdictional claims in published maps and institutional affiliations.

1. Introduction

Ambulatory electrocardiograms (AECG) have evolved greatly from the traditional 24–48 h of Holter monitoring devices. Now, AECGs can last as short as 30 s to as long as 30 days. There is a much wider range for lead selection, with AECGs coming in the form of a small patch with a small vector to a full-scale 12-lead electrocardiogram (ECG) with a wide coverage of heart electrical activity. The clinical applications for AECGs have also expanded from limited arrhythmia analysis to morphology analysis of ST level and QT interval and risk stratification/prediction [1]. Figure 1 is a diagram of the scope for AECG devices and algorithms. In the figure, the most noticeable development of AECG in recent years is innovation in miniaturized devices and applications such as patch devices, wearable devices such as smart watches (Apple Watch), and convenient home use devices such as Alivecor's Kardia Mobile™. These recent innovations are not only convenient to use by patients, they are also FDA-approved medical devices and medical applications [2–4].

Although there have been many new and exciting developments for AECGs involving device type, data capacity, and physical size, the core of AECGs—the basic algorithms, as shown in Figure 1's core circle, which include signal conditioning, beat detection, event detection, and interpretation—have not changed very much. However, key challenges for each of the four processing tasks listed above remain, particularly in noise handling, how to reduce noise during preprocessing, and how to differentiate noise and signals during beat detection and classification. The reason noise continues to be a challenge for AECGs is mainly because most AECGs are recorded outside of hospitals. Therefore, there is no trained medical personnel to monitor the signal quality, patients move around and engage in various activities while carrying the recording devices, and the recording time can also be long. It also needs to be indicated that most recent AECG recordings based on home-use devices such as Kardia Mobile/6L and Apple Watch are short-segment/noncontinuous vs. traditional continuous recordings from Holter recording of 24–72 h. For long Holter recordings, some learning algorithms can be applied for a period to accumulate initial templates, and the learning process can be updated with the longitudinal data to refine the template matching and beat detection. On the other hand, short recordings of AECG cannot afford too long of a learning segment, and thus they are more dependent on pre-learned model performance.

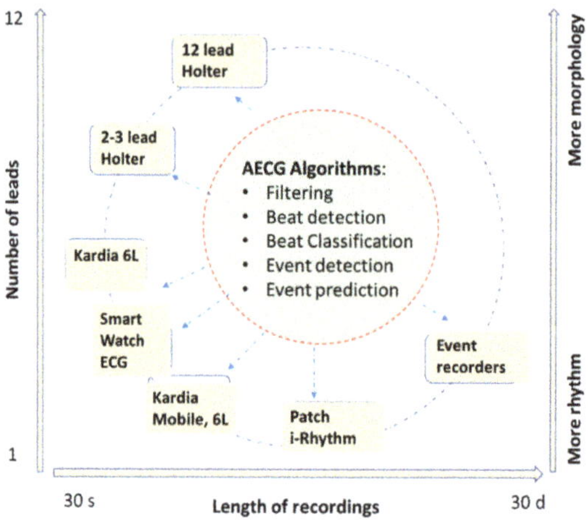

Figure 1. There are a wide range of AECG devices and applications. The recording length can be from 30 s to 30 days, and the number of leads of ECG can be from 1 to 12. At the center of all these lies the AECG algorithms including filtering, beat detection and classifications, and event detection and prediction. For devices with 1 or 2 leads, the events are mainly in rhythm abnormalities, such as sinus, AFIB, or tachycardia/bradycardia. For devices with more leads, some morphology analysis can be added such as ST, QT, LVH, or BBB.

Thirty-five years ago, a limited capacity microprocessor-based AECG algorithm could already achieve AECG beat detection accuracy to around 99% for normal beats and 96% for ectopic beats [5]. It is this remaining 2–3% of AECG signals that remain uncharacterized due to the challenges discussed above. Now the question or challenge is whether modern techniques offer significantly better performances than the 'old' ones.

In recent years, with the rapid development of machine learning (ML), deep learning (DL) in particular, many researchers have applied ML/DL methods to AECG algorithms [6–11]. Unlike the previous wave of interest in ML and neural networks around the 1980s to the 1990s, the recent development in the field has shown more promise, primarily due to the availability of larger training data sets and the maturity of the ML algorithms, especially convolution neural networks (CNN) and recurrent neural networks (RNN) [12,13]. One major advancement of the CNN-based AECG algorithms is that they can directly process ECG waveforms after initial preprocessing, without the need for ECG feature extraction, which the last generation of neural network algorithms relied on. Theoretically, since feature extraction of ECG signals from noisy data is usually difficult and time-consuming, the advantage of not relying on predefined feature extraction is that it can increase the efficiency of building new AECG algorithms from scratch if a large training data set is available.

The traditional 24-48 h Holter AECG might be described as the first attempt at big data analysis, but with less impressive results and with far greater effort compared to some current methods. An average of 100,000 to 200,000 plus heartbeat cycles of ECG data needed to be processed for a single data recording. Either automatic or semi-automatic learning models, which include template matching and clustering, were used during the analysis [14]. These types of learning algorithms were mainly limited to current patient data, instead of using a wide group of patients' data for training sets as the most current DL algorithms do. Most recent applications of DL methods to AECG have used large data sets consisting of multiple patients' data recordings as the training set, followed by the application of automatic pattern matching during analysis. A combination of large data sets at the pre-learning stage and the individual data's continuous learning can be a key to improve the analysis accuracy for such a relatively long AECG analysis [15].

After the initial excitement of experiencing ML/DL's performance of processing AECG, it is important to understand how ML algorithms work and perform in comparison to more traditional algorithms. Instead of a general overview of ML/DL techniques, which have been recently discussed in other reviews, this review focuses on how these new ML/DL algorithms can perform better in recognizing differences between physiologically meaningful signals and noise, and ways in which these new algorithms can be used together with traditional models to achieve even better performance and interpretability. Interpretability is very important for most medical applications, not only because it can help us better understand how the algorithms work, but also because it is more relevant for causality analysis—fundamental to helping physicians find causes and better treatments. We here cover how these points are relevant to each key step of AECG processing.

2. A Summary of Machine Learning Algorithms Used for AECG

Various machine learning algorithms have been used for AECG signal processing and detection for several decades, although DL algorithms have only become more widely used in recent years [13–16]. Therefore, we can list DL vs. non-DL algorithms separately. The purpose of this section is not to introduce each algorithm, rather for the convenience of the discussions in a later section.

2.1. Machine Learning Algorithms without Deep Learning

As shown in Figure 2, there are many ML algorithms in this category. We can divide them into supervised and unsupervised learning [17]. Supervised learning requires an input-label/reference pair for training the algorithm, while unsupervised learning does not need reference/labeling (note: this might not be a complete list of all the algorithms).

- **Fuzzy logic algorithm**—Using fuzzy logic for rule-based classifications. The fuzzy logic uses a 'soft edge' decision boundary to replace the 'crispy' decision boundary on the basis of multiple criteria. It maintains its interpretable nature while being able to adapt to the complicated decision boundaries. However, the algorithm complexity can increase very fast with the increase of the number of input features.
- **Linear regression**—It builds a linear correlation model between inputs and outputs, mostly used for continuous value output.
- **Logistic regression**—It also builds the correlation model between inputs and outputs, converging to binary level output by logistic functions such as sigmoid. It can be used for categorical classifications.
- **Decision tree**—Automatically generates a rule-based classification by a tree-like model. It usually has very intuitive flowchart symbols and rules, and therefore it is simple to understand and interpret. However, it can be relatively inaccurate compared to other predictors with the similar data.

- **SVM (support vector machine)**—A binary classification to maximize the distance between two classes. It has become one of the most robust binary classifiers. It also can perform nonlinear classification with its kernel functions.
- **Bayesian network (naïve Bayesian network)**—Applies simplified Bayesian theorem for classifications. It builds a probabilistic relationship between symptoms and measurements with the causes of diseases, and therefore is usually a better causality model than neural networks' 'black box' models. However, it requires more known knowledge such as conditional/joint probabilities of input variables.
- **k-NN (k-nearest neighbors)**—A very intuitive classification algorithm wherein a sample is classified on the basis of a common majority rule of its k closest neighbors. It is a very effective classification algorithm with limited training samples. It works better with a small number of input features.
- **Random forest**—An ensemble learning method by constructing multiple decision trees. A test sample is classified on the basis of the selections made by most trees. It has some relationship with the decision tree algorithm but is better for avoiding overfitting. It has become one of the most widely used classification algorithms outside of neural networks models.

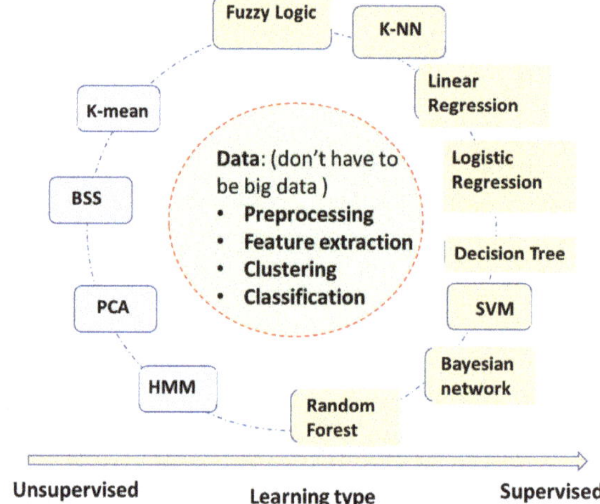

Figure 2. A diagram for the most widely used statistical learning algorithms. Most of the non-deep learning algorithms listed here can work on moderate-sized data since their independent parameters are limited. These algorithms mostly work on previously extracted features, or they can help to identify the best features such as PCA. On the left is unsupervised learning mainly used for clustering and feature optimization; on the right are supervised learning algorithms, requiring pairs of inputs and labels. All these methods have been used widely in AECG applications in the last several decades.

Here is a list for the most widely used **unsupervised algorithms**:

- **K-mean**—A clustering algorithm based on vector quantization that partitions n observations into K clusters, e.g., clustering ECG beats into templates in Holter analysis (not to be confused with k-NN, which is a supervised classification algorithm).
- **BSS (blind signal separation)**—Separates a set of source signals with little information about source signals, e.g., separate signal and noise of AECG.
- **PCA (principal component analysis)**—One of the most popular unsupervised algorithms for data reduction and feature optimization, e.g., determining which ECG features are more important for AFIB detection.

- **HMM (hidden Markov model)**—Assume signals to be a Markov process, current state, $X(n)$ only depend on immediate previous state $X(n\text{-}1)$. e.g., to describe R-R interval sequence of AFIB ECGs.

These groups of algorithms mostly belong to statistical learning and modeling with mathematical equations, instead of distributed weights and layers in neural networks and DL algorithms. Most statistical algorithms listed here can work on moderate-sized data since their independent parameters are limited. These algorithms mostly work on previously extracted features, or they can help to identify the best features.

2.2. Neural Network Deep Learning Algorithms for AECG

Different from the statistical learning algorithms above, neural network (NN) learning algorithms are based on the parallel and layered structure. The optimization is usually done through minimizing cost functions by some sort of backpropagation of gradients [18]. Previous NN learning started in the 1980s with many applications in ECG processing and pattern recognition [19,20]. The NN models were mostly 1–2 hidden layers, so-called 'shallow learning', whose inputs were mostly pre-extracted features or localized waveforms [21]. The current wave of DL models of NN started in around 2012, although a CNN-based DL algorithm for digits was published earlier [22,23]. A list of the most widely used DL algorithms is found below (Figure 3 is a diagram for a summary of DL algorithms).

Supervised Learning:

- **Convolutional neural network deep learning (CNN)**—The most popular DL algorithm with many applications in AECG. The multi-layer structures of CNN act as a filter bank, and the nonlinear activation functions of CNN act as feature extraction. Therefore, the CNN models are capable of handling original ECG waveforms directly.
- **Recursive neural network deep learning (RNN)**—RNN models add links among their hidden units (cells) to add memories for time series signals, similar to the concept of IIR filters used in signal processing. AECG signals are time-series signals whose features such as P-R and R-R intervals can be well fitted to certain time-series models.
- **Transfer learning**—This algorithm is very useful for DL models without enough training data for the current application. A pre-trained DL model for another domain is transferred to the current application. It only re-trains fewer layers of the transferred model or adds fewer new layers for retraining.
- **Ensemble learning**—This algorithm is not only useful for DL models. Random forest learning mentioned above is a type of ensemble learning. General ensemble learning can be extended to a wide range of groups of model collections and voting, suitable for complicated detection and prediction applications. Statistically, ensemble learning can also be viewed as an application of the central limit theorem. However, in practical use, it needs to know if computation power is enough to handle multiple models running.
- **Self (semi)-supervised learning**—This learning method deals with limited labeled data and large amounts of unlabeled data, which almost all applications are facing. It can be very useful in applications of AECG, although there are not too many mentioned thus far.
- **Online learning**—It allows for a pre-trained large model to be updated with the new coming samples, while still maintaining good performance for previously trained samples. This is a particularly good concept but lacks matured algorithms thus far. Some of the transfer learning examples can be viewed as special cases for online learning with only new training samples updated from a pre-trained model, but 'old samples' are most likely forgotten.

Unsupervised learning:

- **Auto-encoder (AE)**—This algorithm is a DL model for encoding the original input, such as ECG waveforms. After training, it can represent ECG waveform with a much-condensed latent variable vector.
- **Variational auto-encoder (VAE)**—It has the same structure and training as AE. However, during the applications, one can adjust its latent vector to form different variation patterns of the original waveforms, for example, to synthesize different noise patterns.

Reinforcement learning (RL):

Reinforcement learning is the third large type of learning algorithm that recently achieved very promising results in many fields, especially in game playing [24,25]. RLs use agents to fit into an environment in order to maximize reward. There are also some attempts of RL for AECG processing but with limited successes [26].

Figure 3 is a diagram of a summary of DL algorithms. In summary, DL models need a large database to train. The rapid development of DL models goes side-by-side with the availability of big data and big computation power. For AECG, a large ECG waveform database with corresponding labels is needed for training purposes.

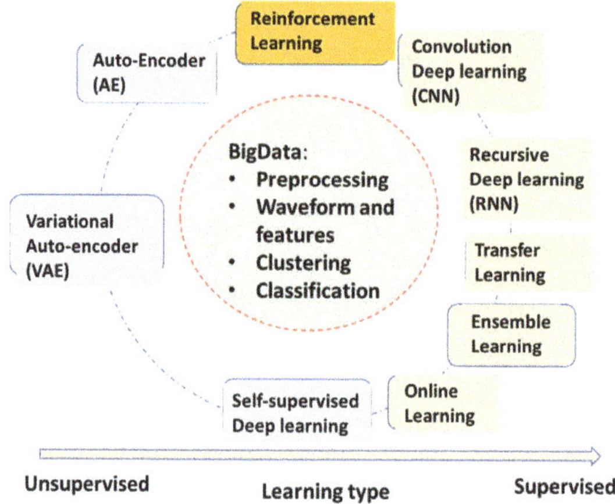

Figure 3. A diagram of a summary of DL algorithms. DL models rely on big data. For AECG, a large ECG waveform database with corresponding labels is needed for training purposes. In the list, CNN and RNN are the most popular DL models. Transfer learning and ensemble learning also become practical for AECG. However, there has been limited use of reinforcement learning and self-supervised learning thus far. AE and VAE are very useful for noise detection and feature extraction.

3. AECG Signal Preprocessing—Noise Filtering

The main purpose of the preprocessing of AECG signals is to reduce noise while bringing minimum distortion to the original signals. Any type of filtering can distort the signals, and therefore we need to make sure that the distortion is tolerable and meets the standards defined for each application.

A typical noise handling task in AECG can be described by a diagram shown in Figure 4, where there are two parallel paths for preprocessing: one (the upper path) is mainly for noise reduction and signal-to-noise ratio (SNR) improvement, but with some signal distortion; the other (the bottom path) is also for noise reduction but with minimal signal distortion. The results from the first path can be used to assist the second path, e.g., signal averaging; as shown in the figure, an average beat is formed with the trigger points detected from the first path.

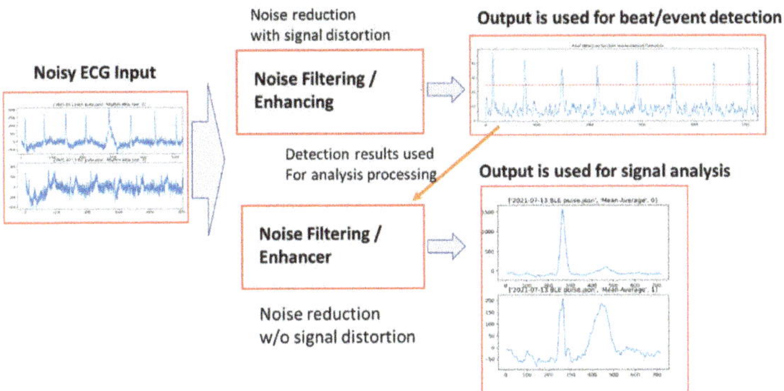

Figure 4. Signal preprocessing can have 2 parallel paths, one (the upper path) is mainly for noise reduction and SNR improvement, but with some signal distortion, while the other (the bottom path) is also for noise reduction but with minimal signal distortion. The results from the first path can be used to assist the second path, e.g., signal averaging; as shown in the figure, an average beat is formed with the trigger points detected from the first path.

3.1. AECG Signal Processing-Noise Reduction

The conventional bandwidth for AECG is 0.5–40 Hz, compared to a diagnostic resting ECGs 0.05–150 Hz [1,27,28]. Since most AECG applications are focused on rhythm and arrhythmia analysis, the main task for preprocessing is to enhance the QRS complex as shown in the upper path of Figure 4. With the applications expanding to other ECG morphology analyses such as ST and QT, preprocessing also needs to consider reducing the signal distortion to other signal segments while enhancing the QRS complex segment, but we here focus on the mainstream application of AECGs.

If the noise's frequency contents are higher than 40 Hz, it is called 'out-band' noise, which can usually be removed or reduced by a bandpass filter. If the noise's frequency contents are within 40 Hz, then it is called 'in-band' noise, which can be caused by motion and muscle contraction; this is almost unavoidable for AECG. It is this in-band noise that needs to be dealt with using more specialized methods. Compared to the preprocessing of diagnostic ECGs with very a small tolerance for signal distortion, AECG signals are allowed to have a larger tolerance for signal distortion. Therefore, there are more choices for filtering methods.

It should be noted that if preprocessing is only to form some type of detection function for the purpose of QRS or P wave detection—not for analyzing ECG morphology—then processing methods can be more flexible as long as signal-to-be-detected can be enhanced while uninterested portions can be attenuated. For example, the QRS detection function applied in this classic paper [5] enhanced QRS complex and attenuated noise, and even P and T waves, such as the output of the upper path of Figure 4.

Usually, there are two parallel paths for ECG signal preprocessing, as shown in Figure 4, one for beat detection (which maximizes the SNR of QRS) and one for more detailed ECG analysis with a minimized distortion to the original waveform shape. If needed, signal averaging or median beats can be formed from the second path signal output.

3.2. Early Stage of ML Filtering of AECG

As most recent studies using ML for AECGs are focused on beat-classification and event detection, it needs to be pointed out that many ML technologies are evolved from the various adaptive filtering algorithms that include the earliest attempts to AECG's powerline interference reduction [29]. In those ECG adaptive filtering applications, the models were learned and converged in real time, and most of the time the model is linear and very compact and therefore it can converge fast. There are some similarities of those adaptive filtering models to the current ML learning model, such as minimizing the error of prediction and target with the steepest descent of gradient of the errors.

The first nonlinear adaptive filtering was proposed and applied to AECG in the early 1990s [19]. This research created a neural network model for real-time filtering, and it achieved better performance for beat detection for noisy data. The reason is that the nonlinear model can adapt to color noise better than linear models so as to improve the accuracy of the match filters.

The wavelet methods are also among the advanced noise filtering methods for AECG [30]. It is also a nonlinear signal processing method, in which different wavelet transformed scales were filtered differently and then inverse transformed back to time-domain signals.

3.3. Using Deep Learning Models for ECG Denoising

The ML algorithms can be used for both the detection function and waveform analysis. The first major task of any ML algorithm is to read and format the ECG data from originally sampled formats, which can be an involved task since data can come from different devices with different formats. For AECG, large data sets are mostly shared with the following three formats: (1) Physionet-MIT data format [31], (2) International Society for Holter and Noninvasive Electrocardiology (ISHNE) format [1], and (3) devices' outputs in Json/XML format with readable ECG data.

The study of [32] applied several DL models to ECG signal denoising, in which a multilayer CNN model and an RNN model were used separately for a supervised regression. For the training set, inputs are noisy ECG signals, and the targets are denoised ECG. The model is more like an auto-encoder where input and target pair have the same signals, with the exception of the target signal, which is to be a leaned version of the input signal [33].

The auto-encoder is designed to extract the independent latent representation, and is also employed in the conditioning of ECG signals [33–36]. The auto-encoder usually has an inversely symmetric architect where the input signal is transformed into a latent representation vector with significantly fewer dimensions. The signal is then reconstructed solely on the basis of the latent representation vector. Therefore, with a clean signal as a target for reconstruction from noisy inputs, an auto-encoder can perform as an ECG signal conditioner. The input of an auto-encoder is usually ECG signal without the need for labels, making it much more achievable than supervised DL methods. Some even applied generative adversarial networks (GAN) in feeding auto-encoders to further reduce the need for large datasets [37]. Due to the dimension reduction nature of the method, auto-encoders can also be applied in other applications such as compression [38–40].

The study of [32] combined conventional filtering with a CNN DL model to obtain a higher SNR improvement. The initial filtering of ECG signals included a bandpass filter (0.1–30 Hz), an IIR notch filter, and a wavelet filter for removing baseline wander. The filtered signals were fed into a 15-layer CNN model.

Unlike arrhythmia analysis or beat classification where ground truth requires medical expertise, for ECG denoising analysis, a training data set can be obtained by adding synthetic noise to a presumed clean ECG signal [41,42].

These types of DL model-based denoising filters have much more computational burden than other types of filters. Therefore, usage is limited for applications with computational and/or battery life limitations.

4. AECG Beat Detection and Classification

Beat detection and classification are the essential tasks of any AECG application. The good news is that the accuracy of beat detection (QRS complex) had already reached more than 99% before any advanced ML algorithms were invented or applied years ago [5]. However, the performances of those beat detections are very much dependent on the test databases. In addition, it needs to be indicated that the performance of beat classifiers is usually not as good as the beat detection only algorithms. The challenges for current ML enthusiasts are not just to prove that ML, especially DL algorithms, can work on AECG beat detection and classification, but also to show improved performance for the remaining 1% of the widely used test databases, and also for improved performance on more challenge databases.

4.1. Conventional Algorithms for Beat Detection and Classification

In a conventional AECG analysis program, after the preprocessing and the detection function is formed, a combination of thresholding and pattern matching algorithms are used for both beat detection and classification. Here, the task of beat detection is to detect any QRS complex, regardless of its source of origin, which can start from sinus, atrial, junctional, ventricle, fusion, etc. The beat classification separates beats into different categories or templates. There are two main challenges for both beat detection and classifications: noise handling and the time-series nature of the beats.

4.1.1. Use Both Thresholding and Template Pattern Matching

The biggest challenge is still noise—electrode motion noise, muscle noise, powerline interference, etc. Noise detection is a major part of any AECG algorithm. After various filtering and techniques are used, as reviewed in the previous section, AECG's SNR is improved so much that the beat detection accuracy can be as good as 99% or even higher. For a 24-h Holter recording, there are about 100,000 beats on average. Thus, a 99% accuracy means there are still about 1000 beats missed. We would hope a Holter ECG reviewer does not have to correct each miss beat-by-beat, otherwise, it could be a time-consuming task. A good thresholding method is to find the optimal threshold between QRS complex and the noise floor. Due to the nature of the AECG's relatively high noise level, sometimes the separation of beat and noise is difficult.

The pattern matching method can work better than using a threshold only, especially under low SNR situations in which a bank of collected templates is matched with an underlying signal beat. If the correlation coefficient is high, then a beat can be identified. If the correlation coefficient is low, then a new template might be added to the template collection. The matching can be feature-based or waveform-based. The feature-based template matching can be more computationally efficient, but it requires a feature extraction process.

4.1.2. Time Series Analysis

By nature, AECG is a time-sequence signal, meaning it has beat cycles. Within each cycle, there are sequences propagating along certain pathways. Therefore, the beat detection and classification also need to take this time-series feature into consideration. For example, when a beat is identified, the following 200–250 ms is called the refractory period when heart tissue cannot be triggered. This logic is easily applied in conventional algorithms, but also needs to be considered for machine learning—deep learning neural network (ML-DNN)-based algorithms.

4.2. ML/DL-Based Beat Detection and Classification

The majority of ML/DL-based AECG studies are beat classifications and arrhythmia detection [7,13]. Among those studies, most DL algorithms are CNN-based supervised learning, while few are RNN-based algorithms trying to fit into time-series features of AECG. Although quite a few papers targeted beat detection and classification, DNN-based algorithms showed a potential of directly classifying arrhythmia events without explicitly marking each beat and its type. In one of the first CNN-based large AECG dataset studies, 12 arrhythmias were classified directly from 1 lead ECG with a selected 30 s segment [10]. Another large study also showed a CNN-based model for classifying multiple arrhythmias and morphology-related abnormalities from multi-lead ECG waveforms directly without beat detection and beat classification [12].

Figure 5 shows a general DL model for beat detection and classifications, where one real value of applying DNN models to AECGs is to possibly skip beat-by-beat analysis altogether, instead obtaining the final event detection from ECG waveform in one large model. This works as long as large training sets are available, and Hannun's study [10] had about 90,000 ECGs while Kashou's study [9] had about 2.4 million ECGs. For the learning data set, one key parameter is the length of the input data for each input-label pair, which can be arranged from 1 s to 10 s, as indicated by the dotted window in Figure 5. A short window with only one QRS beat can be used to teach the model how to recognize various QRS beats, whereas a longer window including 3–4 beats can teach the model for the current QRS beat and its surrounding beats and noise. The longer window is also better for PVC detection by the model since the window includes the current PVC beat and its previous and following beats. However, the longer window for input data usually requires a larger training set since it has more variations of the beat series pattern. Therefore, if the training set is large enough, the optimal input data window is around 3–4 s.

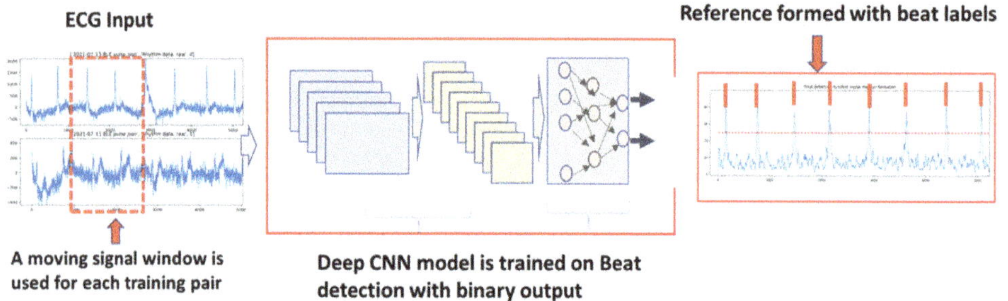

Figure 5. DL models can be trained for beat detection and classification directly from ECG waveform. The key is to define the input/reference pair. The label can indicate beat type and noise segment.

In some AECG applications, especially consumer-based wearable and home-used devices, the one-step 'black box' approach might be more preferred. However, for those AECG applications such as 24 h + Holter and ECG patch, a more comprehensive analysis including detailed beat detection and beat classification might be preferred, since these data are edited and annotated for a final medical report and are also more interpretable.

4.2.1. DL Supervised Learning for Beat Detection and Classification

As shown in the drawing of a general CNN-based ECG learning in Figure 6, the input can be a one-beat or multiple-beat ECG waveform, also being able to have one or multiple leads. Borrowing the idea of image recognition for original CNN applications, we can also call this input an ECG image. For the input of multiple-lead ECG signals, either 2-D CNN or 1-D CNN can be used. However, to handle one lead ECG, 1-D CNN is the better choice. For supervised learning, model structure design also includes selecting a pair of input-targets (reference), in addition to the number of filters, kernel length, and a number of layers. The paper of [6] reviewed several CNN-based beat classifiers by model structure, input data, target class, and other features. All of the studies reviewed took ECG waveforms with different preprocessing as inputs. Some of them also added extracted ECG features such as pre- and post-RR intervals of the current beat. It also stated that the CNN models can perform better for noisy data than conventional feature-based beat detection, and, as shown in the drawing of general CNN-based ECG learning in Figure 6, the input can be a one-beat or multiple-beat ECG waveform, being able to also have one or multiple leads. Borrowing the idea of image recognition for original CNN applications, we can also call this input an ECG image. For the input of multiple-lead ECG signals, either 2-D CNN or 1-D CNN can be used. However, to handle one-lead ECG, 1-D CNN is the better choice. For supervised learning, model structure design also includes selecting a pair of input targets (reference), in addition to the number of filters, kernel length, and the number of layers. Below are several key parameters for CNN model structure on AECG:

- **Input data, length, and dimension**: Some studies used one beat cycle, and some used multiple beat cycles. The advantage of using multiple beat cycles is to add time-series information. It is also beneficial for differentiating signal and noise. However, using multiple segments require larger training sets, since the variation is increased.
- **CNN kernel filter length and number for each layer**: Length of the kernel filter seems not to affect performance by much, but the number of filters usually increases from layer to layer, such as 16, 32, and 64. For AECG processing, the number of filters increased with the complexity of the input signals. For example, if the input signals are rhythm signals in original sampling frequency, the number of kernel filters needed can go up to 128 for deep layers. However, if the input signals are shorter averaged/median beats, the number of filters can be reduced to 32. A large number of filters generate more latent variables. If there are too many latent variables than needed, the model learning speed can be unnecessarily slow.
- **Regular convolution or residual convolution (ResCNN) layer**: ResCNN with a skip layer for each unit can increase training speed. Therefore, most studies adopted the ResCNN model structure.

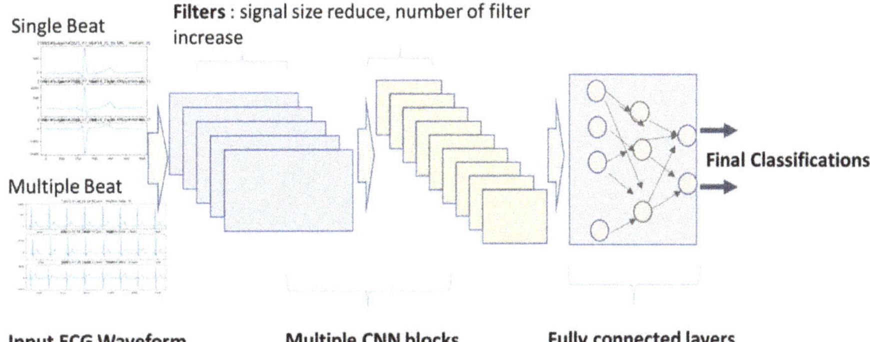

Figure 6. A diagram of CNN model for AECG classifications. The model input can be a single beat or multiple beats. Multiple ECG leads can form an ECG image. If only 1 lead is presented, then 1-D CNN can be used. However, if there are multiple leads as inputs, 2-D CNN can be used. There are multiple CNN blocks to form a deep layer structure. Usually, the number of filters is gradually increased, but the size of the output of each block is reduced by down-sampling after each block of max/average pooling and stride operation. The final block of the model consists of fully connected layers. The final output can be either binary or multiple classifications.

Below are several key strategies for the training of CNN models on AECG:

- **Batch size**: Usually, large batch size is preferred for a big data set and multiple classes, such as 256 or 512. For binary classification and limited data size, a smaller batch size can be used, such as 32 or 64. Another consideration is for GPU memory size, since the whole batch is loaded into GPU memory as one block to speed up the training. The larger batch will take more GPU memory. For AECG processing, the selection of the batch size is also heavily dependent on the available training samples. If the training set is very large, such as 1 million plus, the batch size can be set at 512 to generate a smoother gradient search path by avoiding too much fluctuation of a smaller batch size, and most importantly to avoid being 'trapped' in a local minimum. However, if the training set is relatively small, the batch size has to be reduced, with the reduction of the model layers.
- **Loss function and output function**: These strategies need to be clarified to avoid misuse. The classification task can be divided into three categories (as shown in Table 1):

(1) binary classification, e.g., QRS complex vs. noise; 2) multi-class classification that is mutually exclusive, e.g., classify QRS beats into N, S, V, F, Q beat types; (3) multi-class classification but not mutually exclusive, e.g., morphology-related ECG abnormal: LBBB, ischemia, sinus, etc. Different loss functions and output functions are selected accordingly. The suggestions are also listed in Table 1.

- **Balance of different class types in each batch of training samples**: Very often, we can have a very different distribution of classes. For example, there are many more normal beats than PVC or other abnormal beats. If the same distribution is used in the training batch, it is very likely that the sensitivity of PVC beat detection will be poor. Therefore, the number of PVC beats can be augmented in each training batch, which can be in their original form or with variations. Another method is to use weight balancing. Popular machine learning frameworks, such as Keras, provide a 'class weight' parameter for this purpose [43].

- **Prevent overfitting**: Theoretically, there are concerns for both underfitting and overfitting. However, in most DNN studies, the model size/layer is so large that we might only need to worry about overfitting problems. The overfitting is usually caused by a lack of training samples with too many model parameters that need to be trained. There are several methods that can be used. The first one is to increase the training; for example, by adding certain noise to the original ECG recording to avoid simple repetition of the same data. The second method is to apply drop-off to the training process, which randomly 'disconnects' the weights to the output. The third method is to apply transfer learning [44,45].

Table 1. Selection of loss function and target function for different classification tasks.

Classification Tasks	Binary Classification	Multi-Class Classifications (Mutually Exclusive)	Multi-Class Classifications (Mutually Nonexclusive)
Examples:	QRS beat vs. noise	N, S, V, F, Q beat types	LBBB, ischemia, sinus
Loss function	Binary cross-entropy	Categorical cross-entropy	Binary cross-entropy
Target function	One-hot vector (Softmax)	One-hot vector (Softmax)	Sigmoid, scalar target

4.2.2. Unsupervised Learning for Beat Detection and Classification

Unsupervised learning does not need reference/labels for training, and, therefore, more data are available for building the algorithms or feature extraction from the models. The most widely used non-DL algorithms in this category are blind source separation (BSS) and principal component analysis (PCA).

Blind source separation (BSS) techniques such as ICA and PCA have been widely applied as a key signal condition method for ECG and other bioelectric signals [46–48]. Most bioelectrical signals have a specific source of generation, and noise observed in the sensors is generated from independent sources. ECG can be considered as a single localized electrical signal being observed from the body surface, while typical noise for ECG—movement artifact, EMG, electrode noise, etc.—are either non-localized or localized to different locations, making it spatially independent from ECG signal. By the nature of the ECG signal pattern, the common noise signals are also temporally independent from ECG signals. In sum, the lack of correlation, both spatially and temporally, is the foundation of BSS-based ECG noise reduction.

Hidden Markov model (HMM) is a statistical model where the system is assumed to be a Markovian process—each state entirely depends on its immediately previous states. Cardiac electricity is a well-controlled and organized electrical process and, therefore, largely suits the profile of a Markovian state. The application of HMM usually comprises two stages—the training stage, where the statistical model adapts to the series of events considered to be part of a Markovian process, and the application stage, where the 'trained' HMM is used to provide hidden states as encoded information of the states or provide an estimation of the incoming state [49–52]. The states can be a subcomponent of an ECG beat or segment of ECG rhythms or events. Overall, HMM is a stochastic state model based on a previous state, and the probability distribution of the model depends on the training data. The nature of cardiac electricity is deterministic, unlike other bioelectric signals such as electroencephalogram (EEG), electrogastrogram (EGG), or electromyogram (EMG), making it a suitable target for HMM.

However, one should note that most arrhythmias are overwhelmingly low in prevalence in most populations but very critical for clinical purposes. One good example is high-degree AV blocks. Consequent P waves without QRS or T waves are very unlikely in HMM if such a model is trained on the basis of ECG of the general population. This can lead to false negative detection of P waves that are unsynchronized with QRS waves, arriving in incorrect rhythm analysis. In the process of ECG analysis, beat segmentation after, or as part of QRS detection is a key step in a successful ECG beat classification and analysis. HMM has been applied to ECG segmentation [49,50] to encode each ECG beat detected. Gaussian mixture model (GMM) is also applied in ECG segmentation and delineation processes. With GMM built for P, QRS, T, and ECG baseline, clustering methods can be applied to the ECG signal to increase the resolution of ECG segments [53]. Another unsupervised clustering method, self-organizing neural networks through competitive learning, is also used to improve QRS onset and offset detection [54].

Higher-order HMM (HOHMM) is also applied to beat classification in ECG analysis [51]. HOHMM, with a similar philosophy to RNN, expands the dependency of state to further the immediate previous state to allow for more complex dynamics to be modeled.

For beat classification, k-nearest neighbors (k-NN) or fuzzy C-means (FCM) methods are used very often. Clustering methods have been widely used in ECG beat analysis [55–59] without clearly defined labels of beat types. While the implementation of clustering methods varies, the core concept is to group the items in the targeted dataset on the basis of their 'similarity'. In these clustering methods (KNN, FCM), the said similarity is defined directly through the Euclidean distance in the feature space. The major advantage of the clustering methods is the lack of dependency on labels, a critical bottleneck for all supervised methods. However, knowledge-based inputs are not completely absent, and even item by item labels are no longer needed. For FCM and KCM as examples, the number of clusters is needed for the formulation of the problem, and, therefore, an incorrect number of clusters will result in either duplication of cluster content or unnecessary mixings of different types of beats.

An auto-encoder can also be used for beat classification, which can create a latent representation of ECG with no externally labeled data and, therefore, be able to serve as a feature extraction component for further beat classification [60–62]. Once the auto-encoder reconstructs the ECG beats with proper accuracy, one can assume that the most critical features of the waveform have been captured and consolidated in the latent vector encoded, becoming the ideal feature for either clustering methods or supervised training methods that require fewer data. However, for ECG practices, it is not all sunny. In the training process of auto-encoder methods, the core objective function for reconstruction is usually second-order norm or energy-based. The clear advantage of the energy-based objective function is the simplicity in the generation of first-order derivatives and consequently better efficiency in training. However, this convenience comes at a price. For ECG analysis, the shape and location of P wave, a very small component in amplitude, is critical for the analysis of complicated rhythms such as atrial fibrillation or supraventricular arrhythmias. However, in an energy-based optimization scheme, more efforts will be focused on higher amplitude segments—QRS and T waves due to their size and less on P waves, especially when noise is present. Losing track of P waves may not jeopardize the results of simple tasks such as discriminating normal beats versus ventricular ectopy, but it will affect more complicated analysis involving atrial or junctional activities.

4.2.3. Transfer Learning

Transfer learning is useful when there is not enough training data for a relatively large deep-learning model. The larger-deep model benefits from direct training on ECG waveform instead of extracted features. The main approach for transfer learning is to start with a pre-trained model, adding some padding on the front to adapt to the ECG input, and then adding a couple of layers on the back for final classification. During the training process, most weights from the original model are fixed, while only the weights of newly added layers are trained. In this way, the needed training set is much smaller than if the whole model had been trained. Alternatively, one could use the original deep model for feature extraction or use an additional simple model for final classification with extracted features from the deep model. The study of [45] used a very deep model as automatic feature extraction. The input is the spectrogram of ECG to fit into a 2-D CNN of the original model. The features are then extracted from the deep layer and fed to a support vector machine (SVM) for further training. The performance of classifying normal sinus rhythm, AFIB, VFIB, and ST change is 97%, a very good result considering the limited training samples (a total of 7008 data instances).

4.2.4. Ensemble Learning

Ensemble learning involves combining multiple classification models to form a better performing one. This method has been used for computerized ECG interpretations by statistical learning methods before [63]. Each model can take different or the same input signal/features; the final classification is obtained by a voting method. Ensemble learning is much like multiple experts working on the same problem from different angles, and then the final decision is reached with some 'consultation' of the multiple solutions and proposals. Ensemble learning has been used in AECGs [62,64]. In the studies of [62,64], multiple ML algorithms were assembled to form a 'super' classifier for AFIB detection. The performance of the final ensemble classifier was better than any individual classifier.

The random forest (RF) algorithm is a very powerful and robust ensemble method too. Instead of working on the ECG waveform, it takes ECG features as input. Inside the model, there are several random trees, each of which works on a random group of input variables in parallel. During the test, the sample is classified with the majority of trees agreed upon. One study used the RF algorithm for AFIB detection, employing 150 ECG time and frequency domain features as input, achieving high performance for AFIB detection when compared to other methods [65].

5. AECG Event Detection and Classification

Event detection and classification are the final steps for conventional AECG analysis. Here, the events are the interpretation of rhythm or morphology-related abnormal ECG classifications. For example, in rhythm analysis, ECG events can be classified into normal sinus rhythm, sinus rhythm with sinus arrhythmia, atrial fibrillation (AFIB), atrial flutter (AFLUT), junctional rhythm, ventricular rhythm, etc. Some morphology classifications include left or right bundle branch block (LBBB/RBBB), left ventricular hypertrophy (LVH), ischemia, myocardial infarction (MI), long QT (LQT). An AECG's event detection is mainly focused on rhythm analysis, while resting ECG analysis usually targets more comprehensive rhythm and morphology analysis.

The conventional AECG analysis relies on the early stage of beat detection and beats classification for the final event detection. However, deep-learning models can potentially include all steps into one model, i.e., from ECG waveform input to event classification output, as we mentioned was previously shown by some very good performance studies [9,10]. At the same time, more AECG analyses need to provide more detailed reports on different aspects of beat-related information, such as number of PVC, PVC couplet, short run/long run VT, AFIB burden, and most crucially the trend of RR intervals. Therefore, the analysis modules of beat detection, beat classification, and event detection are still needed, even for ML-DNN-based algorithms.

RNN model is very useful for AECG time series event detection such as AFIB and PVC/VT. Figure 7 shows a diagram of a general structure of RNN event detection. Here, RNN models can include one or multiple RNN layers with multiple cells in each layer. In this case, long short-term memory (LSTM) cells are used. The cells in the same layer are connected with either one-directional links or dual-directional links. The input of RNN is a series of ECG waveforms or ECG parameters, e.g., R-R intervals or P-R-T components. The last block of the model consists of fully connected layers. The final output can be either binary or multiple classifications.

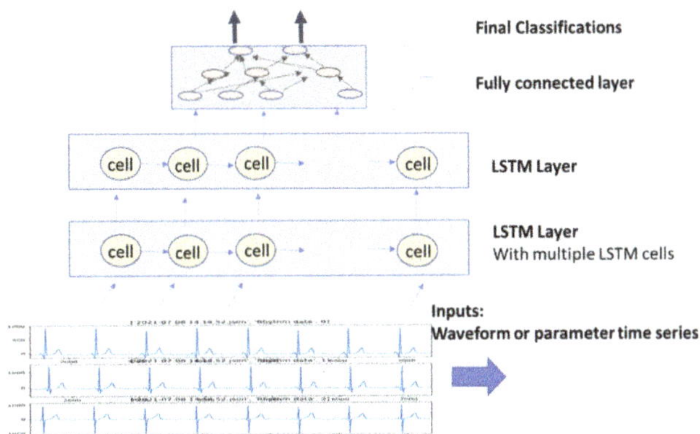

Figure 7. RNN model includes one or multiple RNN layers with multiple cells in each layer. Here, LSTM cells are used. The input of RNN is a series of ECG waveforms or ECG parameters, e.g., R-R intervals or P-R-T components. The last block of the model consists of fully connected layers. The final output can be either binary or multiple classifications.

For the application of unsupervised learning, with proper features extracted, most clustering methods can also be applied to event or rhythm classifications [66–71]. In contrast to beat clustering, rhythm segments contain significantly more information and therefore one of the most critical steps for a successful clustering application is to select proper features. Other than conventional features, DL methods such as auto-encoders have also been applied to provide features for clustering [61] in a similar fashion to ECG beat classifications.

The following are some major event detections of AECG:

5.1. AFIB/AFLUT

AFIB is one of the most prevalent arrhythmias, and missed diagnosis can result in possible stroke and heart failure. Early detection can result in the best treatment options. In the center of a conventional AFIB detection algorithm, there are two features most commonly used: R-R sequence and P wave status. For most AECGs, P wave detection is not reliable due to poor SNR, and therefore R-R sequence becomes the main feature used in most AFIB detection algorithms. It is not so difficult to build a high-sensitivity AFIB detection algorithm with RR sequence analysis alone, especially for longer ECG recordings. It is more challenging to achieve a very high specificity at the same time. Again, in a conventional AFIB detection algorithm, high specificity requires the result P wave status detection. For example, some sinus arrhythmias can have similar RR variations as AFIB, but a regular P wave in front of QRS can rule out AFIB to avoid false positive detection. However, for most one-lead AECGs, especially wearable devices, usually only an equivalent of lead I is available, which is not a very good lead for small P wave detection due to heart vector projection angle. In a standard 12-lead, lead II and v1 usually are the best leads for P wave detection.

Can state-of-art DL models detect AFIB better than conventional algorithms? The Mayo Clinic's study [9] showed that by using 12-lead 10 s ECG as input, the deep-learning model achieved an area under the curve (AUC) of 0.999 for AFIB detection with a 2.4 million ECG data set. Although this model uses the resting ECGs that usually have better SNR, especially for P wave, it still can be a benchmark for other deep-learning models that use ECG waveforms as input directly without explicit feature extraction.

In this AFIB detection challenge [64], there are many interesting AFIB algorithms presented, all of which are machine learning-based. Some of them need explicit feature extraction, while others are deep-learning models with ECG waveform as input. The best performance of the challenge was achieved by the method of feeding time sequence of ECG features to an RNN model [72]. It also provided a high degree of interpretability. These are very encouraging results since interpretability is still very important for medical applications. Most deep-learning models achieved good performance by acting as a 'black box'. Several top performances were achieved with random forest method, a powerful statistical machine learning method that combines features of time and frequency domains [65,73]. Furthermore, an ensemble learning is proposed on the basis of the algorithms joined challenge [64]. It demonstrated that by using top performing algorithms or all algorithms, the AFIB detection accuracy can be higher than any single algorithm can achieve.

5.2. PVC/VT

The detection of premature ventricular contraction (PVC) is one major task of any AECG algorithm. Although a few isolated PVCs might not have any pathological significance, some more severe arrhythmias such as bigeminy, trigeminy, couplet, and ventricular tachycardia are all associated with PVC detection.

Conventional machine learning AECG algorithms use feature-based template matching for PVC detection [74]. The features can include R-R intervals, the pattern of QRS complex, width of QRS complex, and ST-T wave. Noise removal and detection are still the keys for high-accuracy PVC detection. The pattern matching and clustering technique mentioned previously can be used for grouping similar morphology complexes together. Some most often used algorithms include k-nearest neighbors (k-NN) [75], discrete hidden Markov model [76], support vector machine, Bayesian classification [77], and random forest [78].

Deep-learning PVC detection algorithms have been presented in many recent studies [61,79–83]. Many of these algorithms use ECG waveform directly without conducting explicit feature extraction, although some extract features from DL models and feed them into a non-deep model such as SVM. Since PVC detection requires both morphological pattern information and time sequence information, LSTM models that extract time series features from incoming ECG signals perform well when used properly [61].

The potentials of deep-learning models are not only served to improve pattern classification, but can also enhance the causality analysis of PVC. One study applied the CNN model to detect PVC origin [84]. It combined a ventricular computer model into the training scheme. The training datasets were generated by multiplying ventricular current dipoles derived from single pacing at various locations with a patient-specific lead field. The origins of PVC are localized by calculating the weighted center of gravity of classification returned by the CNNs. Although the testing results are limited by the number of cases, it still is a very interesting direction for deep-learning models.

Severe arrhythmias including ventricular and supraventricular tachycardia (VT, SVT) are the critical target events for many AECG algorithms. DL models have shown the potential of detecting VT and SVT already [10,80]. Most of those models are CNN deep-learning with ECG waveform as inputs, while some are RNN-based models. The varying length of signal segments is used ranges between 1 and 30 s. The longer the segment, the larger the data set needed. Again, the key is to let models learn the difference between noise segment and VT/SVT segment. The noise is a major issue for false-positives of

VT/STV/VF detection in conventional algorithms [85], where both time and frequency domain features are extracted from every 5 s segment to form a time sequence.

5.3. QT Analysis

QT interval is one of the most difficult ECG measurements, and therefore most diagnostic QT estimations are based on resting 12-lead ECG. However, AECG can provide continuous QT analysis and trends with their longer recording time. Conventional QT algorithms involve ECG filtering, segmentation, and estimation [86]. The QT algorithm's accuracy has been improved significantly over the previous 12-lead QT algorithm [87].

DL algorithms have been applied for QT estimation. This study takes ECG waveform input directly without feature extraction [88]. It is trained on 2.4 million ECGs from the Mayo Clinic's ECG database, but with only 2 out of 12 leads. The algorithm uses both regression and classification schemes in forming its special cost function. The model consists of three residual layer blocks and two fully connected layers. The test performance against the doctor's annotation is close to that of other major QT algorithms. In addition, the algorithm was also applied to the data collected with the two-lead mobile electrocardiogram device, Kardia 6L^{TM}. The results are also within the error range of the measurement standard (IEC 60601–2-25 (diagnostic electrocardiographs)).

5.4. Noise Segment

It is worth taking a separate section for the discussion of noise in the event detection, since it is so critical in any AECG processing that it is not surprising to know that half of the codes of an AECG program can be devoted to noise detection and the separation of noise from the signals. Another reason for discussing more about noise is that modern machine learning models have made some significant contributions to noise detection, more than any other processing stages for AECG processing.

Using DNN models to detect noise segments has been proven to be very robust. The method of DNN noise detection includes the CNN classification model [89], where training pair is formed with controlled SNR and the noise labels, the auto-encoder method [90]. The differences between these two types of noise detection deep-learning models are that the first type use CNN to classify noisy ECG into noisy/clean directly, and the second type first uses an auto-encoder to train the encoder to form the latent variables and then use this layer as the input for next noise classification model. Both methods do not need to extract features explicitly as conventional noise detection algorithms.

6. ECG Risk Stratification/Prediction

Predictions of severe arrhythmia/cardiac events are always very challenging since the performance of predictions is usually lower than the detections when the events are already happening. The low prevalence of the future event is another reason why the prediction task usually has a very low positive predictive value. Predicting the new onset of AFIB should be meaningful for prevention purposes since adverse outcomes of undetected AFIB can be stroke and heart failure. Previous algorithms for predicting AFIB included P wave averaging analysis [91,92]. The difficulty here is to find the best feature(s) for the prediction.

One study used a large database of 12-lead ECG with 1.6 million ECGs to predict one-year AFIB occurrence probability [93]. The AFIB prediction performance was 0.85 of AUC. A more significant study used a machine learning model to detect the mechanism of AFIB, as well as to guide the best treatment of AFIB ablation [94]. AFIB drives were induced in two computerized atrial models and combined with eight torso models. A total of 103 features were extracted from the signals. A binary decision tree classifier was trained on the simulated data and evaluated using hold-out cross-validation. The classifier yielded 82.6% specificity and 73.9% sensitivity for detecting pulmonary vein drivers on the clinical data.

One study focused on risk stratification of mortality of patients with acute myocardial infarction [95]. It used a large data registry of Korean acute myocardial infarction (AMI), including 22,875 AMI patients from 59 hospitals. The study took 36 hospitals' data for the training set and 23 hospitals' data for the testing set. A DL model with ECG and other clinical parameters was used. The results with the DL model achieved an AUC (area under the curve) of 0.905 vs. the 0.85 GRACE score.

These studies built prediction models for more life-threatening events, cardiac arrest, and ventricular fibrillation through DL. DL-based models extract features from ECG waveforms directly. The advantages might be that the feature extraction is automatic, but in the meantime, it is a non-transparent model approach, and therefore the mechanism under hidden features is not clear. In building the prediction models for risk stratification, understanding the underline mechanism of the risks is also very important as, we learned from T wave alternans (TWA)-related risk stratification for sudden cardiac death (SCD) [96], where researchers found TWA is not only closely associated with an increased risk of SCD, but its mechanism is also connected to perturbations in calcium transport processes. Therefore, TWA may play a role not only in risk stratification but also in the pathogenesis of ventricular tachyarrhythmia events. This is what we hope DL-based models can also contribute to in terms of more such causal prediction and analysis.

7. Discussion

This review covers a very wide range of the AECG application of machine learning-based algorithms, especially DL models. The development and efforts in the AECG algorithms have never been so great and so fast, partially due to the fast advancement of machine learning fields in parallel and interdisciplinary paths.

For conventional AECG algorithms, feature extraction is one of the most important steps, and also one of the most demanding tasks. There are two main reasons to explain why feature extraction is so critical. The first one is that most features used in the algorithm have links to the underlying physiological structures and conditions. For example, P waves are associated with atrial excitation of the heart, QRS is associated with ventricular depolarization, and ST-T segments are related to the repolarization process. The second reason is that most currently used interpretive algorithms are expert system-based, following the same logic and analysis that physicians are using to analyze pathological changes of the heart.

As we have indicated, among the most noticeable advantages of DL models is their ability to handle ECG waveform directly without an explicit feature extraction process. At the same time, this waveform-to-event-detection process of DL models also brings the arguments of the 'black box' model versus a more transparent model of an expert system-based algorithm. Is there a way to take advantage of both approaches to have a more automatic and transparent model? There have been some efforts in this regard. One study used an input sensitivity analysis for trained neural networks for acute MI detection [97]. This feature sensitivity analysis was also extended to DL models, and thus it was made clear as to which part of ECG segments was more significant for their contribution to the final output. For example, for an ML model that differentiates sinus rhythm from atrial fibrillation, it would be helpful to understand if the model makes the detection only on the basis of R-R interval information, or whether it actually also uses P wave information itself. There are some other efforts to make DL models more interpretable [98].

There is already much research to build risk stratification algorithms from AECGs, such as T wave alternans (TWA), heart rate turbulence (HRT), heart rate variability (HRV), and QT interval dynamicity. The center point of those algorithms is to identify the features related to the physiological system. For example, TWA is linked to the substrate of myocardial tissues, and HRT and HRV are related to the control of the autonomous system to the heart condition. Those features help us to understand the mechanism of the changes of ECG. Thus far, most DL models are still only performing a better job on pattern recognition. It is important to improve the accuracy of detecting cardiac events with AECG, but it shall also be meaningful to reveal the causal relationships between surface ECG and underlying mechanisms. With the large data sets and large models, it might help us to find more relevant features.

One topic not discussed here is so-called online learning, or adaptive learning, meaning that an event detection model can be updated with the incoming data and new labels that can be edited by physicians [99]. There have been few to no successful online learning models in AECG. This is partly due to the technical challenge of balancing pre-trained models with the new data and is partly due to the difficulty of meeting regulatory constraints, which require a model to be fully tested and verified to pass certain standards. However, in this era of deep learning and big data, it is reasonable to take more action in this direction. One big advantage of machine learning is that the model can be built quickly and automatically with the available big data. The technical challenge is how to balance the previously trained data set and the new data set. Some methods and ideas from transfer learning can be borrowed, such as fixing most parts of an already trained model, only retraining, or adding a few layers for the new data. However, an automatic verification process also needs to be built into the process to meet the regulator requirements.

It is no secret that DL models heavily rely on the availability of big data sets for training. The model overfitting is caused by not enough training data for a large model so that the performance for independent test data is poor. This is what we called 'model generalization'. To some extent, the test and verification are more important for deep-learning models than conventional detection algorithms when a smaller data set is used. Many papers and research studies use MIT/BME ECG databases on the Physionet [31], classic data sets that have been used for the last 30 years to test various ECG algorithms. The question is whether these databases are still sufficient for building and testing deep-learning models. If not, perhaps work toward building larger public data sets for deep-learning models of AECG is required? We are so glad to see the efforts in this direction in the studies/challenges organized by Computing in Cardiology [100]. In these challenges, large data sets were donated by different institutes, and multiple algorithms were built and shared in the AECG society. What is even more exciting is the effort of putting all these algorithms in the direction of ensemble learning. This type of collaborative learning can also be very promising for targeting some very complicated AECG tasks in the future.

Author Contributions: L.Y. contributed to the portion of unsupervised learning; J.X. contributed to the rest of the paper. All authors have read and agreed to the published version of the manuscript.

Funding: This research received no external funding.

Acknowledgments: The authors would like to acknowledge discussions of AECG and machine learning with Gari Clifford and Bioinformatics group of Emory University and machine learning group of Alivecor Inc.

Conflicts of Interest: Joel Xue is a research fellow with AliveCor Inc.

References

1. Steinberg, J.S.; Varma, N.; Cygankiewicz, I.; Aziz, P.; Balsam, P.; Baranchuk, A.; Cantillon, D.J.; Dilaveris, P.; Dubner, S.J.; El-Sherif, N.; et al. 2017 ISHNE-HRS expert consensus statement on ambulatory ECG and external cardiac monitoring/telemetry. *Heart Rhythm* **2017**, *14*, 55–96, Epub 2017 May 8. [CrossRef] [PubMed]
2. Rosenberg, M.A.; Samuel, M.; Thosani, A.; Zimetbaum, P.J. Use of a Noninvasive Continuous Monitoring Device in the Management of Atrial Fibrillation: A Pilot Study. *Pacing Clin. Electrophysiol.* **2013**, *36*, 328. [CrossRef] [PubMed]
3. Isakadze, N.; Martin, S.S. How useful is the smartwatch ECG? *Trends Cardiovasc. Med.* **2020**, *30*, 442–448. [CrossRef]
4. Hall, A.; Mitchell, A.R.J.; Wood, L.; Holland, C. Effectiveness of a single lead AliveCor electrocardiogram application for the screening of atrial fibrillation: A systematic review. *Medicine* **2020**, *99*, e21388. [CrossRef] [PubMed]
5. Pan, J.; Tompkins, W.J. A Real-Time QRS Detection Algorithm. *IEEE Trans. Biomed. Eng.* **1985**, *32*, 230–236. [CrossRef] [PubMed]
6. Manisha; Dhull, S.K.; Singh, K.K. ECG Beat Classifiers: A Journey from ANN to DNN. *Procedia Comput. Sci.* **2020**, *167*, 747–759. [CrossRef]
7. Somani, S.; Russak, A.J.; Richter, F.; Zhao, S.; Vaid, A.; Chaudhry, F.; De Freitas, J.K.; Naik, N.; Miotto, R.; Nadkarni, G.N.; et al. Deep learning and the electrocardiogram: Review of the current state-of-the-art. *Europace* **2021**, *23*, 1179–1191. [CrossRef]

8. Ribeiro, A.L.P.; Ribeiro, M.H.; Paixão, G.M.M.; Oliveira, D.M.; Gomes, P.R.; Canazart, J.A.; Ferreira, M.P.S.; Andersson, C.R.; Macfarlane, P.W.; Meira, W.; et al. Automatic diagnosis of the 12-lead ECG using a deep neural network. *Nat. Commun.* **2020**, *11*, 1760. [CrossRef]
9. Kashou, A.H.; Ko, W.-Y.; Attia, Z.I.; Cohen, M.S.; Friedman, P.A.; Noseworthy, P.A. A comprehensive artificial intelligence–enabled electrocardiogram interpretation program. *Cardiovasc. Digit. Health J.* **2020**, *1*, 62–70. [CrossRef]
10. Hannun, A.Y.; Rajpurkar, P.; Haghpanahi, M.; Tison, G.H.; Bourn, C.; Turakhia, M.P.; Ng, A.Y. Cardiologist-level arrhythmia detection and classification in ambulatory electrocardiograms using a deep neural network. *Nat. Med.* **2019**, *25*, 65–69. [CrossRef]
11. Lu, W.; Shuai, J.; Gu, S.; Xue, J. Method to Annotate Arrhythmias by Deep Network. In Proceedings of the 2018 IEEE International Conference on Internet of Things (iThings), Halifax, NS, Canada, 30 July–3 August 2018. [CrossRef]
12. Hong, S.; Zhou, Y.; Shang, J.; Xiao, C.; Sun, J. Opportunities and Challenges of Deep Learning Methods for Electrocardiogram Data: A Systematic Review. *Comput. Biol. Med.* **2020**, *122*, 103801. [CrossRef] [PubMed]
13. Ebrahimi, Z.; Loni, M.; Daneshtalab, M.; Gharehbaghi, A. A review on deep learning methods for ECG arrhythmia classification. *Expert Syst. Appl. X* **2020**, *7*, 100033. [CrossRef]
14. Serhani, M.A.; El Kassabi, H.T.; Ismail, H.; Navaz, A.N. ECG Monitoring Systems: Review, Architecture, Processes, and Key Challenges. *Sensors* **2020**, *20*, 1796. [CrossRef] [PubMed]
15. Wasimuddin, M.; Elleithy, K.; Abuzneid, A.S.; Faezipour, M.; Abuzaghleh, O. Stages-based ECG signal analysis from traditional signal processing to machine learning approaches: A survey. *IEEE Access* **2020**, *8*, 177782–177803. [CrossRef]
16. Hoffmann, J.; Mahmood, S.; Fogou, P.S.; George, N.; Raha, S.; Safi, S.; Schmailzl, K.J.; Brandalero, M.; Hubner, M. A Survey on Machine Learning Approaches to ECG Processing. In *Signal Processing—Algorithms, Architectures, Arrangements, and Applications*; IEEE: Piscataway, NJ, USA, 2020; pp. 36–41. [CrossRef]
17. Hastie, T.; Tibshirani, R.; Friedman, J. *The Elements of Statistical Learning—Data Mining, Inference, and Prediction*; Springer Series in Statistics; Springer: New York, NY, USA, 2008.
18. Goodfellow, I.; Bengio, Y.; Courville, A. *Deep Learning*; MIT Press: Cambridge, MA, USA, 2016; Available online: http://www.deeplearningbook.org (accessed on 17 July 2021).
19. Xue, J.Q.; Hu, Y.H.; Tompkins, W.J. Neural-Network-Based Adaptive matched Filtering for QRS Detection. *IEEE Trans. Biomed. Eng.* **1992**, *39*, 317–329. [CrossRef] [PubMed]
20. Xuel, J.Q.; Hu, Y.H.; Tompkins, W.J. Training of ECG signals in nueral network pattern recognition. In Proceedings of the Internaional Conference of the IEEE Engineering in Medicine and Biology Society, Philadelphia, PA, USA, 1–4 November 1990; Volume 12, pp. 1465–1466.
21. Xue, Q.; Reddy, B.R.S. Late Potential Recognition by Artificial Neural Networks. *IEEE Trans. Biomed. Eng.* **1997**, *44*, 132–143. [CrossRef] [PubMed]
22. LeCun, Y.; Haffner, P.; Bottou, L.; Bengio, Y. Object recognition with gradient-based learning. In *Shape, Contour and Grouping in Computer Vision*; Lecture Notes in Computer Science (Including Subseries Lecture Notes in Artificial Intelligence and Lecture Notes in Bioinformatics); Springer: Berlin/Heidelberg, Germany, 1999; Volume 1681, pp. 319–345. [CrossRef]
23. Krizhevsky, A.; Sutskever, I.; Hinton, G.E. ImageNet classification with deep convolutional neural networks. *Commun. ACM* **2017**, *60*, 84–90. [CrossRef]
24. Arulkumaran, K.; Deisenroth, M.P.; Brundage, M.; Bharath, A.A. A Brief Survey of Deep Reinforcement Learning. *IEEE Signal Process. Mag.* **2017**, *34*, 26–38. [CrossRef]
25. Silver, D.; Huang, A.; Maddison, C.J.; Guez, A.; Sifre, L.; Driessche, G.V.D.; Schrittwieser, J.; Antonoglou, I.; Panneershelvam, V.; Lanctot, M.; et al. Mastering the game of Go with deep neural networks and tree search. *Nature* **2016**, *529*, 484–489. [CrossRef]
26. Insani, A.; Jatmiko, W.; Sugiarto, A.T.; Jati, G.; Wibowo, S.A. Investigation Reinforcement learning method for R-Wave detection on Electrocardiogram signal. In Proceedings of the International Seminar on Research of Information Technology and Intelligent Systems, Yogyakarta, Indonesia, 5–6 December 2019. [CrossRef]
27. IEC 60601-2-47:2012 | IEC Webstore. Available online: https://webstore.iec.ch/publication/2666 (accessed on 17 July 2021).
28. IEC 60601-2-25:2011 | IEC Webstore. Available online: https://webstore.iec.ch/publication/2636 (accessed on 17 July 2021).
29. Thakor, N.V.; Zhu, Y.-S. Adaptve filtering to ECG analysis. *IEEE Trans. Biomed. Eng.* **1991**, *38*, 785–794. [CrossRef]
30. Awal, M.A.; Mostafa, S.S.; Ahmad, M.; Rashid, M.A. An adaptive level dependent wavelet thresholding for ECG denoising. *Biocybern. Biomed. Eng.* **2014**, *34*, 238–249. [CrossRef]
31. PhysioBank Databases. Available online: https://archive.physionet.org/physiobank/database/ (accessed on 17 July 2021).
32. Arsene, C. Design of Deep Convolutional Neural Network Architectures for Denoising Electrocardiographic Signals. In Proceedings of the IEEE Conference on Computational Intelligence in Bioinformatics and Computational Biology, Viña del Mar, Chile, 27–29 October 2020. [CrossRef]
33. Xiong, P.; Wang, H.; Liu, M.; Zhou, S.; Hou, Z.; Liu, X. ECG signal enhancement based on improved denoising auto-encoder. *Eng. Appl. Artif. Intell.* **2016**, *52*, 194–202. [CrossRef]
34. Xiong, P.; Wang, H.; Liu, M.; Liu, X. Denoising auto-encoder for eletrocardiogram signal enhancement. *J. Med Imaging Health Inform.* **2015**, *5*, 1804–1810. [CrossRef]
35. Xiong, P.; Wang, H.; Liu, M.; Lin, F.; Hou, Z.; Liu, X. A stacked contractive denoising auto-encoder for ECG signal denoising. *Physiol. Meas.* **2016**, *37*, 2214–2230. [CrossRef] [PubMed]

36. Garus, J.; Pabian, M.; Wisniewski, M.; Sniezynski, B. Electrocardiogram Quality Assessment with Auto-encoder. In Proceedings of the Internarional Conference on Computational Science, 2021 ICCS, Krakow, Poland, 16–18 June 2021; Springer: Cham, Switzerland, 2021; pp. 693–706. [CrossRef]
37. Antczak, K. A Generative Adversarial Approach to ECG Synthesis and Denoising. *arXiv* **2020**, arXiv:2009.02700v1.
38. Yildirim, O.; Tan, R.S.; Acharya, U.R. An efficient compression of ECG signals using deep convolutional auto-encoders. *Cogn. Syst. Res.* **2018**, *52*, 198–211. [CrossRef]
39. Wang, F.; Ma, Q.; Liu, W.; Chang, S.; Wang, H.; He, J.; Huang, Q. A novel ECG signal compression method using spindle convolutional auto-encoder. *Comput. Methods Programs Biomed.* **2019**, *175*, 139–150. [CrossRef]
40. Bekiryazici, T.; Gurkan, H. ECG Compression method based on convolutional auto-encoder and discrete wavelet transform. In Proceedings of the 2020 28th Signal Processing and Communications Applications Conference (SIU 2020), Gaziantep, Turkey, 5–7 October 2020. [CrossRef]
41. Arsene, C. Complex Deep Learning Models for Denoising of Human Heart ECG signals. *arXiv* **2019**, arXiv:1908.10417.
42. Antczak, K. Deep Recurrent Neural Networks for ECG Signal Denoising. *arXiv* **2018**, arXiv:1807.11551v3.
43. Johnson, J.M.; Khoshgoftaar, T.M. Survey on deep learning with class imbalance. *J. Big Data* **2019**, *6*, 27. [CrossRef]
44. Weimann, K.; Conrad, T.O.F. Transfer learning for ECG classification. *Sci. Rep.* **2021**, *11*, 5251. [CrossRef]
45. Salem, M.; Taheri, S.; Yuan, J.-S. ECG Arrhythmia Classification Using Transfer Learning from 2-Dimensional Deep CNN Features. In Proceedings of the IEEE Biomedical Circuits and Systems Conference (BioCAS), Cleveland, OH, USA, 17–19 October 2018. [CrossRef]
46. Xue, J.; Gao, W.; Chen, Y.; Han, X. Study of repolarization heterogeneity and electrocardiographic morphology with a modeling approach. *J. Electrocardiol.* **2008**, *41*, 581–587. [CrossRef]
47. Romeo, I. PCA and ICA applied to noise reduction in multi-lead ECG. In *Computer in Cardiology*; IEEE: Piscataway, NJ, USA, 2011; pp. 613–616.
48. Alickovic, E.; Subasi, A. Effect of Multiscale PCA De-noising in ECG Beat Classification for Diagnosis of Cardiovascular Diseases. *Circuits Syst. Signal Process.* **2015**, *34*, 513–533. [CrossRef]
49. Koski, A. Modelling ECG signals with hidden Markov models. *Artif. Intell. Med.* **1996**, *8*, 453–471. [CrossRef]
50. Andreão, R.V.; Dorizzi, B.; Boudy, J. ECG signal analysis through hidden Markov models. *IEEE Trans. Biomed. Eng.* **2006**, *53*, 1541–1549. [CrossRef]
51. Liao, Y.; Xiang, Y.; Du, D. Automatic Classification of Heartbeats Using ECG Signals via Higher Order Hidden Markov Model. In Proceedings of the IEEE International Conference on Automation Science and Engineering, Hong Kong, China, 20–21 August 2020; pp. 69–74. [CrossRef]
52. Silipo, R.; Deco, G.; Schürmann, B.; Vergassola, R.; Gremigni, C. Investigating the underlying Markovian dynamics of ECG rhythms by information flow. *Chaos Solitons Fractals* **2001**, *12*, 2877–2888. [CrossRef]
53. Hajimolahoseini, H.; Hashemi, J.; Redfearn, D. ECG Delineation for Qt Interval Analysis Using an Unsupervised Learning Method. In Proceedings of the ICASSP, IEEE International Conference on Acoustics, Speech and Signal Processing, Calgary, AB, Canada, 15–20 April 2018; pp. 2541–2545. [CrossRef]
54. Tighiouart, B.; Rubel, P.; Bedda, M. Improvement of QRS boundary recognition by means of unsupervised learning. *Comput. Cardiol.* **2003**, *30*, 49–52. [CrossRef]
55. Lagerholm, M.; Peterson, G. Clustering ECG complexes using hermite functions and self-organizing maps. *IEEE Trans. Biomed. Eng.* **2000**, *47*, 838–848. [CrossRef] [PubMed]
56. Zhang, C.; Wang, G.; Zhao, J.; Gao, P.; Lin, J.; Yang, H. Patient-specific ECG classification based on recurrent neural networks and clustering technique. In Proceedings of the 13th IASTED International Conference on Biomedical Engineering (BioMed 2017), Innsbruck, Austria, 20–21 February 2017; pp. 63–67. [CrossRef]
57. Osowski, S.; Linh, T.H. ECG beat recognition using fuzzy hybrid neural network. *IEEE Trans. Biomed. Eng.* **2001**, *48*, 1265–1271. [CrossRef] [PubMed]
58. Chudáček, V.; Petrík, M.; Georgoulas, G.; Čepek, M.; Lhotská, L.; Stylios, C. Comparison of seven approaches for holter ECG clustering and classification. In Proceedings of the Annual International Conference of the IEEE Engineering in Medicine and Biology, Lyon, France, 22–26 August 2007; pp. 3844–3847. [CrossRef]
59. Korürek, M.; Nizam, A. A new arrhythmia clustering technique based on Ant Colony Optimization. *J. Biomed. Inform.* **2008**, *41*, 874–881. [CrossRef] [PubMed]
60. Nurmaini, S.; Partan, R.U.; Caesarendra, W.; Dewi, T.; Rahmatullah, M.N.; Darmawahyuni, A.; Bhayyu, V.; Firdaus, F. An automated ECG beat classification system using deep neural networks with an unsupervised feature extraction technique. *Appl. Sci.* **2019**, *9*, 2921. [CrossRef]
61. Hou, B.; Yang, J.; Wang, P.; Yan, R. LSTM-Based Auto-Encoder Model for ECG Arrhythmias Classification. *IEEE Trans. Instrum. Meas.* **2020**, *69*, 1232–1240. [CrossRef]
62. Kuznetsov, V.V.; Moskalenko, V.A.; Zolotykh, N.Y. Electrocardiogram Generation and Feature Extraction Using a Variational Auto-encoder. *arXiv* **2020**, arXiv:2002.00254.
63. Rubel, P.; Fayn, J.; Macfarlane, P.W.; Pani, D.; Schlögl, A.; Värri, A. The History and Challenges of SCP-ECG: The Standard Communication Protocol for Computer-Assisted Electrocardiography. *Hearts* **2021**, *2*, 384–409. [CrossRef]

64. Clifford, G.; Liu, C.; Moody, B.; Lehman, L.-W.; Silva, I.; Li, Q.; Johnson, A.; Mark, R. AF Classification from a Short Single Lead ECG Recording: The PhysioNet/Computing in Cardiology Challenge 2017. In *Computing in Cardiology*; IEEE: Piscataway, NJ, USA, 2017. [CrossRef]

65. Zabihi, M.; Rad, A.B.; Katsaggelos, A.K.; Kiranyaz, S.; Narkilahti, S.; Gabbouj, M. Detection of atrial fibrillation in ECG hand-held devices using a random forest classifier. In *Computing in Cardiology*; IEEE: Piscataway, NJ, USA, 2017; Volume 44, pp. 1–4. [CrossRef]

66. Lin, Z.; Ge, Y.; Tao, G. Algorithm for Clustering Analysis of ECG Data. In Proceedings of the Annual International Conference of the IEEE Engineering in Medicine and Biology Society. IEEE Engineering in Medicine and Biology Society. Annual Conference, Shanghai, China, 17–18 January 2006; pp. 3857–3860. [CrossRef]

67. Xia, Y.; Han, J.; Wang, K. Quick detection of QRS complexes and R-waves using a wavelet transform and K-means clustering. *Bio-Med Mater. Eng.* **2015**, *26*, S1059–S1065. [CrossRef]

68. Özbay, Y.; Ceylan, R.; Karlik, B. A fuzzy clustering neural network architecture for classification of ECG arrhythmias. *Comput. Biol. Med.* **2006**, *36*, 376–388. [CrossRef] [PubMed]

69. Stridh, M.; Rosenqvist, M. Automatic screening of atrial fibrillation in thumb-ECG recordings. In Proceedings of the 39th Conference on Computing in Cardiology, Krakow, Poland, 9–12 September 2012; IEEE: Krakow, Poland, 2012; Volume 39, pp. 161–164.

70. Plesinger, F.; Nejedly, P.; Viscor, I.; Halamek, J.; Jurak, P. Automatic detection of atrial fibrillation and other arrhythmias in holter ECG recordings using rhythm features and neural networks. In *Computing in Cardiology*; IEEE: Piscataway, NJ, USA, 2017; Volume 44, pp. 1–4. [CrossRef]

71. Tan, W.W.; Foo, C.L.; Chua, T.W. Type-2 fuzzy system for ECG arrhythmic classification. In Proceedings of the IEEE International Conference on Fuzzy Systems, London, UK, 23–26 July 2007. [CrossRef]

72. Teijeiro, T.; García, C.A.; Castro, D.; Félix, P. Arrhythmia classification from the abductive interpretation of short single-lead ECG records. In *Computing in Cardiology*; IEEE: Piscataway, NJ, USA, 2017; Volume 44, pp. 1–4. [CrossRef]

73. Antink, C.H.; Leonhardt, S.; Walter, M. Fusing QRS detection and robust interval estimation with a random forest to classify atrial fibrillation. In *Computing in Cardiology*; IEEE: Piscataway, NJ, USA, 2017; Volume 44, pp. 1–4. [CrossRef]

74. Yu, J.; Wang, X.; Chen, X.; Guo, J. Automatic Premature Ventricular Contraction Detection Using Deep Metric Learning and KNN. *Biosensors* **2021**, *11*, 69. [CrossRef]

75. Kaya, Y.; Pehlivan, H. Feature selection using genetic algorithms for premature ventricular contraction classification. In Proceedings of the ELECO 2015—9th International Conference on Electrical and Electronics Engineering, Bursa, Turkey, 26–28 November 2015; pp. 1229–1232. [CrossRef]

76. Boublenza, S.; Chikh, A.; Bouchikhi, M.A. Discrete hidden Markov model classifier for premature ventricular contraction detection. *Int. J. Biomed. Eng. Technol.* **2015**, *17*, 371–386.

77. Casas, M.M.; Avitia, R.L.; Gonzalez-Navarro, F.F.; Cardenas-Haro, J.A.; Reyna, M.A. Bayesian Classification Models for Premature Ventricular Contraction Detection on ECG Traces. *J. Healthc. Eng.* **2018**, *2018*, 2694768. [CrossRef]

78. Xie, T.; Li, R.; Shen, S.; Zhang, X.; Zhou, B.; Wang, Z. Intelligent Analysis of Premature Ventricular Contraction Based on Features and Random Forest. *J. Healthc. Eng.* **2019**, *2019*, 5787582. [CrossRef]

79. De Marco, F.; Finlay, D.; Bond, R.R. Classification of Premature Ventricular Contraction Using Deep Learning. In *Computing in Cardiology*; IEEE: Piscataway, NJ, USA, 2020. [CrossRef]

80. Murugesan, B.; Ravichandran, V.; Ram, K.; S.P., P.; Joseph, J.; Shankaranarayana, S.M.; Sivaprakasam, M. ECGNet: Deep Network for Arrhythmia Classification. In Proceedings of the IEEE International Symposium on Medical Measurements and Applications (MeMeA), Rome, Italy, 11–13 June 2018. [CrossRef]

81. Novotna, P.; Vicar, T.; Ronzhina, M.; Hejc, J.; Kolarova, J. Deep-Learning Premature Contraction Localization in 12-lead ECG from Whole Signal Annotations. In *Computing in Cardiology*; IEEE: Piscataway, NJ, USA, 2020. [CrossRef]

82. Javadi, M.; Ebrahimpour, R.; Sajedin, A.; Faridi, S.; Zakernejad, S. Improving ECG classification accuracy using an ensemble of neural network modules. *PLoS ONE* **2011**, *6*, e24386. [CrossRef]

83. Teplitzky, B.A.; McRoberts, M.; Ghanbari, H. Deep learning for comprehensive ECG annotation. *Heart Rhythm* **2020**, *17*, 881–888. [CrossRef]

84. Yang, T.; Yu, L.; Jin, Q.; Wu, L.; He, B. Localization of Origins of Premature Ventricular Contraction by Means of Convolutional Neural Network from 12-Lead ECG. *IEEE Trans. Biomed. Eng.* **2018**, *65*, 1662–1671. [CrossRef] [PubMed]

85. Prabhakararao, E.; Manikandan, M.S. Efficient and robust ventricular tachycardia and fibrillation detection method for wearable cardiac health monitoring devices. *Healthc. Technol. Lett.* **2016**, *3*, 239–246. [CrossRef] [PubMed]

86. Xue, J.Q. Robust QT Interval Estimation—From Algorithm to Validation. *Ann. Noninvasive Electrocardiol.* **2009**, *14* (Suppl. 1), S35. [CrossRef]

87. Hnatkova, K.; Gang, Y.; Batchvarov, V.N.; Malik, M. Precision of QT interval measurement by advanced electrocardiographic equipment. *Pacing Clin. Electrophysiol. PACE* **2006**, *29*, 1277–1284. [CrossRef] [PubMed]

88. Giudicessi, J.R.; Schram, M.; Bos, J.M.; Galloway, C.D.; Shreibati, J.B.; Johnson, P.W.; Carter, R.E.; Disrud, L.W.; Kleiman, R.; Attia, Z.I.; et al. Artificial Intelligence–Enabled Assessment of the Heart Rate Corrected QT Interval Using a Mobile Electrocardiogram Device. *Circulation* **2021**, *143*, 1274–1286. [CrossRef]

89. Ansari, S.; Gryak, J.; Najarian, K. Noise Detection in Electrocardiography Signal for Robust Heart Rate Variability Analysis: A Deep Learning Approach. In Proceedings of the 2018 40th Annual International Conference of the IEEE Engineering in Medicine and Biology Society (EMBC), Honolulu, HI, USA, 18–21 July 2018.

90. Sasmal, B.; Roy, S. ECG Guided Automated Diagnostic Intervention of Cardiac Arrhythmias with Extra-Cardiac Noise Detection using Neural Network. In Proceedings of the 7th International Conference on Optimization and Applications (ICOA), Wolfenbüttel, Germany, 19–20 May 2021. [CrossRef]

91. Dhala, A.; Underwood, D.; Leman, R.; Madu, E.; Baugh, D.; Ozawa, Y.; Kasamaki, Y.; Xue, Q.; Reddy, S. Signal-Averaged P-Wave Analysis of Normal Controls and Patients with Paroxysmal Atrial Fibrillation: A Study in Gender Differences, Age Dependence, and Reproducibility. *Clin. Cardiol.* **2002**, *25*, 525. [CrossRef] [PubMed]

92. Budeus, M.; Hennersdorf, M.; Felix, O.; Reimert, K.; Perings, C.; Sack, S.; Wieneke, H.; Erbel, R. Prediction of atrial fibrillation in patients with cardiac dysfunction: P wave signal-averaged ECG and chemoreflexsensitivity in atrial fibrillation. *Europace* **2007**, *9*, 601–607. [CrossRef] [PubMed]

93. Raghunath, S.; Pfeifer, J.M.; Ulloa-Cerna, A.E.; Nemani, A.; Carbonati, T.; Jing, L.; Vanmaanen, D.P.; Hartzel, D.N.; Ruhl, J.A.; Lagerman, B.F.; et al. Deep Neural Networks Can Predict New-Onset Atrial Fibrillation from the 12-Lead ECG and Help Identify Those at Risk of Atrial Fibrillation-Related Stroke. *Circulation* **2021**, *143*, 1287–1298. [CrossRef] [PubMed]

94. Luongo, G.; Azzolin, L.; Schuler, S.; Rivolta, M.W.; Almeida, T.P.; Martínez, J.P.; Soriano, D.C.; Luik, A.; Müller-Edenborn, B.; Jadidi, A.; et al. Machine learning enables noninvasive prediction of atrial fibrillation driver location and acute pulmonary vein ablation success using the 12-lead ECG. *Cardiovasc. Digit. Health J.* **2021**, *2*, 126–136. [CrossRef] [PubMed]

95. Kwon, J.-M.; Jeon, K.-H.; Kim, H.M.; Kim, M.J.; Lim, S.; Kim, K.-H.; Song, P.S.; Park, J.; Choi, R.K.; Oh, B.-H. Deep-learning-based risk stratification for mortality of patients with acute myocardial infarction. *PLoS ONE* **2019**, *14*, e0224502. [CrossRef] [PubMed]

96. Merchant, F.M.; Sayadi, O.; Moazzami, K.; Puppala, D.; Armoundas, A.A. T-wave Alternans as an Arrhythmic Risk Stratifier: State of the Art. *Curr. Cardiol. Rep.* **2013**, *15*, 398. [CrossRef]

97. Xue, J.; Aufderheide, T.; Wright, R.S.; Klein, J.; Farrell, R.; Rowlandson, I.; Young, B. Added value of new acute coronary syndrome computer algorithm for interpretation of prehospital electrocardiograms. *J. Electrocardiol.* **2004**, *37*, 233–239. [CrossRef] [PubMed]

98. Goh, G.S.W.; Lapuschkin, S.; Weber, L.; Samek, W.; Binder, A. Understanding Integrated Gradients with SmoothTaylor for Deep Neural Network Attribution. *arXiv* **2020**, arXiv:2004.10484.

99. Hoi, S.C.H.; Sahoo, D.; Lu, J.; Zhao, P. Online Learning: A Comprehensive Survey. *Neurocomputing* **2021**, *459*, 249–289. [CrossRef]

100. PhysioNet Index. Available online: https://physionet.org/challenge/ (accessed on 17 July 2021).

Review

Computer Assisted Patient Monitoring: Associated Patient, Clinical and ECG Characteristics and Strategy to Minimize False Alarms

Michele M. Pelter *, David Mortara and Fabio Badilini

School of Nursing, University of California, San Francisco, CA 94143, USA; david.mortara@ucsf.edu (D.M.); fabio.badilini@ucsf.edu (F.B.)
* Correspondence: michele.pelter@ucsf.edu

Abstract: This chapter is a review of studies that have examined false arrhythmia alarms during in-hospital electrocardiographic (ECG) monitoring in the intensive care unit. In addition, we describe an annotation effort being conducted at the UCSF School of Nursing, Center for Physiologic Research designed to improve algorithms for lethal arrhythmias (i.e., asystole, ventricular fibrillation, and ventricular tachycardia). Background: Alarm fatigue is a serious patient safety hazard among hospitalized patients. Data from the past five years, showed that alarm fatigue was responsible for over 650 deaths, which is likely lower than the actual number due to under-reporting. Arrhythmia alarms are a common source of false alarms and 90% are false. While clinical scientists have implemented a number of interventions to reduce these types of alarms (e.g., customized alarm settings; daily skin electrode changes; disposable vs. non-disposable lead wires; and education), only minor improvements have been made. This is likely as these interventions do not address the primary problem of false arrhythmia alarms, namely deficient and outdated arrhythmia algorithms. In this chapter we will describe a number of ECG features associated with false arrhythmia alarms. In addition, we briefly discuss an annotation effort our group has undertaken to improve lethal arrhythmia algorithms.

Keywords: alarm fatigue; annotation of ECG data; arrhythmia alarms; intensive care unit; patient monitoring

Citation: Pelter, M.M.; Mortara, D.; Badilini, F. Computer Assisted Patient Monitoring: Associated Patient, Clinical and ECG Characteristics and Strategy to Minimize False Alarms. *Hearts* **2021**, *2*, 459–471. https://doi.org/10.3390/hearts2040036

Academic Editor: Peter Macfarlane

Received: 29 July 2021
Accepted: 25 September 2021
Published: 1 October 2021

Publisher's Note: MDPI stays neutral with regard to jurisdictional claims in published maps and institutional affiliations.

1. Introduction

Physiologic monitoring, including electrocardiographic (ECG) monitoring, in the intensive care unit (ICU) remains unsatisfactory as evidenced by the well-known alarm fatigue problem. For example, in one study, a single ICU patient generated over 700 alarms per day [1]. In the University of California San Francisco (UCSF) Alarm Study, an average of 187 audible alarms were generated per bed per day during a one-month assessment. Of note, 90% of the ECG arrhythmia alarms were determined to be false [2,3]. Alarm fatigue occurs when nurses and providers are desensitized by frequent alarms, most of which are false or clinically irrelevant (i.e., true, but with no action required). The multitude of alarm sounds become "background noise" that is assimilated into the normal ICU workflow. Over a decade ago, excessive alarm burden was exposed by the press as a significant patient safety concern with the highly publicized death of a patient who was being monitored at a prestigious medical center [4]. Despite multiple heart rate alarms for bradycardia prior to the patient's cardiac arrest, no one working on the unit that day recalled hearing the alarms. In the investigation that ensued, the Centers for Medicare and Medicaid Services reported: "Nurses not recalling hearing low heart rate alarms was indicative of alarm fatigue, which contributed to the patient's death" [4]. The most recent data from the past seven years, show that alarm fatigue was responsible for over 650 hospital deaths [5,6], a number believed to be a substantial underrepresentation due to non- or under-reporting.

Over time, nurses cope with alarm fatigue by: (1) silencing alarms without assessing the patient; (2) lowering the alarm volume; (3) permanently disabling alarms; and/or (4) delay responding by assuming an alarm is false. These actions place patients at risk for adverse events, including death, as true alarms are missed.

A number of federal and national organizations have issued alerts about alarm fatigue. Since 2007, The Emergency Care Research Institute has placed alarm fatigue at or near the top of its list of top ten patient safety hazards [7]. In 2014, The Joint Commission established National Patient Safety Goal 6, Reduce Harm Associated with Clinical Alarms [5], which was updated in 2020 [8]. The American Nurses Association and the American Association of Critical Care Nurses have issued practice alerts regarding alarm fatigue, emphasizing the need for evidence-based approaches to solve this complex problem [9,10]. While a strong desire to solve alarm fatigue exists from these mandates, over a decade has passed with no substantive progress towards a general solution.

Prior clinical studies designed to reduce false alarms have included: daily ECG skin electrode changes [11,12]; customizing alarm parameters and/or alarm settings [11–18]; disposable versus non-disposable ECG lead wires [17,19]; and educational initiatives [12,14,17]. While these strategies have reduced the total number of alarms by 18% [14,19] to 90% [12], these studies do not address the primary problem of false ECG arrhythmia alarms, namely deficient and outdated arrhythmia algorithms. The following review will discuss the research our group has completed to shed light on these deficiencies. In addition, we will briefly discuss an annotation effort our group has undertaken to improve arrhythmia algorithms for asystole, ventricular fibrillation, and ventricular tachycardia.

2. Overview of ECG False Arrhythmia Alarms

False ECG arrhythmia alarms are a major source of alarm fatigue. In the UCSF Alarm study, the frequency, types, and accuracy of ECG monitor alarms were examined in 461 ICU patients [2]. A total of 2,558,760 unique alarms occurred in the 31-day study period: 1,154,201 (45%) arrhythmia; 612,927 (24%) parameter (i.e., too high, too low); and 791,632 (31%) technical (i.e., ECG leads off, artifact, or line/probe disconnect); these are shown in Figure 1. There were 381,560 audible alarms for an audible alarm burden of 187/bed/day. Of 12,671 annotated ECG arrhythmia alarms, 90% were false positive. This study of consecutive ICU patients represents the largest study regarding alarm burden to date, and clearly illustrates the magnitude of alarm fatigue.

As noted in Figure 1, the vast majority of the alarms were for premature ventricular complexes (PVC), followed by technical alarms (i.e., artifact or lead(s) off/fail). In addition to these observational data, the investigator of the UCSF Alarm study annotated six audible arrhythmia alarms as true or false using a standardized protocol [2]. A total of 12,671 audible arrhythmia alarms were annotated. Inter-rater reliability of alarm annotations was tested by randomly selecting 300 alarms that were rated twice by pairs of the annotators. A Cohen's Kappa was run to compare "Rater 1" to "Rater 2". There was 95% agreement as to whether the alarm was a true or false positive (Cohen's Kappa score of 0.86). Table 1 shows the accuracy of the arrhythmia alarms annotated.

Based on these findings, our group has published several studies examining patient, clinical, and ECG factors associated with false arrhythmia alarms. The following sections will describe these studies.

2.1. Patient and Clinical Factors Associated with False Arrhythmia Alarms

In a secondary data analysis of 461 patients enrolled in the UCSF Alarm study with 12,671 audible arrhythmia alarms, 250 (54%) of the patients had at least one of the six alarm types (Table 1) [3]. The number of false alarms per monitored hour for patients' ranged from 0.0 to 7.7, and the duration of the false alarms per hour ranged from 0.0 to 158.8 s. Patient characteristics were compared in relation to: (1) the number and (2) the duration of false arrhythmia alarms per 24-h period, using nonparametric statistics to minimize the influence of outliers. Among the significant associations were the following:

age > 60 years ($p = 0.013$; $p = 0.034$), confused mental status ($p = 0.001$ for both comparisons), cardiovascular diagnoses ($p = 0.001$ for both comparisons), electrocardiographic (ECG) features, including wide QRS complexes due to bundle branch block (BBB, both right or left) ($p = 0.003$; $p = 0.004$) or ventricular paced rhythm ($p = 0.002$ for both comparisons), respiratory diagnoses ($p = 0.004$ for both comparisons), and mechanical ventilation ($p = 0.001$ for both comparisons). In a subsequent study, we found that patients with a left ventricular device also have high rates of false arrhythmia alarms, presumably due to the vibrations and artifact caused by this device [20].

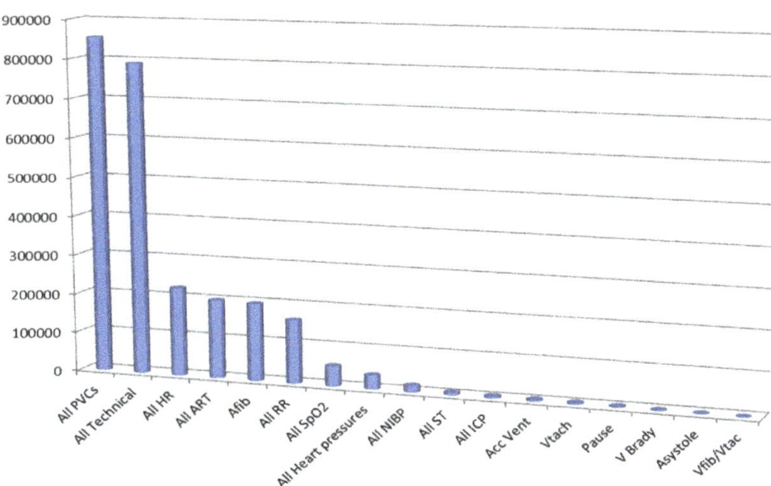

Figure 1. Frequency of all unique physiologic alarms from bedside ICU monitors over a 31-day study period. Figure from Drew, B.J. et al. [2] (open access with permission to use Figure). Abbreviations: ACC Vent = accelerated ventricular rhythm; Afib = atrial fibrillation; ART = arterial blood pressure; HR = heart rate; ICP = intracranial pressure; NIBP = non-invasive blood pressure; PVC = premature ventricular complexes; RR = respiratory rate; Sp02 = saturation of peripheral oxygen; ST = ST-segment; Vtach = ventricular tachycardia; V brady = ventricular bradycardia; and Vfib = ventricular fibrillation.

Table 1. Accuracy of 12,671 audible arrhythmia alarms.

Alarm Type	Number of Alarms	Number of Patients	True Positives	False Positives	False Positive Rate
Asystole	792	113	260	531	67.0%
Ventricular fibrillation	158	19	107	51	32.3%
Ventricular tachycardia	3861	183	502	3352	86.8%
Accelerated ventricular rhythm	4361	99	224	4135	94.8%
Pause	2239	140	272	1963	87.7%
Ventricular bradycardia	1260	39	40	1219	96.7%
Total	12,671			11,251	88.8%

Signal quality as a source of false arrhythmia alarms was examined in the 12,671 audible arrhythmia alarms shown in Table 1 [2]. The following definitions were used: good; fair; or poor signal quality. Good signal quality was defined as a clearly visible P-QRS-T waveform across all available leads with little to no noise, baseline wander, or leads off. Fair signal quality was defined as moderate noise or baseline wander but having identifiable QRS complexes for basic rhythm/rate detection. Poor signal quality was defined as being unanalyzable due to excessive noise, baseline wander, or leads off. Of 12,671 audible arrhythmia alarms, ECG signal quality was rated as good in 73% of the false alarms and 93% of the true alarms, whereas poor signal quality was found in 18% of the false arrhythmia alarms and 6% of the true arrhythmia alarms. Hence, poor signal quality was not the source of the majority of the false arrhythmia alarms.

From our published studies we have found that the vast majority of false alarms occur in a small group of patients. For example, in the UCSF Alarm study, of 461 ICU patients, one patient generated nearly half of the alarms [2], a problem described by others [14,17,21,22]. We found that many of these patients often have ECG features that contribute to false arrhythmia alarms, including: bundle branch block (BBB), ventricular paced rhythms, and low amplitude QRS complexes [3,23,24]. The ECG features associated with false arrhythmia alarms will be described in more detail below.

2.2. Electrocardiographic Features Associated with False Arrhythmia Alarms

2.2.1. Right or Left BBB

Patients with persistent or intermittent right or left BBB generate a high number of false alarms for ventricular tachycardia. This problem occurs as the wide QRSs associated with VT are mistaken as a ventricular rhythm. In a multivariate analysis, we found that patients with right or left BBB were 2.2 times more likely to generate false alarms when compared to patients without this ECG feature (p = 0.020) [3]. Current bedside monitor algorithms lack the ability to recognize right or left BBB. However, this ECG feature, which is present in <10% ICU patients [4], causes a significant number of false alarms, as shown in Figure 2.

2.2.2. Ventricular Paced Rhythms

False alarms for ventricular arrhythmias (i.e., VT or accelerated ventricular rhythm) are common in patients with ventricular paced rhythms. Figure 3 is a false alarm for accelerated ventricular rhythm (AVR; defines as a wide QRS < 100 beats/min) in a patient with a ventricular pacer. Note the pacer spikes before each QRS. The bedside monitor requires the nurse to activate the Pace Mode feature, which adjusts the filter settings in order to detect pacemaker stimuli (i.e., spikes). However, the Pacer Mode feature was not turned on (star) in this patient, which led to non-stop false AVR alarms. We found that the Pacer Mode had been activated in only 33% of the patients with a ventricular pacer [2,25].

2.2.3. Low Amplitude QRS Complexes

We found that low amplitude QRS complexes can cause false asystole alarms [26]. This type of QRS feature can occur in morbidly obese patients, pericardial effusion, and/or BBB. Figure 4 illustrates an outlier patient from the UCSF Alarm Study with low amplitude QRSs who generated 45% of the 12,671 annotated alarms. Note, low amplitude QRS complexes in the limb leads (*), but not in the precordial leads, due to left BBB [2].

The American National Standard (ANS) for cardiac monitors, heart rate meters, and alarms states that ECG devices should not detect a QRS if the waveform is less than 0.15 mV (1.5 mm) in size [27]. This standard was designed to prevent misdiagnosing P waves as QRSs during ventricular standstill. However, some manufacturers use higher QRS detection thresholds (e.g., 0.5 mV or 5 mm) and require that this higher threshold be present in more than one ECG lead. These stricter thresholds result in undercounting the heart rate and cause false asystole alarms [24]. We found that when we examined all seven available ECG leads in patients with these types of alarms, a QRS was readily visible in one or more leads in 91% of false asystole alarms [2].

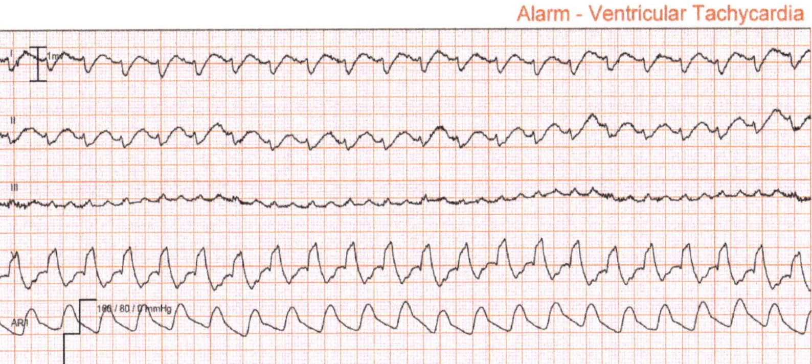

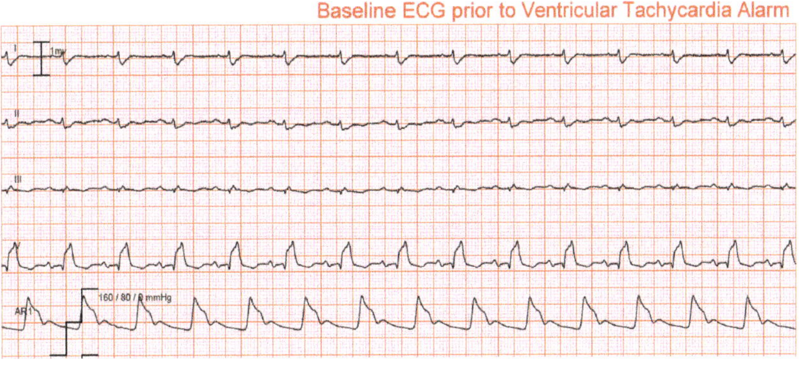

Figure 2. The figures above are rhythm strips showing leads I, II, III, and V (V1—at our hospital) and an arterial blood pressure waveform (AR1) in an intensive care unit patient. The top rhythm strip is an alarm for ventricular tachycardia (VT) during acute respiratory distress. Note the arterial blood pressure is unchanged at 146/89 mmHg and the Sp02 is 96% during the alarm. The bottom figure is a rhythm strip prior to the VT alarm showing normal sinus rhythm, first degree atrioventricular block with right bundle branch block (BBB). The arterial blood pressure is 138/59 and Sp02 is 99%. The QRS morphology in lead V1 in the top strip is identical to that of in the bottom tracings; hence, the top strip in not VT, but rather sinus tachycardia in a patient with right BBB during acute respiratory distress. This patient had a total of 79 false VT alarms (wide QRS heart rate > 100 beats/min) and 120 alarms for accelerated ventricular rhythm (wide QRS < 100 beats/min) during a seven-day ICU stay, illustrating the issue of false alarms due to BBB. Figure from the ECG Monitoring Research Lab, UCSF School of Nursing.

ECG Artifact: Motion artifact is a common cause of false alarms. Typically, the artifact is due to a specific skin electrode on the body surface. The artifact causes pseudo-ectopic beats to be detected, which in turn can trigger a false alarm. Figure 5 illustrates a false V-fib alarm in a patient using their right hand to scratch their head; thus, causing artifact at the right arm (RA) electrode. Artifact in the RA appears in all of the ECG leads that use the RA in their definition. In this example, the leads affected are I, II, and V, the latter as the reference for V is the Wilson terminal [28], which depends in part on the RA electrode. However, lead III (denoted by *) is not affected, as this lead uses the left arm and left leg electrodes, and not the RA.

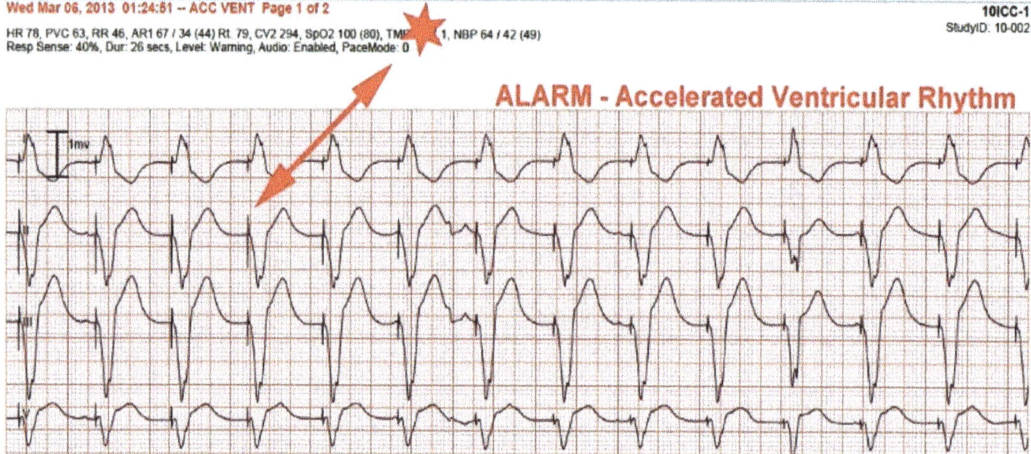

Figure 3. False alarm for accelerated ventricular rhythm defined as >6 ventricular beats with a heart rate between 50 and 100 beats/min in a patient with a ventricular paced rhythm. Figure from the ECG Monitoring Research Lab, UCSF School of Nursing. The Pacer Mode feature was not turned on (star) in this patient, which led to non-stop false AVR alarms.

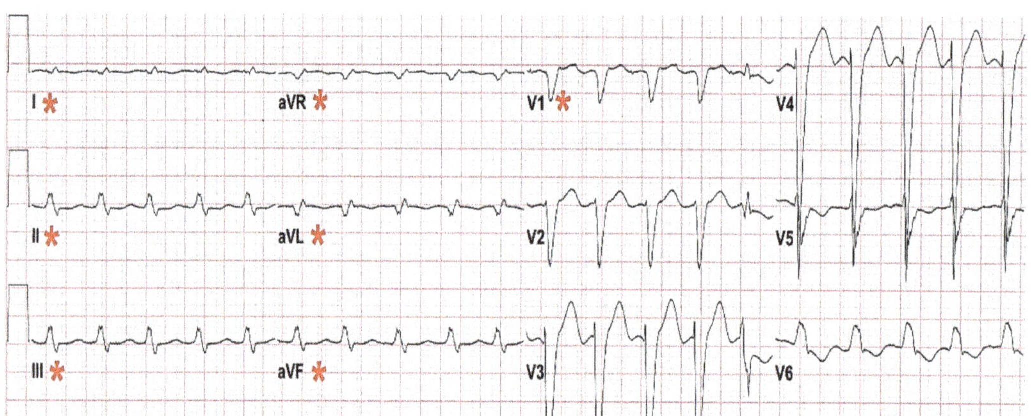

Figure 4. Low amplitude QRS complexes in the limb leads in a patient with left bundle branch block. The * (red) denotes the leads available on the bedside ECG monitor. Reprinted from Ref. [2].

Based on data from the above-mentioned studies, our group at the Center for Physiologic Research in the UCSF School of Nursing, are exploring new algorithms to improve arrhythmia detection of lethal arrhythmias for asystole, ventricular fibrillation (v-fib), and ventricular tachycardia (VT). In order to test the accuracy of these algorithms, our group has established an annotation protocol to examine true versus false alarms for lethal arrhythmias (i.e., asystole, V-fib, and VT) from a dataset of over 6100 ICU patients. Below, we will describe the dataset and the annotation protocol. The final annotation effort has not been published; hence, the focus of discussion in this chapter will be a description of the annotation protocol.

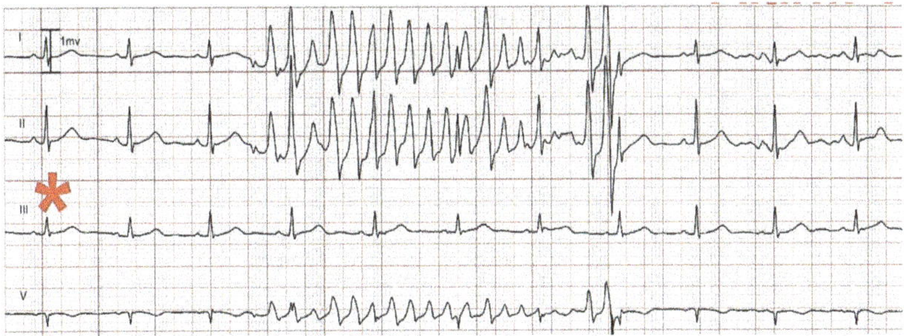

Figure 5. False alarms for ventricular fibrillation. As noted by the *, in lead III, this patient is in a normal sinus rhythm. This particular manufacturer requires a clean ECG signal in at least two ECG leads. Figure from the ECG Monitoring Research Lab, UCSF School of Nursing.

3. Data Capture System

Capturing bedside monitoring data is very challenging, and most frequently limited to a few parametric outputs, such as alarms and vital sign trends. The storage of the actual waveforms (raw data) only occurs under rare, typically research-based circumstances, and requires the implementation of costly and sophisticated data capture infrastructures. In addition, waveform storage typically requires the application of third-party software and hardware, making the workflow even more complex and cumbersome. Only recently have a handful of companies started to provide package solutions that allow seamless data flow from the patient bedside monitors to a hospital-based server and/or data storage center (on or off site).

The consequences of these complexities are that research is easily discouraged or, even worse, is based on bad data (i.e., with gaps or not linked to the correct patient). Indeed, based on the number of false alarms reported in the literature [2,3,14–17,19,21,22,29–33], it is hard to imagine clinically meaningful changes coming out of unsupervised physiologic and alarm data generated by current bedside monitor systems. This is a particularly important point due to the use of artificial intelligence methods and models that are currently very popular approaches used for predicting untoward patient events (i.e., code blue, sepsis). However, these approaches require high quality input data in order to train and test the underlying model being used. Unfortunately, these approaches are being used widely and are being introduced into clinical care.

In the above-described observational study from 2013 [2], all of the available physiologic data were collected using a sophisticated closed network system that connected data from all 77 ICU monitors and the central monitoring station via a gateway system. The waveform data were ultimately sent to a secure server in our research lab for off-line analysis. The following data were collected from each ICU monitor: (1) all available waveforms (e.g., ECG, arterial BP, central venous pressure, intracranial pressure, and SpO2); (2) vital signs (e.g., heart rate, non-invasive BP, and respiratory rate); (3) alarm settings (i.e., crisis, warning or advisory, and message/technical); as well as (4) audible and inaudible alarms. Our hospital's Institutional Review Board approved the study waiving patient consent as bedside monitoring is a standard of care, and we examined the data retrospectively; hence, we collected data from consecutive ICU patients.

While we captured an extremely robust dataset, several challenges had to be overcome in order to use the data in a meaningful way. For example, due to the need for third party hardware to store the data, the data were often not well organized, making it difficult to synchronize, not only the physiologic data, but the electronic health record data with individual patients. In addition, as ICU patients are often in the unit for multiple days and even weeks, the size of the files was very large. The above issues were overcome by our group using several approaches. Our Center has developed an automated procedure that re-organizes all the waveform and alarm data into 24-h time blocks. This standardization makes it much easier to analyze both individual and cohorts of patients based upon the research question. The waveforms are also converted and assembled into a binary public format that is suitable for processing by any algorithm; this fosters data sharing, whenever applicable. In addition, we performed an audit of physiologic alarms captured in an individual patient to that of the electronic health record to verify that our data matched with the patient's electronic health record. As of today, our center has collected and stored bedside monitoring data from all of the adult UCSF ICUs since the pioneering research that started in March 2013; thus, making this the largest database of continuous physiologic data in consecutive ICU patients.

4. Tackling Alarm Fatigue: The Establishment of an Annotated Database Using Bedside Monitoring Data from Consecutive Intensive Care Unit Patients

Alarm fatigue is a substantial problem which is difficult to deal with. Part of the reason is certainly the poor motivation from the industry to invest in and improve existing algorithms. However, even more importantly, the most critical aspect is the lack of a robust database for use as a benchmark for testing of existing and newly designed algorithms. The latter problem has actually raised the attention of regulators, in particular the Food and Drug Administration (FDA), who are seeking an adequate (large, digitized, multi-signal) database to develop, test, and qualify new algorithms. As of today, the benchmark dataset used for medical device testing consists of small number of 30-min 2-channel ECG recordings. In addition, these data are more than 30 years old and from analog Holter recorders with only a few critical arrhythmia events. Therefore, any meaningful answers to alarm fatigue from currently available databases will be limited. Thus, there is a need for an adequate, ideally robust, benchmark dataset from bedside monitors that are currently used in the hospital setting in order to solve this critical problem. Such a dataset should include all of the waveforms typically acquired (i.e., not limited to only ECG) so as to enhance algorithms using multiple signals. To meet this important and yet unmet need, and building on the well-established UCSF Alarm Study that was started in March 2013, our Center is taking on this ambitious challenge.

Using a subset of the existing database, we have assembled 20 months of data (100 patient years) that has been dedicated towards this effort. For this subset of data, we acquired all seven ECG leads (i.e., I, II, III, aVR, AVL, aVF, and a V lead (V1), the latter being the hospital default) and the following waveform data: plethysmograph; transthoracic impedance; and arterial blood pressure. The final sample includes 6143 ICU patients with one or more lethal arrhythmia types (i.e., asystole, V-fib, and VT). The sample was comprised of 46% women and 31% ethnic minorities. Three ICU types were included: Cardiac (16 beds), Medical/Surgical (32 beds), and Neurologic (29 beds). As mentioned previously, the UCSF Institutional Review Board approved the study with waiver of signed patient consent as physiologic monitoring is standard care in the ICU and our data were analyzed off-line and retrospectively; hence, the data are in consecutively enrolled ICU patients.

The seven ECG channels, as well as the other waveform data, were processed by the Center for Physiologic Research lethal arrhythmia algorithms. The physiologic data were converted into a suitable public domain format employing a lossless compression architecture capable of achieving a compression rate of 4:1. For example, the size of a 24 h record inclusive of both ECG and non-ECG waveforms, after compression, is about 100 MB. To make the files more manageable for annotating, multi-day ECG recordings will be separated into 24-h blocks, using midnight to 23:59 PM as the timeframe reference. A HIPAA compliant patient de-identification procedure was applied to the files for annotation, and each record was assigned a unique study ID number. In addition to demographic anonymization, a random date shift was applied to the start date, but not to the start time, for each recording.

The de-identified data were loaded into the Continuous ECG Recording Suite (CER-S, AMPS-LLC) platform for review. A customized version of CER-S has been specifically designed for annotating the database by five PhD prepared ECG expert nurse scientists (protocol described below). Annotators were able to connect remotely to the UCSF network via a dedicated and secure server. Each of the five annotators was assigned arrhythmia alarms to annotate as shown in Figure 6.

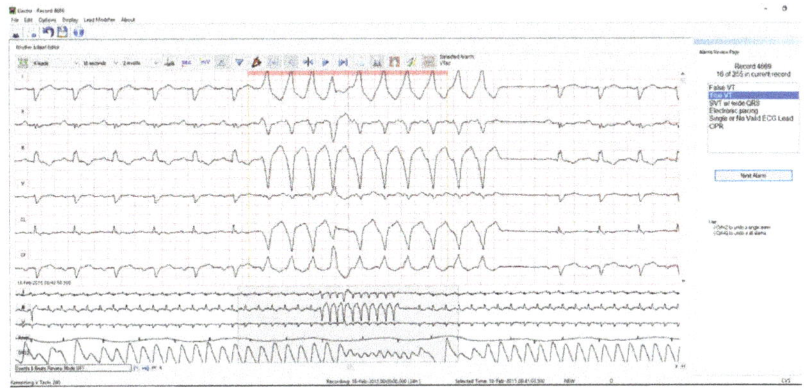

Figure 6. Illustrates an arrhythmia alarm for ventricular tachycardia ready for annotation using the Continuous ECG Recording Suite (CER-S) software program. The annotator selected a response (true, false, etc.) and then selected "Next Alarm" to move onto the next annotation. Note that an Sp02 waveform is shown in the bottom context view for use by the annotator when making a decision. Arterial blood pressure is also available if the patient has this device in place. Figure from the Center for Physiologic Research, UCSF School of Nursing.

5. Annotation Protocol

The annotation protocol was designed as a multi-tiered, multi-expert, ground truth, manual annotation protocol used for triple ascertainment of lethal arrhythmia events with disagreements resolved by a fourth expert. The annotation team (A-Team) was made up of five nurse scientists with decades of experience as bedside clinical nurses and interpreting hospital based-ECG bedside monitoring data. Each of the lethal arrhythmia alarms was randomly assigned to three of the five annotators. Alarms with consensus (all three annotators agree), were finalized. Alarms for which only two annotators agreed were re-assigned to a fourth annotator. If the disagreement was resolved (i.e., fourth annotators concurred with the two initial annotators) the annotation was finalized. If the disagreement was not resolved, the alarm was reviewed by two independent reviewers for a final decision.

Assessment of False Negatives

In addition to the annotation effort described above using the Center's lethal arrhythmia algorithm, the incidence of false-negative alarms (i.e., undetected alarms) will be assessed in a subsequent study. For each 24-h data block, a 15-min data segment will be randomly selected (i.e., regardless of whether it contained any true alarm already identified). The same five annotators reviewed each of the 15-min segments, marking the presence of any lethal cardiac arrhythmia.

The annotation effort using the 20-month database (both true versus false and false negatives) is still underway, and the final data will be published at a later date. The goal of this effort is to improve current algorithms used in bedside physiologic monitors by reducing false arrhythmia alarms that are associated with alarm fatigue in clinicians and the associated patient hazards related to missed true events. Furthermore, both monitoring manufacturers and regulatory agencies will benefit from this work, as they work towards solutions to this complex problem.

6. Discussion

In this review we discuss the issue of false arrhythmia alarms that contribute to alarm fatigue in clinicians working in hospitals. While nurses/physicians experience alarm fatigue from repeated exposure to alarms, patients are subjected to stress, both psychological (i.e., fear, anxiety) [34,35] and physiological (i.e., increased heart rate and blood pressure, sleep deprivation, and delirium) [36,37] from alarm noise. Patients report being frightened by frequent alarms that often go unanswered [34]. Sentinel events have also been reported due to missed true events that are buried among high numbers of false alarms, with resultant mortality and morbidity among hospitalized patients. As such, a number of federal and professional organizations have issued alerts about alarm fatigue, yet very few interventions have had a substantial and sustained impact on this problem [7–10,13].

A widely held belief is that poor skin electrode contact causes false arrhythmia alarms. However, we found that only 9% of false arrhythmia alarms were due to poor signal quality (i.e., unanalyzable due to excessive noise, baseline wander, or leads off). Rather, our group has shown that false arrhythmia alarms are more common in patients with ECG features such as bundle branch block, a ventricular pacemaker, and/or low amplitude QRS complexes [3,24,25,38]. Of note, these features, present in only a few patients, were shown to be responsible for 60% of false alarms. Our published research suggests that clinical interventions (i.e., skin electrode changes, alarm adjustments, and education) will not address the vast majority of false arrhythmia alarms as the central problem are deficiencies in currently used arrhythmia algorithms. Importantly, algorithms used in current monitors were developed using three existing databases created in the mid 1970s (i.e., AHA-ECRI, the CUBD, and the MIT/BIH Databases). None of these databases include digitally acquired ECGs, or any other signals currently acquired in modern patient monitors (Sp02 or arterial blood pressure). Additionally, only a small number of patients and arrhythmias were captured during only hours of ECG monitoring. Therefore, new algorithm development will remain stagnant until more robust data sets are available to test and improve arrhythmia algorithms.

Alternative Approaches and Future Directions

In this chapter, we describe an effort to improve the detection of lethal arrhythmias (i.e., asystole, V-fib, and VT) during in-hospital ECG monitoring by annotating a large database ($n > 6100$ ICU patients) using currently available bedside monitors (i.e., digitized multi-channel ECG and all available physiologic signals). The ultimate goal of this effort is to create a database that can be used to develop and test new algorithms that incorporate not only ECG, but existing physiologic signals (i.e., SpO2 or arterial blood pressure). It should be noted that a number of studies have been published using varied algorithm-based approaches, mostly machine learning, to address false lethal arrhythmia alarms [39–49]. While several of these studies have shown improved detection of VT, all of these studies have used existing databases, which have limitations as stated above (i.e., decades old, non-digitized, two-channel ECG and one or two physiologic signals, small sample of patients and arrhythmias, sampling bias, and recordings of short duration). Therefore, future studies are needed to examine these novel approaches using contemporary data.

In addition to algorithm-based solutions, there is a need to develop and test new technologies (wearables and sensors) as well as "smart alarms" that integrate multiple physiologic parameters [50]. Clinicians would also benefit from the integration of multiple data elements (i.e., ECG, physiologic data, electronic health record, laboratory and pharmacologic data) displayed in such a manner that the interpretation of complex multi-layered data elements can be completed promptly. Lastly, there is a need to pair algorithm-based solutions with patient outcomes, which could be used to guide default settings in bedside monitors to differentiate actionable from non-actionable arrhythmias.

7. Conclusions

In this chapter a number of research studies that have examined false arrhythmia alarms during in-hospital ECG monitoring in the intensive care unit were reviewed. The central problem that is created from false arrhythmia alarms is alarm fatigue (i.e., desensitization), which directly impacts nurses and providers who are exposed to high numbers of false alarms but also patient and their families who hear the alarms and wonder if something is wrong. Discussed were a number of patient, clinical and ECG waveform factors associated with false arrhythmia alarms. Our research group at the UCSF Center for Physiologic Research is using a very large database of over 6100 ICU patients with multi-lead and multi-parameter data to develop and test new algorithms to identify lethal arrhythmias, namely asystole, V-fib, and VT. We described the protocol that will be used to annotate these data which we anticipate will serve as a benchmark database that will move the science forward towards meaningful change.

Author Contributions: Conceptualization, M.M.P., D.M. and F.B.; writing—original draft preparation, M.M.P.; and writing—review and editing, D.M. and F.B. All authors have read and agreed to the published version of the manuscript.

Funding: This research received no external funding.

Institutional Review Board Statement: The studies reviewed in this chapter were conducted according to the guidelines of the Declaration of Helsinki, and approved by the Institutional Review Board of The University of California, San Francisco California (IRB #12-09723 original approval 08/29/2012 with ongoing approval).

Informed Consent Statement: For all of the published studies presented in this review, patient consent was waived because physiologic monitoring is used as a standard of care in all of the intensive care units and the data were analyzed off-line and retrospectively and therefore, were not used for clinical decision making.

Data Availability Statement: No new data were created or analyzed in this study. Data sharing is not applicable to this article.

Conflicts of Interest: The authors declare no conflict of interest.

References

1. Cvach, M. Monitor alarm fatigue: An integrative review. *Biomed. Instrum. Technol.* **2012**, *46*, 268–277. [CrossRef] [PubMed]
2. Drew, B.J.; Harris, P.; Zegre-Hemsey, J.K.; Mammone, T.; Schindler, D.; Salas-Boni, R.; Bai, Y.; Tinoco, A.; Ding, Q.; Hu, X. Insights into the problem of alarm fatigue with physiologic monitor devices: A comprehensive observational study of consecutive intensive care unit patients. *PLoS ONE* **2014**, *9*, e110274. [CrossRef] [PubMed]
3. Harris, P.R.; Zegre-Hemsey, J.K.; Schindler, D.; Bai, Y.; Pelter, M.M.; Hu, X. Patient characteristics associated with false arrhythmia alarms in intensive care. *Ther. Clin. Risk Manag.* **2017**, *13*, 499–513. [CrossRef] [PubMed]
4. Kowalczyk, L. MGH Death Spurs Review of Patient Monitors. Available online: http://www.boston.com/news/health/articles/2010/02/21/mgh_death_spurs_review_of_patient_monitors/ (accessed on 6 January 2016).
5. The Joint Commission. National Patient Safety Goal. Available online: http://www.jointcommission.org/assets/1/18/JCP0713_Announce_New_NSPG.pdf (accessed on 15 July 2021).
6. Ruskin, K.J.; Hueske-Kraus, D. Alarm fatigue: Impacts on patient safety. *Curr. Opin. Anaesthesiol.* **2015**, *28*, 685–690. [CrossRef]
7. Keller, J.P., Jr. Clinical alarm hazards: A "top ten" health technology safety concern. *J. Electrocardiol.* **2012**, *45*, 588–591. [CrossRef] [PubMed]
8. The Joint Commission. National Patient Safety Goals. Available online: https://www.jointcommission.org/-/media/tjc/documents/standards/national-patient-safety-goals/2020/npsg_chapter_hap_jul2020.pdf (accessed on 9 June 2021).
9. American Nurses Association. Medical Alarm Safety in Hospitals. Available online: http://www.nursingworld.org/MainMenuCategories/ThePracticeofProfessionalNursing/Improving-Your-Practice/One-Strong-Voice-Clinically-Speaking/Medical-Alarm-Safety-in-Hospitals.html (accessed on 15 July 2021).
10. American Association of Critical Care Nurses. Alarm Fatigue. Available online: https://www.aacn.org (accessed on 15 July 2021).
11. Phillips, J.; Barnsteiner, J.H. Clinical alarms: Improving efficiency and effectiveness. *Crit. Care Nurs. Q.* **2005**, *28*, 317–323. [CrossRef] [PubMed]
12. Whalen, D.A.; Covelle, P.M.; Piepenbrink, J.C.; Villanova, K.L.; Cuneo, C.L.; Awtry, E.H. Novel approach to cardiac alarm management on telemetry units. *J. Cardiovasc. Nurs.* **2014**, *29*, E13–E22. [CrossRef]

13. A Siren Call to Action: Priority Issues from the Medical Device Alarms Summit. Available online: https://www.aami.org/anthology-alarm-management-solutions/alarm-anthology-siren-call (accessed on 15 March 2021).
14. Graham, K.C.; Cvach, M. Monitor alarm fatigue: Standardizing use of physiological monitors and decreasing nuisance alarms. *Am. J. Crit. Care* **2010**, *19*, 28–34, quiz 35. [CrossRef]
15. Gross, B.; Dahl, D.; Nielsen, L. Physiologic monitoring alarm load on medical/surgical floors of a community hospital. *Biomed. Instrum. Technol.* **2011**. [CrossRef] [PubMed]
16. Ruppel, H.; Funk, M.; Whittemore, R. Measurement of Physiological Monitor Alarm Accuracy and Clinical Relevance in Intensive Care Units. *Am. J. Crit. Care* **2018**, *27*, 11–21. [CrossRef] [PubMed]
17. Sendelbach, S.; Wahl, S.; Anthony, A.; Shotts, P. Stop the Noise: A Quality Improvement Project to Decrease Electrocardiographic Nuisance Alarms. *Crit. Care Nurse* **2015**, *35*, 15–22, quiz 11p following 22. [CrossRef]
18. Sowan, A.K.; Vera, A.G.; Fonseca, E.I.; Reed, C.C.; Tarriela, A.F.; Berndt, A.E. Nurse Competence on Physiologic Monitors Use: Toward Eliminating Alarm Fatigue in Intensive Care Units. *Open Med. Inform. J.* **2017**, *11*, 1–11. [CrossRef]
19. Cvach, M.M.; Biggs, M.; Rothwell, K.J.; Charles-Hudson, C. Daily electrode change and effect on cardiac monitor alarms: An evidence-based practice approach. *J. Nurs. Care Qual.* **2013**, *28*, 265–271. [CrossRef]
20. Watanakeeree, K.; Suba, S.; Mackin, L.; Badinili, F.; Pelter, M.M. ECG Arrhythmia and Tehcnical Alarms during Left Ventricular Assist Device (LVAD) Therapy in the ICU. *Heart Lung* **2021**, in press. [CrossRef] [PubMed]
21. Cvach, M.; Kitchens, M.; Smith, K.; Harris, P.; Flack, M.N. Customizing Alarm Limits Based on Specific Needs of Patients. *Biomed. Instrum. Technol.* **2017**, *51*, 227–234. [CrossRef] [PubMed]
22. Cvach, M.; Rothwell, K.J.; Cullen, A.M.; Nayden, M.G.; Cvach, N.; Pham, J.C. Effect of altering alarm settings: A randomized controlled study. *Biomed. Instrum. Technol.* **2015**, *49*, 214–222. [CrossRef]
23. Nguyen, S.C.; Suba, S.; Hu, X.; Pelter, M.M. Double Trouble: Patients With Both True and False Arrhythmia Alarms. *Crit. Care Nurse* **2020**, *40*, 14–23. [CrossRef]
24. Pelter, M.M.; Fidler, R.; Hu, X. Research: Association of Low-Amplitude QRSs with False-Positive Asystole Alarms. *Biomed. Instrum. Technol.* **2016**, *50*, 329–335. [CrossRef]
25. Suba, S.; Sandoval, C.P.; Zegre-Hemsey, J.K.; Hu, X.; Pelter, M.M. Contribution of Electrocardiographic Accelerated Ventricular Rhythm Alarms to Alarm Fatigue. *Am. J. Crit. Care* **2019**, *28*, 222–229. [CrossRef]
26. Pelter, M.M.; Adams, M.G.; Drew, B.J. Computer versus manual measurement of ST-segment deviation. *J. Electrocardiol.* **1996**. [CrossRef]
27. Association for the Advancement of Medical Instrumentation Center for Radiological Devices & Health U S A Food & Drug Administration. *American National Standard: Cardiac Monitors, Heart Rate Meters, and Alarms*; Association for the Advancement of Medical Instrumentation Arlington: Virginia, VA, USA, 2002; corrected 2008; Volume EC13:2002/(R)2007.
28. Gargiulo, G.D. True unipolar ECG machine for Wilson Central Terminal measurements. *Biomed. Res. Int.* **2015**, *2015*, 586397. [CrossRef] [PubMed]
29. Bonafide, C.P.; Localio, A.R.; Holmes, J.H.; Nadkarni, V.M.; Stemler, S.; MacMurchy, M.; Zander, M.; Roberts, K.E.; Lin, R.; Keren, R. Video Analysis of Factors Associated With Response Time to Physiologic Monitor Alarms in a Children's Hospital. *JAMA Pediatr.* **2017**, *171*, 524–531. [CrossRef] [PubMed]
30. Dandoy, C.E.; Davies, S.M.; Flesch, L.; Hayward, M.; Koons, C.; Coleman, K.; Jacobs, J.; McKenna, L.A.; Olomajeye, A.; Olson, C.; et al. A team-based approach to reducing cardiac monitor alarms. *Pediatrics* **2014**, *134*, e1686–e1694. [CrossRef] [PubMed]
31. Feder, S.; Funk, M. Over-monitoring and alarm fatigue: For whom do the bells toll? *Heart Lung* **2013**, *42*, 395–396. [CrossRef]
32. Khademi, G.; Roudi, M.; Shah Farhat, A.; Shahabian, M. Noise pollution in intensive care units and emergency wards. *Iran. J. Otorhinolaryngol.* **2011**, *23*, 141–148.
33. Siebig, S.; Kuhls, S.; Imhoff, M.; Gather, U.; Scholmerich, J.; Wrede, C.E. Intensive care unit alarms—How many do we need? *Crit. Care Med.* **2010**, *38*, 451–456. [CrossRef]
34. Johansson, L.; Bergbom, I.; Lindahl, B. Meanings of being critically ill in a sound-intensive ICU patient room—A phenomenological hermeneutical study. *Open Nurs. J.* **2012**, *6*, 108–116. [CrossRef] [PubMed]
35. Johansson, L.; Bergbom, I.; Waye, K.P.; Ryherd, E.; Lindahl, B. The sound environment in an ICU patient room—A content analysis of sound levels and patient experiences. *Intensive Crit. Care Nurs.* **2012**, *28*, 269–279. [CrossRef]
36. Berglund, B.; Lindvall, T. Guidlines for Community Noise. Available online: http://www.wholint/docstore/peh/noise/Noiseold.html (accessed on 20 May 2021).
37. Figueroa-Ramos, M.I.; Arroyo-Novoa, C.M.; Lee, K.A.; Padilla, G.; Puntillo, K.A. Sleep and delirium in ICU patients: A review of mechanisms and manifestations. *Intensive Care Med.* **2009**, *35*, 781–795. [CrossRef]
38. Nguyen, S.C.; Suba, S.; Hu, X.; Pelter, M.M. Double trouble: Patients with both true and false ECG arrhythmia alarms and the impact on alarm fatigue. *Crit. Care Nurse* **2019**, in press.
39. Aboukhalil, A.; Nielsen, L.; Saeed, M.; Mark, R.G.; Clifford, G.D. Reducing false alarm rates for critical arrhythmias using the arterial blood pressure waveform. *J. Biomed. Inform.* **2008**, *41*, 442–451. [CrossRef]
40. Baumgartner, B.; Rodel, K.; Knoll, A. A data mining approach to reduce the false alarm rate of patient monitors. *Conf. Proc. IEEE Eng. Med. Biol. Soc.* **2012**, *2012*, 5935–5938. [CrossRef]

41. Desai, K.; Lexa, M.; Matthews, B.; Genc, S. Hemodynamic-impact-based prioritization of ventricular tachycardia alarms. *Conf. Proc. IEEE Eng. Med. Biol. Soc.* **2014**, *2014*, 3456–3459. [CrossRef]
42. Lehman, E.P.; Krishnan, R.G.; Zhao, X.; Mark, R.G.; Lehman, L.H. Representation Learning Approaches to Detect False Arrhythmia Alarms from ECG Dynamics. *Proc. Mach. Learn. Res.* **2018**, *85*, 571–586.
43. Li, Q.; Clifford, G.D. Signal quality and data fusion for false alarm reduction in the intensive care unit. *J. Electrocardiol.* **2012**, *45*, 596–603. [CrossRef] [PubMed]
44. Salas-Boni, R.; Bai, Y.; Harris, P.R.; Drew, B.J.; Hu, X. False ventricular tachycardia alarm suppression in the ICU based on the discrete wavelet transform in the ECG signal. *J. Electrocardiol.* **2014**, *47*, 775–780. [CrossRef]
45. Clifford, G.D.; Behar, J.; Li, Q.; Rezek, I. Signal quality indices and data fusion for determining clinical acceptability of electrocardiograms. *Physiol. Meas.* **2012**, *33*, 1419–1433. [CrossRef] [PubMed]
46. Mousavi, S.; Fotoohinasab, A.; Afghah, F. Single-modal and multi-modal false arrhythmia alarm reduction using attention-based convolutional and recurrent neural networks. *PLoS ONE* **2020**, *15*, e0226990. [CrossRef]
47. Rodrigues, R.; Couto, P. Detection of false arrhythmia alarms with emphasis on ventricular tachycardia. *Physiol. Meas.* **2016**, *37*, 1326–1339. [CrossRef] [PubMed]
48. Au-Yeung, W.M.; Sahani, A.K.; Isselbacher, E.M.; Armoundas, A.A. Reduction of false alarms in the intensive care unit using an optimized machine learning based approach. *NPJ Digit. Med.* **2019**, *2*, 86. [CrossRef]
49. Sadr, N.; Huvanandana, J.; Nguyen, D.T.; Kalra, C.; McEwan, A.; de Chazal, P. Reducing false arrhythmia alarms in the ICU using multimodal signals and robust QRS detection. *Physiol. Meas.* **2016**, *37*, 1340–1354. [CrossRef] [PubMed]
50. Winters, B.D.; Cvach, M.M.; Bonafide, C.P.; Hu, X.; Konkani, A.; O'Connor, M.F.; Rothschild, J.M.; Selby, N.M.; Pelter, M.M.; McLean, B.; et al. Technologic Distractions (Part 2): A Summary of Approaches to Manage Clinical Alarms With Intent to Reduce Alarm Fatigue. *Crit. Care Med.* **2017**. [CrossRef]

Article

Body Surface Potential Mapping: Contemporary Applications and Future Perspectives

Jake Bergquist [1,2,3,*], Lindsay Rupp [1,2,3], Brian Zenger [1,2,3,4], James Brundage [4] and Anna Busatto [1,2] and Rob S. MacLeod [1,2,3]

1. Department of Biomedical Engineering, University of Utah, Salt Lake City, UT 84112, USA; lindsay.rupp@utah.edu (L.R.); brian.zenger@utah.edu (B.Z.); anna.busatto@utah.edu (A.B.); macleod@sci.utah.edu (R.S.M.)
2. Scientific Computing and Imaging Institute, University of Utah, Salt Lake City, UT 84112, USA
3. Nora Eccles Harrison Cardiovascular Research and Training Institute, University of Utah, Salt Lake City, UT 84112, USA
4. School of Medicine, University of Utah, Salt Lake City, UT 84112, USA; James.Brundage@hsc.utah.edu
*   Correspondence: jbergquist@sci.utah.edu

Abstract: Body surface potential mapping (BSPM) is a noninvasive modality to assess cardiac bioelectric activity with a rich history of practical applications for both research and clinical investigation. BSPM provides comprehensive acquisition of bioelectric signals across the entire thorax, allowing for more complex and extensive analysis than the standard electrocardiogram (ECG). Despite its advantages, BSPM is not a common clinical tool. BSPM does, however, serve as a valuable research tool and as an input for other modes of analysis such as electrocardiographic imaging and, more recently, machine learning and artificial intelligence. In this report, we examine contemporary uses of BSPM, and provide an assessment of its future prospects in both clinical and research environments. We assess the state of the art of BSPM implementations and explore modern applications of advanced modeling and statistical analysis of BSPM data. We predict that BSPM will continue to be a valuable research tool, and will find clinical utility at the intersection of computational modeling approaches and artificial intelligence.

Keywords: body surface mapping; electrocardiographic imaging; image processing; clinical applications

Citation: Bergquist, J.; Rupp, L.; Zenger, B.; Brundage, J.; Busatto, A.; MacLeod, R.S. Body Surface Potential Mapping: Contemporary Applications and Future Perspectives. *Hearts* **2021**, 2, 514–542. https://doi.org/10.3390/hearts2040040

Academic Editor: Peter Macfarlane

Received: 2 September 2021
Accepted: 5 October 2021
Published: 5 November 2021

Publisher's Note: MDPI stays neutral with regard to jurisdictional claims in published maps and institutional affiliations.

1. Background

Body surface potential mapping (BSPM) has a long and rich history as a noninvasive technique used to sample the heart's electrical activity by sampling over the entire surface of the thorax. There are numerous reviews that cover the history and utility of BSPM [1–7]. BSPM differentiates from other forms of electrocardiograms (ECGs) by its comprehensive acquisition of bioelectric potentials with the goal of capturing all that is available from the body surface, at the cost of substantial redundancy in information. BSPM was first reported widely by Taccardi et al. as a tool to demonstrate the inadequacies of single-dipole source models to describe cardiac electric sources [8] and many other investigators have demonstrated its superior ability to reveal a wide range of pathologies [6]. Despite its advantages, BSPM is rarely available as part of routine clinical management. However, it remains a useful tool for both research studies and as the input for an imaging modality that seeks to reconstruct cardiac electrical activity noninvasively. In this report, we provide a contemporary view of BSPM and its value in exploring mechanisms of electrocardiography as well as its clinical potential in the setting of electrocardiographic imaging and emerging applications of machine learning.

1.1. BSPM Analysis Approaches

Once the raw electrocardiograms are acquired (see Section 2), there are three main pathways of processing and interpreting BSPM measurements: (1) signal-based analyses,

(2) mapping approaches, and (3) reconstructions of the cardiac sources. Figure 1 illustrates the three approaches. Signal-based analysis encompasses the same set of temporal analysis techniques used for standard ECG signals, applied to many more channels under the assumption that richer sampling will generate richer information. Even though there are multiple signals, each is typically analyzed separately—essentially a one-dimensional approach. The results of the analysis across leads are, of course, also combined to yield a diagnosis or interpretation of events. Mapping approaches are two-dimensional because they leverage both the temporal (signal) and explicitly the spatially dense sampling of BSPM to create maps of bioelectricity. The resulting mapping analysis approaches resemble other two-dimensional imaging modalities, e.g., X-rays or, even more accurately, fluoroscopic images that contain both space and time. Another similarity to medical imaging lies in the distortion intrinsic to these modalities. Medical images are smeared by tissues of differing densities and are captured as shadows of the objects of interest. BSPM signals and the maps constructed from them are also smeared and attenuated compared to potentials measured invasively from the heart, in this case because of the variable electrical resistance of the torso. The third approach in BSPM encompasses additional information in order to remove these distortions and reconstruct the underlying cardiac sources. A medical imaging parallel is computed tomography (CT), which encompasses multiple views and explicit knowledge of the projection direction of each view to reconstruct three-dimensional anatomy. Electrocardiographic reconstruction is known as ECG imaging (ECGI) and like CT, it removes the attenuating and distorting effects of interfering tissues by applying laws of physics, implemented with numerical methods in a computer, to estimate the desired sources or objects of interest.

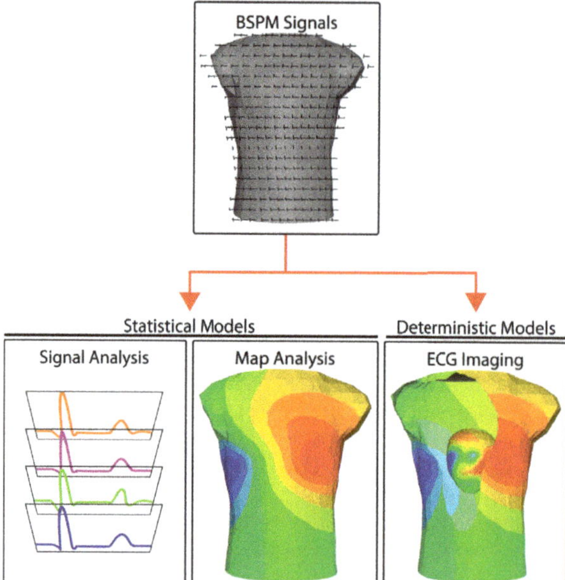

Figure 1. BSPM analysis approaches. BSPM signals are analyzed using one of three different pathways and using two types of mathematical models. Signal analysis methods generally operate on the BSP signals isolated from their geometry. Map analysis extends signal analysis by including the geometry of the torso from which the BSP are recorded. Both signal analysis and map analysis usually rely predominately on statistical models. ECG imaging is based predominately on a deterministic model to reconstructing the cardiac activity at the heart (see the cutaway in the last panel) using the BSP signals and the geometry of the thorax.

1.1.1. Signal-Based Approaches

These methods generally focus on features embedded in the time signals, either single instants during the heartbeat, e.g., ST-segment deviations or peak values of the QRS or T wave, or signal shapes, e.g., P, QRS, or T-wave shapes, durations, and symmetry. Most such features are represented in any type of ECG and their use in BSPM assumes that enhanced spatial sampling will yield improved diagnostics, a hypothesis supported by many clinical applications [4,6,9–13]. Signal-based analysis can also be based on transformations of the time signals into the frequency domain, again similar to frequency domain approaches first reported in standard ECGs. The key feature of all these approaches is that there is no explicit attention to the spatial component of placement or organization of the electrodes, but rather to the signals and the parameters included in the signals. From the image analysis perspective, such approaches resemble using one or more single beams of ultrasound to estimate cardiac motion.

1.1.2. Mapping Approaches

Incorporating explicitly the spatial organization of BSPM measurement into the analysis approaches yields more value from BSPM at the same time as it presents challenges for clinical interpretation. Clinicians are well versed in signal-based approaches from their training and experience with standard ECG. However, there is very little training available to learn how to read the maps that BSPM yields. Maps can take the form of static images in which the body surface voltage distribution at a particular time instant is captured in the form of isovalues, often mapped to color for clarity (see the center panel of Figure 1). The isovalues can be raw electric potentials or parameters derived from the potentials, e.g., specific amplitudes or features from each individual ECG, integral values taken over durations such as the QRS complex, ST segment, or the T wave [6,14]. With the advent of computing and scientific visualization tools [15,16], interactive displays are now common and allow for time-evolving maps under interactive control. When geometric models of the torso surface are available, it is possible to localize the recording electrodes on the surface, even achieve a subject-specific rendering, and display BSPM data over the entire thorax. From this spatial perspective arise new metrics that have been used to identify features of cardiac bioelectricity and disease, both in normal subjects at rest [17,18] and during exercise [19] as well as pathophysiological states such as acute ischemia [7,10,20–23], infarction [24–27], coronary artery disease [28–34], sudden infant death [35], pre-excitation of the ventricles [36], vulnerability to ventricular arrhythmias [37–40], ventricular repolarization [41,42], or the effects of cardiac resynchronization therapy [43].

1.1.3. Reconstruction Approaches

As with all image-reconstruction processes, ECG imaging (ECGI) requires additional information to estimate the sources from remote measurements. Medical image reconstructions, e.g., CT, require multiple views from known directions as well as information about the X-ray opacity of tissues to estimate the anatomy of the hidden objects. ECGI takes raw BSPM signals and augments them with knowledge of the torso anatomy, the locations of the BSPM electrodes, and the electrical tissue conductivities within the thorax to estimate the bioelectric sources in the heart. All reconstruction approaches are based on physical laws; for ECGI, these laws come from electrostatics and lead to solutions to Laplace's and Poisson's equations [44–46]. While these equations are straightforward, solving them in ways that reveal bioelectric sources leads to ill-posed problems, which further results in ill-conditioned numerical problems [47–49]. Thus, every application of ECGI requires a series of related choices, including the bioelectric source model for the heart, the form of the thoracic geometric model and its passive conductivity values, the numerical techniques used to implement the underlying equations describing the electrostatics, and the approaches to deal with the ill-posed nature of the problem. We outline some of these choices in subsequent sections of this paper and refer to excellent reviews for details [45,47,48,50].

Despite the many required choices and challenges, ECGI has been applied to a wide range of clinical situations and is the only BSPM approach implemented commercially. Following a 20 year series of groundbreaking validation studies in animals [51–58] and then in humans [59,60], ECGI has been applied to explore ischemia and myocardial infarction [61], dispersion of repolarization [49], localization and characterization of ventricular [62–65] and atrial arrhythmias [66–68], and the effects of cardiac resynchronization therapy [69]. In addition to diagnostic and mechanistic insights, ECGI can also provide essential guidance for ablation therapies [70–73] and novel, noninvasive treatments of severe arrhythmias by means of radiation therapy [74,75].

1.2. Deterministic versus Statistical Models

A second perspective on the analysis of BSPM is based on the underlying assumptions about cardiac sources (and their reflection on the body surface) and the associated mathematical models. We can re-contextualize the modes of analysis explored above based on what type of model they rely on. The first model is statistical, based on the notion that robust relationships exist between features in the signals and associated aspects of cardiac function or dysfunction. These relationships have little physical basis or physiological origins but instead are gleaned from linking signal features—in time and space—to behaviors of the heart obtained by other means. Once such relationships are identified and verified, they can serve as diagnostic indicators or part of a clinical differential. These statistical models can be based on linear regression or other well-established correlation methods, or they may also derive from the rapidly emerging fields of machine learning and neural networks, in which the statistical model comes from the data. We will focus below on these machine learning approaches as they clearly have enormous potential for new insights that make maximal use of the rich data the BSPM provides.

The second family of models in BSPM analysis is deterministic and is based on physical laws and physiology. Each deterministic approach starts with an explicit model of the cardiac sources and then incorporates the effects of the volume conductor between the heart and body surface. The goal then becomes to manipulate this model in order to reconstruct the cardiac source(s) from the recorded BSPM signals. Deterministic models enable the reconstruction approaches described above and are most common in the setting of ECGI.

Because of their complementary nature, these two families of models are not exclusive, but can rather both be incorporated into a complete analysis technique. Both the signal-based and the mapping analyses rely heavily on the statistical model, while reconstruction-based analyses typically have a greater deterministic component associated to a small statistical one. However, crossover between models exists throughout, e.g., deterministic models assume quasi-static conditions [76], which dictates assumptions in the statistical models that use features of the ECG wave to diagnose disease states [10,11]. Similarly, statistical approaches guide many of the parameter choices required for ECGI, an approach based on deterministic assumptions [77–79].

2. Technical Requirements

The recording, processing, storing, and visualizing of up to hundreds of individual body surface electrograms is a daunting task, with each step between body and computer representing years of progressive innovations and improvements in technology and methodology. In order to understand the contemporary applications and novel breakthroughs in BSPM, it is necessary to appreciate the workflow and technical requirements for acquiring, processing, and visualizing BSP maps. We first assess the added requirements BSPM introduces over standard clinical 12 lead ECG techniques. We then cover the technical requirements for leveraging the spatial information provided by BSPM.

BSPM requires increased sampling over standard 12 lead clinical ECG and many of technological advances have been directed to this goal. Researchers recorded some of the earliest body surface maps using custom silver or steel 5 mm diameter electrodes [8].

Researchers placed these electrodes three to five at a time at pre-determined locations across the entire torso, front and back, and displayed the electrogram traces on cathode ray tube oscilloscope screens. The researchers then took photographs of each set of ECG tracings in sequence and time aligned the resulting 200+ signals using a standard reference signal recorded concurrently. Visualization and evaluation of the resulting signals took place by projecting the films, selecting discrete timepoints across all of the signals, and hand drawing isopotential curves across a flattened map of the torso surface. To capture the time course of the BSP maps required overlaying the isopotential contours on photos of the subject's torso for each time instant and concatenating the resulting frames to form a movie. This process was clearly laborious, with the initial recordings requiring an average of three hours [8], driving a crucial assumption of stability of signal morphology so that subsequent time alignment was possible. These initial BSP maps, with their sparse temporal sampling and laborious acquisition requirements, laid the foundation for decades of technical improvement and a growing understanding of the rich volume of information available. Modern BSPM systems have benefited from decades of advances in recording, storage, processing, and visualization techniques, but still follow roughly the same steps as they did in the 1960s.

2.1. Electrodes

Recording of BSPM differs from recordings of clinical 12 lead ECGs in large part due to the number and placement of electrodes. BSPM recordings usually utilize a leadset consisting of 64 to 300 electrodes spread across the torso surface. These hundreds of leads represent a marked increase from standard clinical ECG systems that commonly use ten physical electrodes (nine recording electrodes and one ground electrode). Modern electrodes are made of either silver, silver-silver chloride, or in some cases graphite and often utilize conductive gel to increase the skin-electrode conductivity, improving signal to noise ratio. For BSPM, these same electrodes can be stitched into strips or even entire vests to aid in rapid and reproducible electrode placement. There is a split in preference between single-use electrodes (higher per-use cost, but ease of manufacture and use) versus reusable electrodes (lower overall cost but more complex design and more constrained placement due to the arrangement of the electrodes and other medical equipment). A common problem is finding a reliable source of electrodes for use in both research and clinical settings. The standard clinical ECG systems utilize single-use electrodes that are often not suitable for BSPM applications due to the high costs and complexity of placement of all 100–300 leads individually. There is a unsolved need to produce high-quality torso surface electrodes for use in BSPM applications such that many institutions turn to internal production. Such custom electrode manufacturing across multiple institutions leads to heterogeneous—and often incompatible—electrode specifications, configurations, and interfaces.

2.2. Leadsets

The selection of lead numbers and locations varies across institutions and scenarios, and a common problem when working with BSPM is the conversion among different lead sets. These differences in number and placement are motivated by various concerns such as diverse philosophies of optimal lead placement, limited acquisition capacity, electrode packaging (individual, strips, or vests), and the need to integrate with other clinical or research equipment (Figure 2). Fortunately, there exist schemes to convert between lead sets and this enables comparisons across institutions [80,81]. Modern lead set placements depend heavily on the use case with a general rule that electrodes should be placed evenly across the thorax, with modifications to increase density in the precordial area [82]. Several studies have sought to identify reduced leadsets based on specific clinical targets [83,84] or reconstruction of full BSPM using linear predictions trained from a database of full-resolution recordings [85–87]. Leadsets such as the Medtronic "ECG Belt" or even the standard 12-lead ECG can drive estimation algorithms that either generate the full BSP maps numerically or may provide enough information for BSPM analysis approaches that are focused on specific diagnostic questions. Researchers developed these techniques to save time and storage space; however, computational resources and storage improvements have made the latter a less relevant concern. Overall, the number and placement of leads are primarily driven by the design of the electrodes themselves, and the use case for the BSP maps.

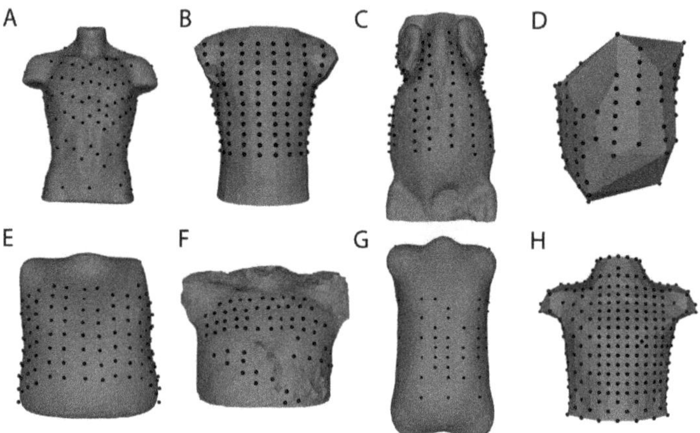

Figure 2. Body surface mapping lead arrangements and torso geometry examples. Bordeaux Torso Tank array (**A**) [88]. Utah Torso Tank array (**B**) [89]. Utah Large Animal Body Surface Map (**C**) [90]. Maastricht Dog Torso Map (**D**) [58]. EP Solutions patient 24 (**E**) [91]. KIT 20 PVC torso (**F**) Karlsruhe Institute of Technology. Nijmegen Human Torso 2004-12-09 (**G**) University of Nijmegen. Dalhausi Human Torso (**H**) ([82]). These geometries and associated body surface data (except C) can be found on EDGAR, a cardiac electrophysiology open database (edgar.sci.utah.edu) (accessed date 29 October 2021) [92].

2.3. Analog Signal Processing

With the electrodes in place on the torso, the main concern becomes maintaining the quality of the recorded signals. Torso surface electrogram voltages are on the order of a few (1–3) millivolts therefor electrical noise is a constant challenge, due mainly to the long connecting wires and the noise generated by other nearby electrical equipment. Analog processing of the recorded signals usually consists of buffering, amplification by a factor of 100–1000, filtering to limit the bandwidth to the physiologically relevant range, e.g., 0.2–300 Hz, and optionally to remove power line noise [93,94]. This analog processing ideally occurs as close to the electrode-tissue interface as possible, with some systems using amplifier circuits built into electrode strips such as the "Active Electrodes" from BioSemi (https://www.biosemi.com/strip_electrode.htm (accessed date 29 October 2021)) [95]. Clinical ECG systems often use a bandpass filter with an upper cutoff as low as 200 Hz. However, BSPM systems have a higher cutoff frequency in the 200 to 1000 Hz region to capture more complex high-frequency components of the ECG signals [93,94]. The cost of higher cutoff frequencies is susceptibility to noise, often requiring special care during acquisition to limit noise at the source, e.g., proper grounding of recording equipment, electrical isolation of the subject, shielding of sensitive devices, and minimizing the use of other electrical equipment during BSP signal acquisition. A further requirement of this stage is subject safety, often requiring optical isolation to limit the maximum ground current to a few μamps even in the face of defibrillation voltages of 600 V. Similarly, protective circuitry is also necessary to protect the recording equipment, including over-voltage protection, electrical isolation via optical coupling, and in some cases, manual disconnects between recording leads and the recording system.

2.4. Signal Acquisition and Digitization

Modern BSPM recording equipment captures the output of the analog stage and converts the signals to digital form to enable subsequent, processing, display and quantitative analysis. Such acquisition occurs with highly multiplexed systems that allow for simultaneous capture of up to 1000 channels at suitable temporal sampling, typically 1000–2000 samples/s. of recordings to be made at a high-temporal and -voltage resolution while also allowing for further digital processing of the signals to improve their quality and aid in analysis. Once digitized, the signals can be processed on a computer using techniques such as simple digital noise filters and complex algorithms to remove baseline drift and other noise sources as well as to segment recorded signals into single beats or other time periods of interest [96]. Ongoing studies comparing the resulting techniques and their impact on subsequent evaluation of BSPM findings suggest that even subtle variations in approaches and parameters can impact the results of BSPM analysis such as ECGI [79,97,98].

2.5. Map Construction

Once the digitized signals have been recorded and processed, the next step is to visualize the signals. Traditionally, recording systems visualize ECG leads as time signals, similar to a standard ECG, often by selecting groups of leads and displaying them in ways that provide either a comprehensive overview of the entire body surface or focused views covering limited areas. However, displaying more than a handful of time series signals becomes overly complex (see the top image in Figure 1). A more common approach is to create spatial maps of body-surface potential that incorporate spatial information from the lead placement into the display of the electrical information. The creation of such a body surface potential map requires a geometric representation of the torso, such that the recorded signals can be arranged relative to each other in space. Such a torso geometry can be generic or patient specific and can either be two-dimensional, based on some form of flattening or unwrapping of the torso surface, or truly three-dimensional such as those in Figure 2. Subject-specific torso geometries are derived from measurements of the electrode locations, typically using a 3D camera, CT or MRI imaging, or a mechanical digitizer. In each case, the electrode positions define a surface mesh, usually via triangulation of the points. Visualization software maps voltage values at the mesh nodes to color and then interpolates the colors over the triangular elements, resulting in a potential map that displays the potential (or other derived quantities) across space. By stepping through time instants, it is possible to create controlled, often interactive, animations of the temporal progression of the spatial patterns of cardiac bioelectricity. Individual timepoints of interest can then be highlighted as in Figure 3. A wide array of visualization options, such as scaling, cropping, or adjusting the color-mapping function provide a rich toolset to interpret the measured values [15,16,99–101].

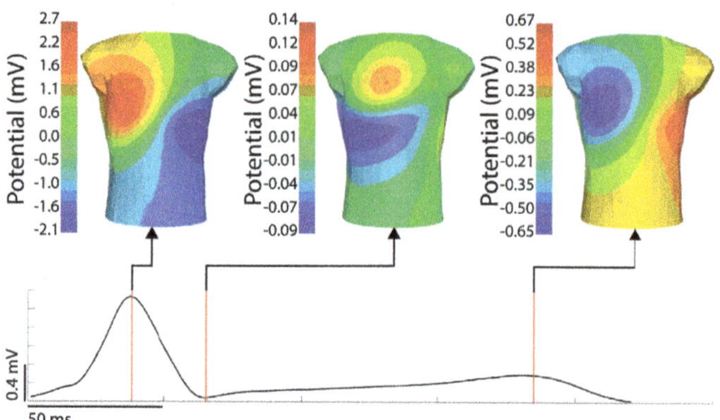

Figure 3. An example BSP map with timepoint of interest visualized. The signal shown is from a stimulated ventricular activation from the anterior left ventricle. The time singnal (bottom) is the RMS of the torso surface signals. The time instances shown are the peak of the RMS QRS, the end of the QRS, and the peak of the T-wave.

2.6. Current State of Mapping Systems

Modern BSPM recording systems can be broken generally into the electrode configuration and the back-end recording hardware and software. This distinction highlights the ability of multiple electrode configurations to be used with a single acquisition system. However, the interface between any given electrode configuration and acquisition system depends heavily on the design of both components, and some electrode systems are custom built for their acquisition hardware. Contemporary electrode configurations typically consist of 100 to 300 electrodes spread across the torso surface. The CardioInsight commercial system, for example, utilizes 242 electrodes stitched into a wearable vest, allowing for rapid and reproducible electrode placement [102]. However, a challenge with such vest systems is ensuring inadequate fitting to different torso shapes and maintaining adequate contact between the electrodes and skin. Other systems such as that from EP Solutions utilize strips of electrodes placed around the thorax up to a maximum of 224 electrodes. The strips allow for more flexible placement at the expense of a more complicated procedural setup. Another innovation to improve signal quality is to include signal amplifiers for each electrode integrated into the strips, called 'active electrodes' by the authors [95]. Studies by other groups describe multiple strips consisting of between 4 and 12 electrodes each that connect to a variable interface system before analog processing [17,86,94,103,104]. An example recording system for experimental use is shown in Figure 4.

Signal acquisition hardware and software are based on either commercial products such as the Biosemi active two system (biosemi.com), or are custom-built systems with integrated analog processing, analog to digital conversion, digital processing, and acquisition [105–107]. While older systems were limited in their recording resolution (both in time and voltage level) and number of simultaneous recording electrodes, modern systems can often record up to a thousand or more inputs at a sampling frequency of over 1000 Hz and voltage resolution in the milli to micro volt range, which is more than sufficient for most BSPM applications. Zenger et al. described a hybrid system (Figure 4) based on custom-designed hardware to interface BSP electrode systems with a commercial recording system (intantech.com) initially designed for neural signals [94]. Such custom systems built for experimental use are designed to be flexible and can often record from electrode arrays on the heart surface or embedded within the cardiac tissue. Commercial systems such as Medtronic, CardioInsight, or EP Solutions use purpose-built hardware and software to process, record, save, and visualize the BSPM, and any post-processed outputs.

Processing pipelines pass the recorded BSP through a range of signal cleaning and visualization software. Rodenhauser et al. describes a software pipeline for filtering, baseline correcting, and segmenting the BSP signals [96]. Visualizing these time series signals as they are recorded is often a feature of the recording system software, allowing for live interrogation of individual or groups of leads. Reconstruction of the torso geometry is typically accomplished via computed tomography, magnetic resonance imaging, or mechanical digitizer. In some cases, a generic torso shape is assumed instead of a patient-specific geometry. In the experimental torso-tank system, the torso geometry is fixed and known ahead of time. The creation of the torso surface meshes is accomplished using segmentation and meshing software. Mapping the processed electrogram data onto the torso meshes to create BSPM is done either by proprietary software, custom algorithms, or open-source software such as map3d [16] and SCIRun [99]. Researchers or clinicians can construct the torso maps either during recording, or after all the recordings have been finished and processed. Since BSPM systems save the recorded signals, they can be reviewed later and post-processed using various tools and approaches to aid in interpretation.

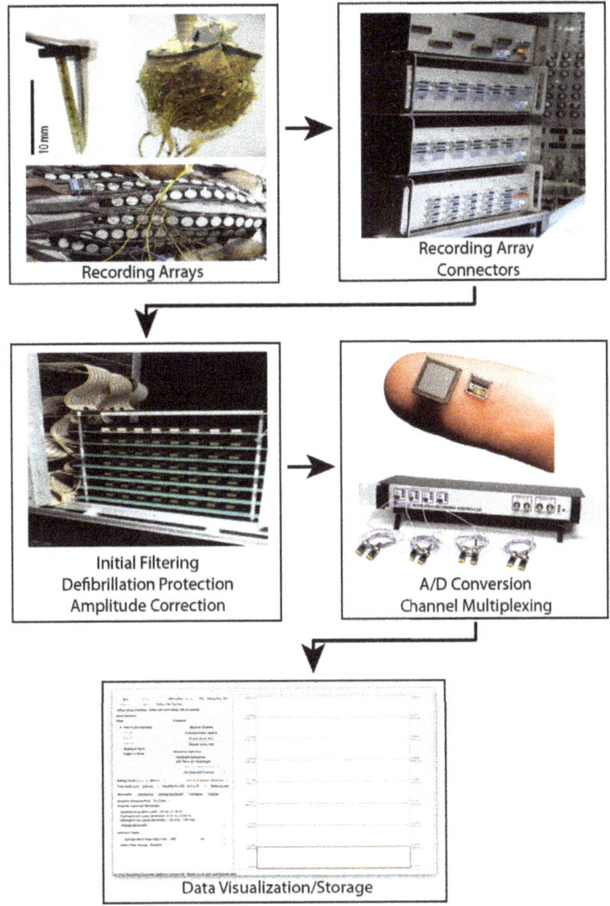

Figure 4. Custom signal acquisition system described by Zenger et al. [90]. This system includes custom electrode arrays, a front-end interface for connecting various electrode array configurations, analog processing, analog to digital conversion by a commercial intan recording system, and data visualization and saving software. The ADC and display software are designed by Intan Technologies (intantech.com) (accessed date 29 October 2021).

3. Technical Extensions

Beyond their inherent information content, BSP maps can serve as inputs for further analysis. BSP maps can be challenging to interpret in their original form and postprocessing and advanced analysis techniques are often necessary to leverage the abundance of information present to provide both clinical and research insights. A natural division among these modes of analysis is between deterministic models, based on physical systems of bioelectric sources and simulations, and data-driven statistical models, without rigid underlying assumptions of physical systems. We will explore here promising and insightful implementations of such analyses of BSP maps, beginning with their technical descriptions.

3.1. Deterministic Modeling: Electrocardiographic Imaging

The first step in a deterministic modeling approach is to devise a method to simulate the potentials on the body surface. In the context of cardiac bioelectricity and BSP mapping, this class of modeling and the associated problems are collectively known as the "forward problem" of electrocardiography. The forward problem relies on assuming an underlying model that expresses the activity of the cardiac bioelectric source (the source model) and a way to project that activity to the measurement location, the torso surface. Source models in cardiac bioelectric activity vary in complexity depending on the specific aspect of interest. For example, a single time-varying current dipole can approximate the bulk activation of the heart, the basis for the Einthoven lead system [108] and still a common source used for clinical interpretation [109]. At the other extreme are models that consider the cellular membrane potentials of each myocyte in the cardiac tissue [14,50]. Across this range of source models, there are specialized methods to project these cardiac source representations to the torso surface. This forward projection takes the geometries and electrical conductivities of the heart and torso along with the source model as inputs to predict the BSP signals. Different implementations cater to different source models, and can include additional details such as the geometries and conductivities of other organs in the torso. Each one of these forward methods implements a physics-based model of bioelectricity to accomplish the projection from cardiac source to BSP signal. The forward problem leads naturally to the more clinically relevant scenario where BSPM is an input to reconstruct the cardiac source itself, known as the "inverse problem" of electrocardiography or electrocardiographic imaging.

Electrocardiographic imaging (ECGI) is an approach that falls into a broad category of inverse problems which focus on reconstructing a source given distant measurements. Figure 5 shows an example ECGI implementation to estimate the cardiac source given BSP recordings via the inverse problem. Inverse problems are generically ill conditioned and ill posed, meaning there is rarely a single unique solution, and small perturbations in the input, such as noise, can have nonlinear effects on the output [48]. A technique known as regularization addresses these numerical instabilities by enforcing additional assumptions on the inverse reconstructions, thereby limiting the solution space to meaningful solutions [48,110]. The ECGI inverse problem takes the mathematical relationship established by the forward problem and applies numerical methods to achieve an estimate of the cardiac source that would give rise to the observed body surface potentials [48]. This reconstruction converts the BSPM into a form that has more direct clinical interpretations than the original BSPM, displaying clinically relevant aspects of the cardiac activity such as sites of early or abnormal activation, abnormal repolarization, re-entrant circuits, or regions of myocardial ischemia [50]. ECGI also enables researchers to noninvasively characterize the heart electrical activity under various experimental scenarios, providing a record of the response of the heart to stimuli without the need for invasive measurements. Such insights come at a cost, a set of additional constraints and considerations above what BSPM alone require.

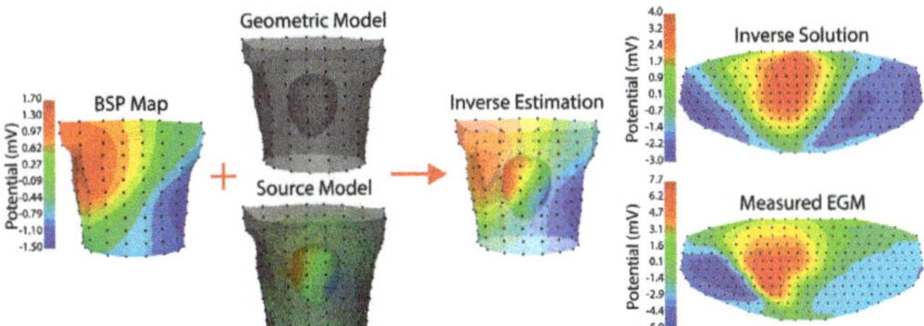

Figure 5. Example ECGI implementation. Recorded body surface potentials (left) are combined with a geometric model and a source model. The geometric model is made up of the relative positions of the cardiac geometry and torso geometry. The source model in this case is extracelular potentials, and the relationship used for the forward model is the boundary element method. The resulting inverse estimation is extracellular potentials on the cardiac surface. The final column shows a comparison between the inverse solution and the measured extracellular potentials on a flattened version of the cardiac geometry. The cardiac and torso geometries were generated as described in Bergquist et al. [111] where the cardiac geometry is a 256 electrode pericardiac cage array and the torso geometry is a 192 electrode torso tank. Tikhonov 2nd order regularization with L curve was used. The peak of the RMS of the QRS was visualized in all steps.

The costs associated with ECGI are diverse and represent a set of compromises associated with each key aspect of the reconstruction process. The need for precise torso and cardiac geometries is often a complicated addition to a BSPM procedure. Imaging techniques such as CT or MRI usually provide this geometric relationship. However, these modalities bring additional procedural complexities and costs to the ECGI setup. Recent studies have investigated reconstructing the cardiac geometry location using only body surface recordings, which would allow for so-called "imageless ECGI" that does not rely on costly CT/MRI [89,112]. Such "imageless" techniques rely instead on iterative optimization frameworks that estimate cardiac position on a beat by beat basis. The selection of a source model and forward/inverse framework is an additional consideration that depends on the cardiac activity of interest. Extracellular potential source models can be susceptible to line-of-block artifacts in subsequent activation sequence estimations, whereas transmembrane potential source models can avoid this problem via smoothing of the reconstruction [113]. However, transmembrane potential source models require a more complex cardiac geometry and can be difficult to interpret, where as extracellular potential source models are readily interpretable. The regularization of the inverse problem remains one of the most challenging aspects of ECGI. A range of regularization techniques have been developed to enforce constraints on the inverse solution based on various assumptions about the cardiac source. Common regularization methods target features of the cardiac source such as its amplitude, spatial gradients, smoothness in space, and presence of edges. Assumptions about these features are formulated into constraints that for example enforce low amplitude (Tikhonov 0 order), enforce small spatial gradients (Tikhonov 1st order), enforce spatial smoothness (Tikhonov 2nd order), or preserve edges (Total Variation) which are all assumed characters of the cardiac source. Other regularization approaches leverage assumptions about the nature of the inverse problem to avoid unstable solutions such as truncated singular value decomposition which removes small singular value components of the forward matrix to avoid the exaggerated effects of noise in these singular vectors when the inverse solution is computed [110]. The amount of influence the regularization has on the inverse solution is determined by a weight associated with the regularization. This weight can be determined in a myriad of ways but always represents a trade off between an over-regularized solution (a solution biased by the assumptions of the regularization) and an under-regularized solution (a solution vulnerable to numerical instabilities of the inverse estimation). Techniques to develop novel regularization methods, implementations, weight selections, and even combination of regularization techniques are an area of active research. Commercial and open-source tools have been developed to address the need to integrate all of the many steps associated with ECGI and allow for extensive exploration of the underlying methods and parameters [50,114,115].

3.2. Uncertainty Quantification

Error and uncertainty are often inherent when identifying parameter values for cardiac simulations such as ECGI, and understanding how variation in the input parameters affects the model output is a field of study known as Uncertainty Quantification (UQ). Techniques for assessing model output uncertainty given variation in input parameters come with a range of complexity and computational cost. Modern UQ methods leverage sophisticated mathematical approaches such as polynomial chaos expansion (PCE), which leverages assumptions about the nature of the stochastic field or process to minimize the number of samples necessary for the computation of accurate statistics [116–120]. ECGI can benefit from such approaches as they allow for the robust exploration of parameter choices along every step of the ECGI pipeline. Previous studies have applied UQ to investigate the effects of organ and tissue conductivity, cardiac position, parameters of ion channel models, and cardiac fiber orientation on various aspects of cardiac electrophysiological models [121–126].

Most UQ studies focus on forward problems; however, there is a need to apply the same UQ techniques to the inverse problem more directly. By understanding how parameter variation affects inverse solutions, both clinicians can better understand the variability of ECGI solutions and researchers can explore which parameters are more or less important when designing new ECGI approaches. Application of UQ to the inverse problem is an area of active research, for example Tate et al. examined heart shape variability and its effect on ECGI inverse solutions [127].

3.3. Statistical Modeling: Machine Learning

In contrast to deterministic approaches such as ECGI, statistical approaches derive their output from features identified and learned in example data. Under these statistical models, the relationship between model input and output does not need to explicitly rely on an underlying physical model. BSPM analysis has benefited from statistical approaches, based either on linear regression approaches [83,128] or decomposition techniques [39,86,129–132]. Machine learning (ML) encompasses a related set of statistical approaches where the data provided to train the model dictates how the model converts the input into the output. These techniques tend to perform best on large datasets, which have become increasingly available in many domains. ML techniques are becoming a vital part of the research and clinical landscape of cardiac electrophysiology [133–136]. They have the advantage of providing concise and interpretable outputs using complex inputs such as BSPM, making them attractive in both clinical and research settings. Here, we will lay out some of the foundational principles of ML and how they may apply to the analysis of BSPMs.

3.3.1. Supervised Approaches

ML approaches can be used to simplify BSPM analysis by answering specific clinical and research questions such as absence or presence of a disease using BSPM as an input. The generation of such ML models occurs in a process called supervised learning, in which the parameters of the underlying ML model, often called 'weights', are generated via a data-driven mechanism termed 'training', which optimizes values of these parameters to produce accurate label outputs on the training data. Training data consists of a set of inputs paired with desired outputs such as labels, quantities of interest, or transformations of the input. In the case of BSPM analysis, these target outputs could be any clinical or research value of interest, including presence of a disease, value of some physiological parameters, or location of a feature of cardiac activity or anatomy to name a few. After training, the resulting algorithm can then receive new BSPM recordings as inputs and output an estimate of the target variable, a process termed 'inference'.

There is an enormous variety in the structure of supervised learning techniques, also called architectures, each with trade-offs regarding task performance, ability to detect relevant signal features, and computational cost. Many architectures were developed in the context of other problems, such as estimation on tabular data (where each column in the dataset represents some descriptive quantity or categorization about a phenomenon), computer vision, and natural language processing [137]. There is not yet a consensus on which architectures perform best in the context of BSPM; thus, it is crucial to consider how to use BSPMs as inputs to a variety of model architectures and to consider the trade-offs associated with these architectures.

Traditional machine learning models were designed for use with vector data, where each input is an n-dimensional vector of real numbers. These architectures include logistic regression, support vector machines, k-nearest neighbors, decision trees, and ensembled forests such as AdaBoost and XGBoost [133]. An intuitive way to apply these architectures to BSPM analysis would be to extract relevant features from the BSPMs (e.g., ST-segment potential, QRS amplitude, T wave integral) and use those features as inputs to the model, as shown in Figure 6. This deliberate feature extraction ensures that the algorithm uses only information that the designer of the architecture has already deemed relevant *a priori*. However, there is evidence that these architectures can perform well when the entire BSPM recordings are used rather than relying on extracted features [138]. Such an approach requires that the BSPM, which is typically represented as an $n \times m$ matrix of n electrodes by m time instances, is linearized into a single $n * m \times 1$ vector (Figure 6). While this approach does not represent a traditional usage of these algorithms, it does allow traditional ML models to determine relevant signal features rather than relying exclusively on features already assumed to be relevant. Such traditional models usually have fewer parameters and thus require less computational resources at inference, increasing the feasibility of clinical deployment but perhaps limiting the scope of problems they can address. Additionally, these traditional model architectures neglect the availability of spatial information that BSPM provide in the form of relative electrode locations and spatial-potential maps.

One can also consider BSPM as images in the context of computer vision ML. Computer vision tasks focus on identifying features or labels given inputs that are images. Tensors represent these images with a shape $c \times w \times h$ where c is the number of channels (usually three for RGB images), w and h are the image width and height, respectively. In the same way, a stack of BSPM leads could be structured as a $c \times m \times n$ tensor where c is 1 (analogous to a single channel grayscale image), m is the number of leads, and n is the number of timepoints in each recording Figure 6. Convolutional neural networks (CNN) are a foundational family of architectures for computer vision that have promise for BSPM. These architectures help detect features with variable locations within an image, and utilize filters that span across all dimensions of the input images. Thus, CNN architectures can leverage the spatial information implied in the BSPM input via the ordering of the electrodes in tensor form. Such an approach, however, does not fully realize the available spatial information, as physical distances between electrodes are not explicitly embedded in the tensor ordering. Additionally, by ordering the BSP recordings into a BSPM tensor, there will inevitably be some electrodes that are physically distant but close to each other along the spatial dimension and other that are physically close but distant in the tensor. Despite this limitation, the incorporation of spatial information implied by tensor ordering can be facilitated with careful architecture design, particularly with respect to the shape and stride of the convolutional filters used. Application of CNNs to 12-lead ECG analysis is perhaps the most common supervised learning approach implemented successfully using these concepts [139], and for this reason, a CNN approach is likely to have success when applied to BSPM. Furthermore, the spatial information present in BSPM could be explicitly included in a CNN-based ML analysis via graph analysis. Graph analysis allows for the encoding of complex spatial relationships into a structured and regular sized input such as the tensors used as inputs for CNN. Dhamala et al. demonstrated such a graph-based approach to characterize spatially heterogeneous scar tissue in a 3D cardiac model via a CNN architecture [140]. Such a graph input could be constructed for the BSPM torso geometry in order to leverage all of the advantages provided by the spatial information encoded in BSPM recordings.

Other models, initially developed in natural language processing (NLP), might also be helpful for BSPM analysis. In NLP, the model input is often a series of embedded words in a space that stores relationships between words. A model then takes all the word as inputs and learns parameters that relate each word to the words before it. Popular architectures in this domain include recurrent neural networks (RNN), long short-term memory (LSTM), and transformers. BSPMs can be restructured for these models by splitting each BSPM recording into different subsets representing the word tokens of the signal. Each BSPM would take the shape of a $w \times m \times s$ matrix, where w is the number chunks (words), m is the number of leads, and s is the number of timepoints per word (calculated as $s = n/w$ where n is the number of timepoints per recording). It is worth noting that of the common architectures, transformers are generally the cheapest computationally and have become a favorite in many NLP applications [141,142]. Transformer networks have also been used with some success in 12-lead ECG analysis tasks, indicating that they may be suited for BSPM analysis as well [139]. NLP architectures carry many of the same benefits of image-based networks when the words are constructed as described above, by splitting the time domain. In this way, each input still carries the spatial relationship of the electrodes implied by their ordering in the inputs. A separate approach would be to break up the BSPM into words based on the electrode configuration across the torso, with similar positions grouped into words.

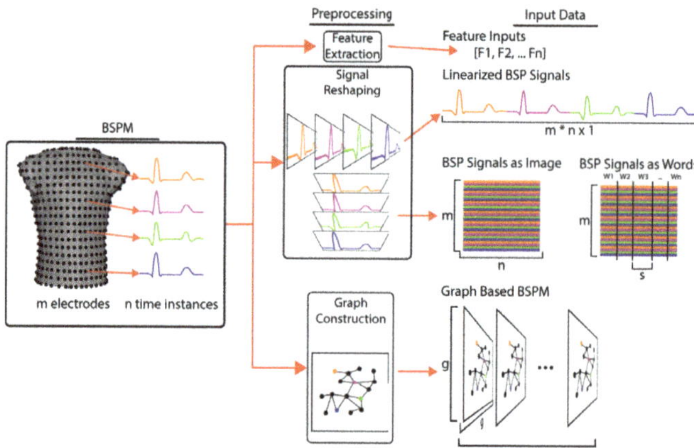

Figure 6. Transformation of BSP maps into inputs for various types of machine learning. BSPM signals are first preprocessed, which varies depending on the type of ML model. For feature-based models, characteristics of the BSP signals (QRS integral, T wave peak, activation time, etc.) are calculated and provided as the input signals. For simple linear neural networks and other vector-based ML models the BSP signals are linearized, concatenating the signal form each electrodes into a single vector. For image- and natural language-based ML models, the BSP signals are arranged into a matrix of m leads by n electrodes, which can then be spilt into s length words. For graph-based ML models, the torso geometry is used to create a computational graph that relates the BSP signals to each other based on their spatial relationships.

Finally, it is possible to combine these families of models. One example, the winner of the PhysioNet 2020 challenge using 12-lead ECGs as inputs, used CNN layers to embed ECG signal features which fed into a transformer architecture [143]. The final layer of this transformer then combined with the top-ranked extracted features as determined by a random forest model. There are many opportunities for creativity and innovation in architecture design and selection so that BSPM ML applications will likely grow. When selecting a model architecture or combination of architectures for a BSPM supervised learning task, it is likely easiest to start with simpler model architectures and increase complexity as needed.

3.3.2. Unsupervised ML Approaches

Machine learning can also simplify BSPM analysis by learning characteristics of a BSPM dataset without explicit labels. The goal of this form of ML is usually to reduce the dimensionality of BSPMs such that they are more easily interpretable in a clinical or research context. Because these algorithms explore relationships in the data without explicit labels related to a specific problem, this type of ML is described as unsupervised. The lower-dimensional BSPMs may be directly visualized or labeled as input to a supervised learning architecture predicting some target label. Some examples of unsupervised learning techniques have been used previously for BSPM and include principal component analysis (PCA) [129,131,132] spectral clustering, and k-means clustering [144–146]. In each case, the BSPM recordings can be restructured to fit the input requirements of the technique. While the lack of a requirement for explicit predefined labels lowers the requirement for use of unsupervised methods to only requiring the data to train on, the resulting lower dimensional outputs can be difficult to interpret. These outputs often require further processing such as in Good et al., where Laplacian eigenmaps was used detect myocardial ischemia [147]. Good et al. computed a secondary metric based on the lower dimensional representation of the electrogram data which they leveraged to detect myocardial ischemia. Additionally, unsupervised networks do not always have an explicit method for incorporating the spatial information that BSP maps provide. Unsupervised methods are often considered to be able to learn any necessary spatial relationships without direct enforcement.

4. Contemporary Applications of Body Surface Mapping

As outlined in Section 1, the interpretation of signals acquired at the body surface can be accomplished using various techniques with a wide range of complexity. These techniques include direct manual examination of the BSP signals and BSP maps, extraction of key features via signal processing, and applying models to convert the input BSP data to readily interpretable outputs such as reconstructions of the cardiac source or clinical variables such as presence or absence of a disease.

4.1. Direct Interpretation of BSP Signals

One goal of research applications of BSPM is on detecting and localizing ischemic cardiomyopathy through experimental models, cardiac stress testing, and acute MI events. Research from our group has focused on interpreting body surface potential changes during hyperacute episodes of ischemia [90,148,149]. Zenger et al. recorded electrical activity on the torso surface using a large-animal experimental model while simultaneously inducing partial occlusion of a coronary artery and applying cardiac stress to create controlled acute myocardial ischemia. We have used these experimental recordings to explain complex physiological differences between various types of cardiac stressors, e.g., pharmacological versus exercise cardiac stress [148]. Furthermore, we identified a complex epicardial shielding phenomenon of ischemic potentials visible within the heart that propagated only partially to the torso surface [149]. Other groups have focused on understanding and improving exercise stress test diagnosis of myocardial ischemia by leveraging BSPM. Kania et al. showed improved sensitivity and specificity of BSPM over 12-lead ECG, when used during an exercise stress test [150]. Specifically, they observed over 20% increase in sensitivity when using BSPM compared to standard 12-lead ECGs. Other groups have replicated these results [151,152]. The reason for the improved sensitivity and specificity is associated with an improved overall coverage of body surface potentials, allowing for capture and examination of ST-segment changes that may not appear initially within the region observed by 12-lead ECGs [152]. Others have validated these results across multiple patient groups, and with gold-standard SPECT imaging [152].

Another major focus for the research uses of BSPM has been the detection and localization of acute myocardial infarction (AMI). Despite many decades of research, detecting AMI from 12-lead ECGs has low sensitivity and specificity. Therefore, further development has targeted new avenues for analysis. Daly et al. showed that BSPM could significantly improve the detection of left main coronary artery stenosis and left circumflex stenosis [153,154]. They found that BSPM demonstrated 89% sensitivity at identifying left main coronary AMI compared to 49% using 12-lead ECG, which the authors attributed to increased coverage of the BSPM system. Other groups have replicated results across broader populations of patients treated at different tertiary care centers [23,155]. Wang et al. identified changes to the U-wave as seen on BSPMs, which correlated to specific locations of AMIs throughout the heart [156]. Finally, other groups have used BSPM to predict overall MI size and severity, and identified changes to the Q-wave on BSPM as moderately correlated with the overall AMI chronic scar formation [157].

Researchers have applied BSPM to examine both the diagnosis and mechanistic development of atrial fibrillation. Recent developments include examining BSPMs in AF patients to understand the role of the dominant frequency. The motivating hypothesis is that sites where AF electrical abnormalities anchor, or dominant frequency sites, could be a target for AF treatment with catheter ablation. Guillem et al. found that the highest dominant frequency sites on the torso showed a significant correlation with dominant frequencies found in the nearer atrium ($\rho = 0.96$ for the right atrium and $\rho = 0.92$ for the left atrium) [158]. Other groups have targeted AF analysis using principal component analysis and wavelet transform on BSPM data to reduce the dimensionality of the signals and gather further insights into the mechanistic underpinnings of AF [159,160]. One study found maps of patients with atrial fibrillation had atrial activity that was dispersed around standard ECG leads and out of recording range, suggesting that noninvasive assessment of AF complexity by the standard 12-lead ECG is not adequate [160].

Ultra high frequency (UHF) has seen applications to ECG analysis that could also readily be transferred to BSP map analysis. UHF has been applied to assess a range of cardiac conditions such as dyssynchrony and myocardial ischemia [161–163]. Extensions of UHF analysis to BSPM could leverage the increased spatial sampling of BSP maps in combination with the increased temporal sampling incorporated into UHF.

BSPM analysis has also been applied in several other research settings that all target electrical changes during complex cardiac disease. These examinations include the development of ventricular fibrillation, ideal parameters and device design for cardiac resynchronization therapies, better understanding the nuances of Brugada syndrome, and investigations into heterogeneity of repolarization times [164–170].

4.2. BSPM Simplification and Interpretation Techniques

The contemporary application of BSPM to clinical needs is primarily focused on simplification. The overwhelming amount of spatial and temporal data available during body surface mapping makes meaningful clinical quantification difficult. Furthermore, in the rush of routine clinical care, a test result must be quickly and easily assessed with high sensitivity and specificity to be incorporated into a standard clinical workflow. Reviewing complex spatial and temporal data of BSPM is essentially impossible without years of training in addition to standard clinical experience. Therefore, advances in BSPM in the clinical setting have primarily targeted the analysis of BSPM data for interpretability.

4.2.1. Deterministic Approach: ECGI

Both the forward and inverse problems of electrocardiography have been used in BSP map analysis. Researchers have used simulation of BSP maps via the forward problem to investigate the presence of different ventricular ectopic activation on BSPM using simulation models to produce the cardiac source [171].

The emergence and development of ECGI as a deterministic tool for the interpretation of BSPM has driven several advances and innovations for BSPM recording and processing. ECGI researchers have developed a range of signal processing techniques to improve BSPM signal quality and address the ill-conditioned nature of the inverse problem [48,110]. These include techniques such as signal averaging across multiple heartbeats, baseline correction, and various noise filters and noise reduction algorithms. The ill-posed nature of the inverse problem has also driven the exploration of the optimal number, and placement of BSPM recording electrodes [83,85,86,172]. Studies by Dogrusoz et al. have investigated various interpolation techniques to reconstruct BSPM measurements lost due to either poor signal quality or obstruction of the leads by other equipment [98]. The notion of reconstructing an entire BSPM from a limited set of leads has been explored by Lux et al. to both reduce data size and enable BSPM to be more readily recorded [85]. ECGI also adds a requirement to measure the cardiac geometry in addition to the torso geometry. Thus, the emergence of ECGI has driven the development of BSPM procedures that include torso and cardiac imaging via CT or MRI. Recent studies by Bergquist et al. and others have investigated reconstructing the cardiac position using only BSPM, opening the avenue for "imageless ECGI" that does not require costly and complex MRI/CT imaging studies [173]. The output of an ECGI system represents the electrical activity of the heart as activation times or electrograms. These reconstructions can be projected onto the cardiac anatomy and allow physicians and researchers to interrogate activity occurring at the heart itself, rather than making inferences from body surface recordings.

ECGI has demonstrated clinical success in several scenarios, including premature ventricular contraction localization, cardiac resynchronization therapy, atrial fibrillation, and ventricular scar-related tachycardia. Contemporary results of these applications have shown success; however, there are still ample avenues for improvement [174–176]. PVC localization via ECGI has shown variable performance, with some approaches reporting 95% accuracy in correctly identifying the site of PVC origination within the AHA 17-segment anatomy [174]. Other studies have shown less promising results with localization accuracy [175,176]. These studies have explored the technical achievements and limitations of PVC localization by ECGI. However, the utility of ECGI through reduced procedure times and better patient outcomes has yet to be thoroughly validated. We hypothesize that as ECGI techniques continue to mature, this transition into meaningful clinical application and assessment of clinical importance will be an area of active research and development.

ECGI has also been used to explore and improve treatment methods of cardiac resynchronization therapy (CRT) [177–179]. Ghosh et al. utilized ECGI to characterize the activation and repolarization patterns of CRT patients [178]. Berger et al. utilized ECGI to explore endocardial and epicardial activation patterns in CRT patients. These authors suggest that ECGI can be used to guide CRT lead placement and pacing strategies [177]. Ploux et al. showed that analysis of ECGI reconstructions in CRT patients could better predict the therapeutic success of CRT than traditional ECG-based metrics [179]. In each of these studies, ECGI provided additional insight into the cardiac bioelectric behavior that allowed for better therapeutic planning or better mechanistic understanding than is possible with standard ECG.

Clinicians and researchers have used ECGI to visualize the electrical patterns of ventricular tachycardia (VT) that occur after significant regions of myocardial scar have formed through other disease processes such as myocardial infarction or ischemic cardiomyopathy [61,115,180–182]. Some studies have investigated VT formation in humans studies and reported accurate results in reconstructing epicardial circuits for re-entry [182] and location of the epicardial and endocardial circuits [183]. Others have used ECGI to detect regions of scar tissue within the myocardial wall and compared these ventricular scar patterns to the gold standard of MRI imaging [181]. Specifically, Horacek et al. examined an improved method for identifying intramural myocardial scar compared to epicardial only approaches [181]. This study showed good agreement to gold-standard MRI imaging with the intramural detection approach. Finally, Cuclich et al. demonstrated that ECGI could be used in a clinical workflow to aid in the identification of re-entrant tachycardic circuits and allow for more precise preoperative planning, reducing overall treatment time [61,115]. These studies demonstrate a promising clinical application of ECGI that has shown promise and progress in recent years. We theorize that the continued development of these techniques will allow clinicians to leverage BSPM via ECGI to improve their clinical workflows.

Atrial fibrillation has also become a focus of ECGI research efforts. ECGI allows for the noninvasive construction of dominant frequency maps on the atria, aiding in driver localization and ablation planning. ECGI has been used to identify regions of dominant frequency and assist physicians in targeting these regions for treatment. Recent studies in this domain have explored the use of ECGI to identify atrial fibrillation drivers using both simulated data and real-world clinical data [159,184–186]. While these methods are still in development, the results show promise for ECGI in the context of AF.

A handful of companies have developed commercial systems to translate ECGI into clinical workflows. The most long-standing of these systems is CardioInsight by Medtronic, primarily used to target ventricular and atrial arrhythmia mapping [71,72]. The Amy-card 01c by EP Solutions is another ECGI system that targets both ventricular mapping approaches (CRT planning and ventricular arrhythmia mapping) as well as atrial mapping [187]. Finally, the Acorys system by Corify is a relatively new system that targets arrhythmia localization and pre-ablation/operative planning applications.

4.2.2. Statistical Approach: Machine Learning/Artificial Intelligence

There has been a steadily growing interest in machine learning application (ML) techniques to analyze cardiac bioelectric signals. To date, most AI/ML studies addressing electrocardiography data utilize 12-lead ECG recordings because of the large volume of data needed across large populations to create accurate and representative ML models. However, ML may provide new and unique insights into using BSPM data to significantly improve diagnostic accuracy without adding significant user training. Applications of ML tools in BSPM are primarily focused on easing interpretation by lowering the dimensionality of BSP maps or by training models to estimate specific clinical variables.

Dimensionality reduction in BSPM via ML represents the complex phenomena captured by the hundreds of body surface recordings in a more compact form. Good et al. demonstrated this by using Laplacian Eigenmaps, a method of unsupervised ML dimensionality reduction, to more quickly and accurately identify myocardial ischemia from recorded electrograms when compared to traditional metrics [147]. While this study primarily focused on electrogram data recorded at the heart surface, Good et al. also explored a limited use case with BSPM data [188]. The output of the Laplacian Eignemap model is a 3D representation of the electrogram time series recorded from hundreds of leads. While this method does compress the data into a smaller dimension, interpretation of the resulting Eigenmap manifold is not straightforward and requires custom metrics and modes of analysis. Interpretation of the output of an unsupervised ML method such as Laplacian Eigenmaps can often be an additional layer of complexity, making direct analysis unsuitable for clinical implementation. Supervised ML models, however, can be designed to output a clinical variable of interest directly.

Supervised ML models have been applied with increasing frequency to identify clinical variables using ECGs as an input. Many of these studies focus on 12-lead ECG data and are associated with competitions such as the Computing in Cardiology Physionet 2020 Challenge, which challenged contestants to detect 26 common cardiac pathologies using ML approaches [189]. Common deep learning architectures used include convolutional neural networks (CNN), long short-term memory models (LSTM), and transformer networks. However, there is some early evidence indicating that combinations of CNNs with Transformers or LSTMs seem to perform well on a variety of tasks [139,143]. Despite the differences in data, we hypothesize that many of the same architectures and innovations developed by these studies can be readily applied to BSPM data. We also predict that the additional information content provided by the increased spatial sample of BSPM data may provide additional benefits such as reduced model complexity, improved performance and allow for the investigation of more complex tasks. Brundage et al. applied both logistic regression (a simple ML architecture) and XGBoost (a low parameter decision tree-based architecture) models to detect myocardial ischemia using 96 lead BSPM animal data which was collected according to Zenger et al. [90]. These models demonstrate >0.95 and receiver operator area under the curve, indicating excellent model performance despite being less complex than the architectures commonly applied to 12-lead ECG classification problems [138].

Others have used ML as a personalization tool rather than strictly for increasing the ease of interpretation. Giffard et al. investigated the use of ML tools which take baseline BSPM recordings as input in predicting two critical parameters of a forward model of cardiac electrocardiography, that estimated the response of the heart to various pacing conditions [190]. Personalization-focused ML tools could be used clinically to tune CRT to patients and improve outcomes. The same group later demonstrated that after training an initial model from scratch on simulated data, including data accounting for various pacing and heart positions, this model could be fine-tuned on patient-specific torso geometries and used to predict the same key forward modeling parameters with greater accuracy and less compute at inference [191]. The significant decrease in computational time achieved by transfer learning on patient-specific geometries makes clinical deployment much more feasible. These results suggest that data from heterogeneous sources, including simulations and animal models, may help develop ML models that could be personalized on a subject-specific basis. Given the lack of large publicly available BSPM datasets for humans, an opportunity to use simulated and animal model data would ease the difficulty of data acquisition.

5. Conclusions and Prospective View

Body surface potential maps represent a wealth of information to serve both clinical and research goals that can be easily and safely acquired from any patient or subject. BSP maps provide increased sampling of cardiac electrical activity over the standard 12-lead ECG. This increased coverage and density have been exploited in a number of research applications to explore the effects of various pathophysiological process and develop diagnosis tools. Another advantage of BSPM is that the analysis of maps can follow a range of diverse pathways. The signals can be directly analyzed to assess cardiac bioelectric response under a variety of conditions, or used as inputs to computational models designed to identify key features of physiology including noninvasively reconstructing cardiac activity on or within the heart. No other ECG modality allows for such detailed examination while also remaining completely noninvasive.

Contemporary BSPM systems see the bulk of their use in research settings and only to a lesser extent in clinical practice, despite the existence of commercial ECGI devices. Research BSPM systems consist almost entirely of custom built acquisition hardware and software, coupled with custom built electrode systems. This heterogeneity between implementations makes comparisons between datasets difficult, but not insurmountable, another byproduct of the rich coverage and sampling density. One challenge that all mapping systems face is the shortage of sources of high-quality electrodes that can readily interface with a BSPM system. Most researchers turn to in-house production of electrodes, leading to further heterogeneity among BSPM datasets. We see the need for a consistent source of high-quality, reasonably priced body surface electrode arrays purpose built for body surface mapping as a major challenge preventing BSPM from being a common research tool.

Implementing BSPM in clinical practice is also hindered by the complexity of BSP map interpretation, the heterogeneity of acquisition and electrode systems, and the lack of established diagnostic tests and routines based on BSP maps. Most contemporary applications of BSPM approach some of these challenges by leveraging methods that take BSP maps as inputs to produce an output that is more readily interpretable than the raw signals. For example, ECG imaging (ECGI) transforms the BSP map, a representation of the cardiac bioelectric activity measured from distant sensors, into a direct representation of the bioelectric activity at the heart. Using ECGI, relevant clinical features such as paths of re-entry, sites of abnormal activation or conduction, and tissue heterogeneities can be directly observed rather than inferred. Acquiring the same information from direct measurements would require invasive procedures that carry significant extra risk and cost when compared to the relative ease of a BSPM measurement.

ECGI has been successfully integrated into some clinical workflows; however, it is far from a common, clinical procedure, despite the apparent advantages. This paucity of use is likely because ECGI still lacks a specific application or a use case that makes a dramatic improvement to clinical practice. Additionally, ECGI still suffers from a number of limitations and common difficulties that likely inhibits its uptake into common clinical workflows. The ill-posed nature of the ECGI inverse problem can lead to erroneous solutions even with sufficient regularization. These errors result in misleading features in reconstructed signals such as errant lines of conduction block. Contemporary research in the development of ECGI seeks to address such errors by a combination of approaches that include advances in pre-processing of BSP maps, improvements to regularization and ECGI implementations, and advances in post-processing of the resulting ECGI reconstructions [97,98,113,192]. Such methodologies have shown promise in resolving these difficulties in ECGI. We hypothesize that as ECGI is developed, honed, and applied to more pathologies in research settings, these improvements will result in a clinical ECGI system that is more fit for a common clinical workflow.

Machine learning models also present an opportunity to use BSP maps as an input to generate a clinically useful diagnostic output. The design and specific output of an ML model is more flexible than ECGI, as ML does not traditionally rely on specific underlying physical relationships but rather learns a highly nonlinear mapping between the inputs and desired outputs. This flexibility in design allows for the development of ML models to address very specific clinical questions, with model outputs such as presence or absence of a specific disease, or value of some relevant physiological measure that would assist in patient diagnosis and treatment planning. ML models can be tuned to answer a much larger set of specific questions than ECGI, so long as there is sufficient data available to train the model. ML present an opportunity to utilize BSPM for diagnosis of more than just cardiac-related pathophysiology. Clinical outcomes such as likelihood of an adverse cardiac event after a procedure may be predictable using ML with BSP maps as the input. While ML has yet to be extensively applied to BSPM data, based on successes of ML models in other biomedical fields and in 12-lead ECG analysis, we anticipate that ML applications leveraging BSPM data will open the door to a wide variety of clinical applications.

Across all instances of BSPM implementation and subsequent interpretation, there has been a common restriction based on the computational power and storage needed to acquire and manage BSP maps. Once a BSPM system is set up, recording additional heartbeats is relatively trivial except for the need for additional computational resources to store and process the additional signals. This restriction based on computational resources, which in early analyses constrained investigations to single heartbeat recordings or even sub-sections of single heartbeats, has rapidly become less of a limiting factor in BSP map analysis. Modern computational resources greatly facilitate the incorporation of entire BSP maps from multiple heartbeats due to dramatically increased storage capacity and processing power. This additional capacity allows for more comprehensive investigations into scenarios with highly dynamic cardiac behavior such as response to pharmaceuticals, acute disease progression, and response to exercise. Modes of analysis, whether traditional signal processing, ECGI, or AI, have the potential to leverage information from multiple heartbeats to better explore the utility of BSP maps and further develop our understanding of the underlying physiology of dynamic beat-to-beat change.

We have so far considered a division between models that are deterministic (e.g., ECGI) and models based on statistical data driven relationships (e.g., ML). However, such a distinction does not preclude the opportunity to combine aspects of each of these modeling approaches. For example, recent studies have demonstrated the ML techniques can account for geometric inaccuracies in the ECGI forward model. Additionally, physics informed neural networks (PINN), which incorporate physics-based deterministic models into ML neural network architectures, have emerged recently as a promising bridge between ML techniques and deterministic models. Machine learning based approaches are well suited for addressing complex nonlinear relationships in a form that can be less computationally taxing and easier to establish than a deterministic model of the same phenomena. However, ML approaches can often neglect physiologically relevant constraints, leading to models that produce non-meaningful outputs under certain circumstances. The combination of ML and deterministic models could offer the best of both worlds, allowing for both accurate and efficient models that are constrained in part by physiological and physical models. We suspect that this intersection between statistical and deterministic models will provide the most fertile ground for the development of future BSPM analysis methods and models.

Author Contributions: J.B. (Jake Bergquist): Conceptualization, Project Management, Primary Writing, Editing and Proofreading; L.R.: Conceptualization, Primary Writing, Editing and Proofreading; B.Z.: Conceptualization, Primary Writing, Editing and Proofreading; J.B. (James Brundage): Conceptualization, Primary Writing, Editing and Proofreading; A.B.: Conceptualization, Primary Writing, Editing and Proofreading; R.S.M.: Conceptualization, Primary Writing, Editing and Proofreading, Funding Acquistion. All authors have read and agreed to the published version of the manuscript.

Funding: This work was supported by NIH NHLBI grant no. F30HL149327; NIH NIGMS Center for Integrative Biomedical Computing (www.sci.utah.edu/cibc), NIH NIGMS grants P41GM103545 and R24 GM136986; the NSF GRFP; the Utah Graduate Research Fellowship; and the Nora Eccles Treadwell Foundation for Cardiovascular Research.

Institutional Review Board Statement: Not applicable.

Informed Consent Statement: Not applicable.

Data Availability Statement: Data used for example figures and visulizations can be found on the EDGAR (edgar.sci.utah.edu) (accessed date 29 October 2021).

Conflicts of Interest: The authors declare no conflict of interest.

References

1. Taccardi, B.; Ambroggi, L.D.; Viganotti, C. Body-Saurface Mapping of Heart Potentials. In *The Theoretical Basis of Electrocardiology*; Nelson, C., Geselowitz, D., Eds.; Claredon Press: Oxford, UK, 1976; pp. 436–466.
2. Taccardi, B. Future Prospects and Applications. In *Body Surface Electrocardiographic Mapping*; Mirvis, D., Ed.; Kluwer Academic Publishers: Boston, MA, USA; Dordrecht, The Netherlands; London, UK, 1988; pp. 193–200.
3. Taccardi, B. Body surface mapping and the cardiac electric sources: A historical survey. *J. Electrocardiol.* **1990**, *23*, 150–154. [CrossRef]
4. Flowers, N.; Horan, L. Body Surface Potential Mapping. In *Cardiac Electrophysiology: From Cell to Bedside*, 2nd ed.; Zipes, D., Jalife, J., Eds.; W.B. Saunders Company: Newberg, OR, USA, 1995; Volume 93, pp. 1049–1067.
5. Taccardi, B.; Punske, B.; Lux, R.; MacLeod, R.; Ershler, P.; Dustman, T.; Vyhmeister, Y. Useful lessons from body surface potential mapping. *J. Cardiovasc. Electrophysiol.* **1998**, *9*, 773–786. [CrossRef] [PubMed]
6. Ambroggi, L.D.; Musso, E.; Taccardi, B. Body Surface Potential Mapping. In *Comprehensive Electrocardiology*; Macfarlane, P., Veitch Lawrie, T., Eds.; Springer: Berlin/Heidelberg, Germany, 2005; Volume 32, pp. 1375–1413.
7. Robinson, M.; Phil, D.; Curzen, N. Electrocardiographic Body Surface Mapping: Potential Tool for the Detection of Transient Myocardial Ischemia in the 21st Century? *Ann. Noninvasive Electrocardiol.* **2009**, *14*, 201–210. [CrossRef] [PubMed]
8. Taccardi, B. Distribution of Heart Potentials on the Thoracic Surface of Normal Human Subjects. *Circ. Res.* **1963**, *4*, 341–351. [CrossRef] [PubMed]
9. Sridharan, M.; Horan, L.; Hand, R.; Orander, P.; Killam, H.; Flowers, N. Use of body surface maps to identify vessel site of coronary occlusions. *J. Electrocardiol.* **1989**, *22*, 72–81. [CrossRef]
10. Kornreich, F.; Montague, T.; Rautaharju, P. Location and magnitude of ST changes in acute myocardial infarction by analysis of body surface potential maps. *J. Electrocardiol.* **1992**, *25*, 15. [CrossRef]

11. Kornreich, F.; Montague, T.; Rautaharju, P. Best ECG leads for diagnosing acute myocardial infarction by multivariate analysis of body surface potential maps. In Proceedings of the IEEE Computers in Cardiology, IEEE Computer Society, Durham, NC, USA, 11–14 October 1992; pp. 439–442.

12. Kornreich, F.; MacLeod, R.; Dzavik, V.; Kornreich, A.; Stoupel, E.; de Almeida, J.; Walker, D.; Montague, T. Body surface potential mapping of QRST changes during and after percutanesous transluminal coronary angioplasty. *J. Electrocardiol.* **1994**, *27*, 113–117. [CrossRef]

13. Kornreich, F.; Lux, R.; MacLeod, R. Map representation and diagnostic performance of the standard 12-lead ECG. *J. Electrocardiol.* **1995**, *28*, 121–123. [CrossRef]

14. Lux, R. Mapping Techniques. In *Comprehensive Electrocardiology*; Macfarlane, P., Veitch Lawrie, T., Eds.; Pergamon Press: Oxford, UK, 1989; Volume 2, Chapter 26, pp. 1001–1014.

15. MacLeod, R.; Johnson, C.; Matheson, M. Visualizing Bioelectric Fields. *IEEE Comp. Graph. Applic.* **1993**, *13*, 10–12. [CrossRef]

16. MacLeod, R.; Johnson, C. Map3d: Interactive scientific visualization for bioengineering data. In Proceedings of the IEEE Engineering in Medicine and Biology Society 15th Annual International Conference, San Diego, CA, USA, 31 October 1993; pp. 30–31.

17. Montague, T.; Smith, E.; Cameron, D.; Rautaharju, P.; Klassen, G.; Flemington, C.; Horacek, B. Isointegral Analysis of Body Surface Maps: Surface Distribution and Temporal Variability in Normal Subjects. *Circulation* **1981**, *63*, 1167–1172. [CrossRef]

18. Green, L.; Lux, R.; Haws, C.; Williams, R.; Hunt, S.; Burgess, M. Effects of age, sex, and body habitus on QRS and ST-T potential maps of 1100 normal subjects. *Circulation* **1985**, *71*, 244–253. [CrossRef] [PubMed]

19. McPherson, D.; Horacek, B.; Sutherland, D.; Armstrong, C.; Spencer, A.; Montague, T. Exercise Electrocardiographic Mapping in Normal Subjects. *J. Electrocardiol.* **1985**, *18*, 351–360. [CrossRef]

20. Kornreich, F.; Montague, T.; Rautaharju, P. Body surface potential mapping of ST segment changes in acute myocardial infarction: Implications for ECG enrollment criteria for thrombolytic therapy. *Circulation* **1993**, *87*, 773. [CrossRef] [PubMed]

21. McPherson, D.; Horacek, B.; Spencer, C.; Johnstone, D.; Lalonde, L.; Cousins, C.; Montague, T. Indirect Measurements of Infarct Size, Correlative Variability of Enzyme, Radionuclear, Angiographic and Body Surface Map Variables in 34 Patients during acute Phase of First Myocardial Infarction. *Chest* **1985**, *8*, 841–848. [CrossRef] [PubMed]

22. Franks, M.; Lawson, L. Body surface mapping improves diagnosis of acute myocardial infarction in the emergency Department. *Adv. Emerg. Nurs. J.* **2012**, *34*, 32–40. [CrossRef] [PubMed]

23. Ornato, J.P.; Menown, I.B.A.; Peberdy, M.A.; Kontos, M.C.; Riddell, J.W.; Higgins, G.L.; Maynard, S.J.; Adgey, J. Body surface mapping vs 12-lead electrocardiography to detect ST-elevation myocardial infarction. *Am. J. Emerg. Med.* **2009**, *27*, 779–784. [CrossRef]

24. Montague, T.; Smith, E.; Johnstone, D.; Spencer, C.; Lalonde, L.; Bessoudo, R.; Gardner, M.; Anderson, R.; Horacek, B. Temporal Evaluation of Body Surface Mapping Patterns Following Acute Inferior Myocardial Infarction. *J. Electrocardiol.* **1984**, *17*, 319–328. [CrossRef]

25. Montague, T.; Johnstone, D.; Spencer, A.; Lalonde, L.; Gardner, M.; O'Reilly, M.; Horacek, B. Non-Q-Wave Acute Myocardial Infarction: Body Surface Potential Map and Ventriculographic Patterns. *Am. J. Cardiol.* **1986**, *58*, 1173–1180. [CrossRef]

26. McPherson, D.; Horacek, B.; Johnstone, D.; Lalonde, L.; Spencer, C.; Montague, T. Q-Wave Infarction: Pathophysiology of Body Surface Potential map and Ventriculographic Patterns in Anterior and Inferior Groups. *Can. J. Cardiol.* **1986**, *Suppl A*, 91A–98A.

27. Ambroggi, L.D.; Bertoni, T.; Breghi, M.; Marconi, M.; Mosca, M. Diagnostic value of body surface potential mapping in old anterior non-Q myocardial infarction. *J. Electrocardiol.* **1988**, *21*, 321–329. [CrossRef]

28. Green, L.; Lux, R.; Haws, C. Detection and Localization of Coronary Artery Disease with Body Surface Mapping in Patients with Normal Electrocardiograms. *Circulation* **1987**, *76*, 1290–1297. [CrossRef]

29. Montague, T.; Johnstone, D.; Apencer, A.; Miller, R.; MacKenzie, B.; Gardner, M.; Horacek, B. Quantitation of Myocardial Ischemia by Body Surface Potential Mapping: Exercise Maps in Patients with Isolated left Anterior Descending Coronary Artery Disease. *Am. J. Cardiol.* **1988**, *61*, 273–282. [CrossRef]

30. Montague, T.; Johnstone, D.; Miller, R.; MacKenzie, B.; Gardner, M.; Horacek, B. Quantitative Body Surface Mapping: Exercise Maps in Patients with Single and Multiple Coronary Artery Obstructions. In *Canadian Cardiovascular Society Annual Meeting*; Canadian Cardiovascular Society: Ottawa, ON, Canada, 1987.

31. Montague, T.; Macdonald, R.; Henderson, M.; Miller, R.; Horacek, B. Quantitative Body Surface Mapping: Resting Maps Before and After Successful Angioplsty. In *Canadian Cardiovascular Society Annual Meeting*; Canadian Cardiovascular Society: Ottawa, ON, Canada, 1987.

32. Montague, T.; Johnstone, D.; Spencer, A.; Miller, R.; MacKenzie, B.; Gardner, M.; Horacek, B. Body Surface Potential Maps with Low-Level Exercise in Isolated Left Anterior Descending Coronary Artery Disease. *Am. J. Cardiol.* **1988**, *61*, 273–282. [CrossRef]

33. Montague, T.; Witkowski, F.X. The Clinical Utility of Body Surface Potential Mapping in Coronary Artery Disease. *Am. J. Cardiol.* **1989**, *64*, 378–383. [CrossRef]

34. Montague, T.; Witkowski, F.; Miller, R.; Johnstone, D.; Mackenzie, R.; Spencer, C.; Horacek, B. Exercise Body Surface Potential mapping in single amd multiple coronary artery disease. *Chest* **1990**, *97*, 1333–1342. [CrossRef]

35. Montague, T.; Finley, J.; Mukelabai, K.; Black, S.; Rigby, S.; Spencer, A.; Horacek, B. Cardiac Rhythm, Rate and Ventricular Repolarization Properties in Infants at Risk for Suddent Infant Death Syndrome: Comparison with Age- and Sex-Matched Control Infants. *Am. J. Cardiol.* **1984**, *54*, 301–307. [CrossRef]

36. Ambroggi, L.D.; Taccardi, B.; Macchi, E. Body Surface Maps of Heart Potential: Tentative Localization of Preexcited Area of Forty-two Wolff-Parkinson-White patients. *Circulation* **1976**, *54*, 251. [CrossRef]
37. Ambroggi, L.D.; Bertoni, T.; Locati, E.; Stramba-Badiale, M.; Schwartz, P. Mapping of body surface potentials in patients with the idiopathic long QT syndrome. *Circulation* **1986**, *74*, 1334–1345. [CrossRef] [PubMed]
38. Gardner, M.; Montague, T.; Armstrong, C.; Horacek, M.; Smith, E. Vulnerability to ventricular arrhythmia: Assessment by mapping of body surface potential. *Circulation* **1986**, *73*, 684–692. [CrossRef]
39. Hubley-Kozey, C.; Mitchell, B.; Gardner, M.; Warren, J.; Penney, C.; Smith, E.; Horáček, B. Spatial features in body surface potentials maps can identify patients with a history of sustained ventricular tachycardia. *Circulation* **1995**, *92*, 1825–1838. [CrossRef]
40. Ambroggi, L.D.; Aime, E.; Ceriotti, C.; Rovida, M.; Negroni, S. Mapping of Ventricular Repolarization Potentials in Patients With Arrhythmogenic Right Ventricular Dysplasia: Principal Component Analysis of the ST-T Waves. *Circulation* **1997**, *96*, 4314–4318. [CrossRef]
41. Lux, R.; Green, L.; MacLeod, R.; Taccardi, B. Assessment of spatial and temporal characteristics of ventricular repolarization. *J. Electrocardiol.* **1994**, *27*, 100–104. [CrossRef]
42. Lux, R.; Fuller, M.; MacLeod, R.; Ershler, P.; Punske, B.; Taccardi, B. Noninvasive indices of repolarization and its dispersion. *J. Electrocardiol.* **1999**, *32*, 153–157. [CrossRef]
43. Bank, A.; Gage, R.; Curtin, A.; Burns, K.; Gillberg, J.; Ghosh, S. Body surface activation mapping of electrical dyssynchrony in cardiac resynchronization therapy patients: Potential for optimization. *J. Electrocardiol.* **2018**, *51*, 534–541. [CrossRef] [PubMed]
44. Gulrajani, R.; Savard, P.; Roberge, F. The Inverse Problem in Electrocardiography: Solutions in Terms of Equivalent Sources. *Crit. Rev. Biomed. Eng.* **1988**, *16*, 171–214.
45. Gulrajani, R. *Bioelectricity and Biomagnetism*, 1st ed.; John Wiley & Sons: New York, NY, USA, 1998.
46. MacLeod, R.; Buist, M. The Forward Problem of Electrocardiography. In *Comprehensive Electrocardiology*; Macfarlane, P., van Oosterom, A., Pahlm, O., Kligfield, P., Janse, M., Camm, J., Eds.; Springer: London, UK, 2010; pp. 247–298.
47. Messinger-Rapport, B.; Rudy, Y. Regularization of the Inverse Problem in Electrocardiography: A Model Study. *Math. Biosci.* **1988**, *89*, 79–118. [CrossRef]
48. Pullan, A.; Cheng, L.K.; Nash, M.; Brooks, D.; Ghodrati, A.; MacLeod, R. The Inverse Problem of Electrocardiography. In *Comprehensive Electrocardiology*; Macfarlane, P., van Oosterom, A., Pahlm, O., Kligfield, P., Janse, M., Camm, J., Eds.; Springer: London, UK, 2010; pp. 299–344.
49. Ghanem, R.; Burnes, J.; Waldo, A.; Rudy, Y. Imaging dispersion of myocardial repolarization, II: Noninvasive reconstruction of epicardial measures. *Circulation* **2001**, *104*, 1306–1312. [CrossRef]
50. Cluitmans, M.; Brooks, D.H.; MacLeod, R.; Dössel, O.; Guillem, M.S.; van Dam, P.M.; Svehlikova, J.; He, B.; Sapp, J.; Wang, L.; et al. Validation and opportunities of electrocardiographic imaging: From technical achievements to clinical applications. *Front. Physiol.* **2018**, *9*, 1305. [CrossRef]
51. Franzone, P.C.; Gassaniga, G.; Guerri, L.; Taccardi, B.; Viganotti, C. Accuracy Evaluation in Direct and Inverse Electrocardiology. In *Progress in Electrocardiography*; Macfarlane, P., Ed.; Pitman Medical: London, UK, 1979; pp. 83–87.
52. Oster, H.; Taccardi, B.; Lux, R.; Ershler, P.; Rudy, Y. Noninvasive electrocardiographic imaging: Reconstruction of epicardial potentials, electrograms, and isochrones and localization of single and multiple electrocardiac events. *Circulation* **1997**, *96*, 1012–1024. [CrossRef] [PubMed]
53. Burnes, J.; Taccardi, B.; MacLeod, R.; Rudy, Y. Noninvasive electrocardiographic imaging of electrophysiologically abnormal substrates in infarcted hearts: A model study. *Circulation* **2000**, *101*, 533–540. [CrossRef]
54. Burnes, J.; Taccardi, B.; Rudy, Y. A Noninvasive Imaging Modality for Cardiac Arrhythmias. *Circulation* **2000**, *102*, 2152–2158. [CrossRef]
55. Jia, P.; Punske, B.; Taccardi, B.; Rudy, Y. Electrophysiologic endocardial mapping from a noncontact nonexpandable catheter: A validation study of a geometry-based concept. *J. Cardiovasc. Electrophysiol.* **2000**, *11*, 1238–1251. [CrossRef] [PubMed]
56. Burnes, J.; Ghanem, R.; Waldo, A.; Rudy, Y. Imaging dispersion of myocardial repolarization, I: Comparison of. *Circulation* **2001**, *104*, 1299–1505. [CrossRef]
57. Ramanathan, C.; Rudy, Y. Electrocardiographic Imaging: I. effect of torso inhomgeneties on body surface electrocardiographic potentials. *J. Cardiovasc. Electrophysiol.* **2001**, *12*, 229–240. [CrossRef] [PubMed]
58. Cluitmans, M.; Bonizzi, P.; Karel, J.; Das, M.; Kietselaer, B.; de Jong, M.; Prinzen, F.; Peeters, R.; Westra, R.; Volders, P. In Vivo Validation of Electrocardiographic Imaging. *JACC Clin. Electrophysiol.* **2017**, *3*, 232–242. [CrossRef] [PubMed]
59. Ramanathan, C.; Ghanem, R.; Jia, P.; Ryu, K.; Rudy, Y. Noninvasive electrocardiographic imaging for cardiac electrophysiology and arrhythmia. *Nat. Med.* **2004**, *10*, 422–428. [CrossRef]
60. Ghanem, R.; Jia, P.; Ramanathan, C.; Ryu, K.; Markowitz, A.; Rudy, Y. Noninvasive electrocardiographic imaging (ECGI): Comparison to intraoperative mapping in patients. *Heart Rhythm J.* **2005**, *2*, 339–354. [CrossRef]
61. Cuculich, P.; Zhang, J.; Wang, Y.; Desouza, K.; Vijayakumar, R.; Woodard, P.; Rudy, Y. The electrophysiological cardiac ventricular substrate in patients after myocardial infarction noninvasive characterization with electrocardiographic imaging. *J. Am. Coll. Cardiol.* **2011**, *58*, 1893–1902. [CrossRef]

62. Intini, A.; Goldstein, R.; Jia, P.; Ramanathan, C.; Ryu, K.; Giannattasio, B.; Gilkeson, R.; Stambler, B.; Brugada, P.; Stevenson, W.; et al. Electrocardiographic imaging (ECGI), a novel diagnostic modality used for mapping of focal left ventricular tachycardia in a young athlete. *Heart Rhythm J.* **2005**, *2*, 1250–1252. [CrossRef]
63. Cuculich, P.; Wang, Y.; Lindsay, B.; Vijayakumar, R.; Rudy, Y. Noninvasive real-time mapping of an incomplete pulmonary vein isolation using electrocardiographic imaging. *Heart Rhythm J.* **2010**, *7*, 1316–1317. [CrossRef]
64. Vijayakumar, R.; Silva, J.; Desouza, K.; Abraham, R.; Strom, M.; Sacher, F.; Hare, G.V.; Haissaguerre, M.; Roden, D.; Rudy, Y. Electrophysiologic substrate in congenital Long QT syndrome: Noninvasive mapping with electrocardiographic imaging (ECGI). *Circulation* **2014**, *130*, 1936–1943. [CrossRef]
65. Rudy, Y.; Lindsay, B. Electrocardiographic imaging of heart rhythm disorders: From bench to bedside. *Card. Electrophysiol. Clin.* **2015**, *7*, 17–35. [CrossRef] [PubMed]
66. Wang, Y.; Schuessler, R.; Damiano, R.; Woodard, P.; Rudy, Y. Noninvasive electrocardiographic imaging (ECGI) of scar-related atypical atrial flutter. *Heart Rhythm J.* **2007**, *4*, 1565–1567. [CrossRef] [PubMed]
67. Wang, Y.; Cuculich, P.; Zhang, J.; Desouza, K.; Vijayakumar, R.; Chen, J.; Faddis, M.; Lindsay, B.; Smith, T.; Rudy, Y. Noninvasive electroanatomic mapping of human ventricular arrhythmias with electrocardiographic imaging. *Sci. Transl. Med.* **2011**, *3*, 98ra84. [CrossRef] [PubMed]
68. Cuculich, P.; Wang, Y.; Lindsay, B.; Faddis, M.; Schuessler, R.; RJ, J.D.; Li, L.; Rudy, Y. Noninvasive characterization of epicardial activation in humans with diverse atrial fibrillation patterns. *Circulation* **2010**, *122*, 1364–1372. [CrossRef]
69. Jia, P.; Ramanathan, C.; Ghanem, R.; Ryu, K.; Varma, N.; Rudy, Y. Electrocardiographic imaging of cardiac resynchronization therapy in heart failure: Observation of variable electrophysiologic responses. *Heart Rhythm J.* **2006**, *3*, 296–310. [CrossRef] [PubMed]
70. Latacha, M.; Memon, N.; Cuculich, P.; Hertel, J.; Wang, Y.; Rudy, Y.; Smith, T. Pathologic examination after epicardial ablation of ventricular tachycardia in cardiac sarcoidosis. *Heart Rhythm J.* **2010**, *7*, 705–707. [CrossRef]
71. Haissaguerre, M.; Hocini, M.; Shah, A.; Derval, N.; Sacher, F.; Jais, P.; Dubois, R. Noninvasive panoramic mapping of human atrial fibrillation mechanisms: A feasibility report. *J. Cardiovasc. Electrophysiol.* **2013**, *24*, 711–717. [CrossRef]
72. Cochet, H.; Dubois, R.; Sacher, F.; Derval, N.; Sermesant, M.; Hocini, M.; Montaudon, M.; Haissaguerre, M.; Laurent, F.; Jais, P. Cardiac arrythmias: Multimodal assessment integrating body surface ECG mapping into cardiac imaging. *Radiology* **2014**, *271*, 239–247. [CrossRef]
73. Hocini, M.; Shah, A.; Neumann, T.; Kuniss, M.; Erkapic, D.; Chaumeil, A.; Copley, S.; Lim, P.; Kanagaratnam, P.; Denis, A.; et al. Focal Arrhythmia Ablation Determined by High Resolution Non-invasive Maps: Multicenter Feasibility Study. *J. Cardiovasc. Electrophysiol.* **2015**, *26*, 754–760. [CrossRef] [PubMed]
74. Cuculich, P.; Schill, M.; Kashani, R.; Mutic, S.; Lang, A.; Cooper, D.; Faddis, M.; Gleva, M.; Noheria, A.; Smith, T.; et al. Noninvasive Cardiac Radiation for Ablation of Ventricular Tachycardia. *N. Engl. J. Med.* **2017**, *377*, 2325–2336. [CrossRef]
75. Robinson, C.; Samson, P.; Moore, K.; Hugo, G.; Knutson, N.; Mutic, S.; Goddu, S.; Lang, A.; Cooper, D.; Faddis, M.; et al. Phase I/II Trial of Electrophysiology-Guided Noninvasive Cardiac Radioablation for Ventricular Tachycardia. *Circulation* **2019**, *139*, 313–321. [CrossRef] [PubMed]
76. Plonsey, R. *Bioelectric Phenomena*; McGraw-Hill: New York, NY, USA, 1969.
77. Ghodrati, A.; Brooks, D.; MacLeod, R. Methods of solving reduced lead systems for inverse electrocardiography. *IEEE Trans. Biomed. Eng.* **2007**, *54*, 339–343. [CrossRef]
78. Figuera, C.; Suarez-Gutierrez, V.; Hernandez-Romero, I.; Rodrigo, M.; Liberos, A.; Atienza, F.; Guillem, M.; Barquero-Perez, O.; Climent, A.; Alonso-Atienza, F. Regularization Techniques for ECG Imaging during Atrial Fibrillation: A Computational Study. *Front. Physiol.* **2016**, *7*, 466. [PubMed]
79. Dogrusoz, Y.; Bear, L.; Svehlikova, J.; Coll-Font, J.; Good, W.; Dubois, R.; van Dam, E.; MacLeod, R. Reduction of Effects of Noise on the Inverse Problem of Electrocardiography with Bayesian Estimation. *IEEE Comput. Cardiol.* **2018**, *45*, 1–4.
80. MacLeod, R.; Lux, R.; Taccardi, B. Translation of body surface maps between different electrode configurations using a three-dimensional interpolation scheme. In *Electrocardiology '93: Proceedings of the International Congress on Electrocardiology, XXth Annual Meeting*; MacFarlane, P., Ed.; World Scientific: Singapore, 1993; pp. 179–182.
81. Hoekema, R.; Uijen, G.; Stili, D.; van Oosterom, A. Lead system transformation and body surface map data. *J. Electrocardiol.* **1998**, *31*, 71–82. [CrossRef]
82. MacLeod, R.; Miller, R.; Gardner, M.; Horacek, B. Application of an electrocardiographic inverse solution to localize myocardial ischemia during percutaneous transluminal coronary angioplasty. *J. Cardiovasc. Electrophysiol.* **1995**, *6*, 2–18. [CrossRef]
83. Kornreich, F.; Rautaharju, P.; Warren, J.; Montague, T.; Horacek, B. Identification of Best Electrocardiographic Leads for Diagnosing Myocardial Infarction by Statistical Analysis of Body Surface Potential Maps. *Am. J. Cardiol.* **1985**, *56*, 852–856. [CrossRef]
84. Kornreich, F.; Montague, T.J.; Rautaharju, P.; Kavadias, M.; Horacek, B. Identification of Best Electrocardiographic Leads for Diagnosing Left Ventricular Hypertrophy by Statistical Analysis of Body Surface Potential Maps. *Am. J. Cardiol.* **1988**, *62*, 1285–1291. [CrossRef]
85. Lux, R.; Smith, C.; Wyatt, R.; Abildskov, J. Limited lead selection for estimation of body surface potential maps in electrocardiography. *IEEE Trans. Biomed. Eng.* **1978**, *25*, 270–276. [CrossRef]
86. Lux, R.; Burgess, M.; Wyatt, R.; Evans, A.; Vincent, G.; Abildskov, J. Clinically practical lead systems for improved electrocardiography: Comparison with precordial grids and conventional lead systems. *Circulation* **1979**, *59*, 356–363. [CrossRef]

87. Lux, R.; Evans, K.; Burgess, M.; Wyatt, R.; Abildskov, J. Redundancy Reduction for Improved Display and Analysis of Body Surface Potential Maps: I. Spatial Compression. *Circ. Res.* **1981**, *49*, 186–196. [CrossRef]
88. Bear, L.R.; Cheng, L.K.; LeGrice, I.J.; Sands, G.B.; Lever, N.A.; Paterson, D.J.; Smaill, B.H. Forward problem of electrocardiography: Is it solved? *Circ. Arrhythmia Electrophysiol.* **2015**, *8*, 677–684. [CrossRef] [PubMed]
89. Bergquist, J.A.; Zenger, B.; Good, W.W.; Rupp, L.C.; Bear, L.R.; MacLeod, R.S. Novel Experimental Preparation to Assess Electrocardiographic Imaging Reconstruction Techniques. *Comput. Cardiol.* 2020, 1–4. [CrossRef]
90. Zenger, B.; Good, W.W.; Bergquist, J.A.; Burton, B.M.; Tate, J.D.; Berkenbile, L.; Sharma, V.; MacLeod, R.S. Novel experimental model for studying the spatiotemporal electrical signature of acute myocardial ischemia: A translational platform. *Physiol. Meas.* **2020**, *41*, 15002. [CrossRef]
91. Potyagaylo, D.; Chmelevsky, M.; van Dam, P.; Budanova, M.; Zubarev, S.; Treshkur, T.; Lebedev, D. ECG Adapted Fastest Route Algorithm to Localize the Ectopic Excitation Origin in CRT Patients. *Front. Physiol.* **2019**, *10*, 183. [CrossRef]
92. Aras, K.; Good, W.; Tate, J.; Burton, B.; Brooks, D.; Coll-Font, J.; Doessel, O.; Schulze, W.; Patyogaylo, D.; Wang, L.; et al. Experimental Data and Geometric Analysis Repository: EDGAR. *J. Electrocardiol.* **2015**, *48*, 975–981. [CrossRef]
93. Ershler, P.; Steadman, B.; Moore, K.; Lux, R. Systems for measuring and tracking electrophysiologic distributions. *Proc. IEEE Eng. Med. Biol. Soc. Annu. Int. Conf.* **1998**, *17*, 56–61. [CrossRef]
94. Zenger, B.; Bergquist, J.A.; Good, W.W.; Rupp, L.C.; MacLeod, R.S. High-capacity cardiac signal acquisition system for flexible, simultaneous, multidomain acquisition. *Comput. Cardiol.* **2020**, 1–4. [CrossRef]
95. Dessel, P.V.; Hemel, N.M.V.; Bakker, J.M.D.; Linnenbank, A.C.; Potse, M.; Jessurun, E.R.; Sippensgroenewegen, A.; Wever, E.F. Relation Between Body Surface Mapping and Endocardial Spread of Ventricular Activation in Postinfarction Heart. *J. Cardiovasc. Electrophysiol.* **2001**, *12*, 1232–1241. [CrossRef] [PubMed]
96. Rodenhauser, A.; Good, W.; Zenger, B.; Tate, J.; Aras, K.; Burton, B.; MacLeod, R. PFEIFER: Preprocessing Framework for Electrograms Intermittently Fiducialized from Experimental Recordings. *J. Open Source Softw.* **2018**, *3*, 472. [CrossRef]
97. Bear, L.; Svehlikova, J.; Bergquist, J.; Good, W.; Rababah, A.; Coll-Font, J.; MacLeod, R.; van Dam, E.; Dubois, R. Impact of baseline drift removal on ECG beat classification and alignment. *Comput. Cardiol.* **2021**, in press.
98. Dogrusoz, Y.S.; Bear, L.R.; Bergquist, J.; Dubois, R.; Good, W.; MacLeod, R.S.; Rababah, A.; Stoks, J. Effects of Interpolation on the Inverse Problem of Electrocardiography. In Proceedings of the 2019 Computing in Cardiology (CinC), Singapore, 8–11 September 2019; pp. 1–4. [CrossRef]
99. Parker, S.; Johnson, C. SCIRun: A scientific programming environment for computational steering. In Proceedings of the ACM IEEE Supercomputing Conference IEEE, San Diego, CA, USA, 8 December 1995; Volume 2, pp. 1419–1439.
100. Hansen, C.; Johnson, C. *The Visualization Handbook*; Elsevier: Amsterdam, The Netherlands, 2005.
101. Institute, S. SCIRun: A Scientific Computing Problem Solving Environment, Scientific Computing and Imaging Institute (SCI). 2015. Available online: http://www.scirun.org (accessed on 29 October 2021).
102. Cheniti, G.; Puyo, S.; Martin, C.A.; Frontera, A.; Vlachos, K.; Takigawa, M.; Bourier, F.; Kitamura, T.; Lam, A.; Dumas-Pommier, C.; et al. Noninvasive Mapping and Electrocardiographic Imaging in Atrial and Ventricular Arrhythmias (CardioInsight). *Card. Electrophysiol. Clin.* **2019**, *11*, 459–471. doi: 10.1016/j.ccep.2019.05.004 . [CrossRef] [PubMed]
103. Essen, R.V.; Hinsen, R.; Louis, R.; Merx, W.; Silny, J.; Rau, G.; Effert, S. On-line monitoring of multiple precordial leads in high risk patients with coronary artery disease—A pilot study. *Europ. Heart J.* **1985**, *5*, 203–209. [CrossRef] [PubMed]
104. SippensGroenewegen, A.; Spekhorst, H.; Hauer, R.; van Hemel, N.; Broekhuijsen, P.; Dunning, A. A Radiotransparent Carbon Electrode Array for Body Surface Mapping During Cardiac Catheterization. In Proceedings of the IEEE Engineering in Medicine and Biology Society 9th Annual International Conference, Boston, MA, USA, 13–16 November 1987; pp. 1782–1793.
105. Ershler, P.; Lux, R.; Steadman, B. A 128 Lead online Intraoperative Mapping System. In Proceedings of the IEEE Engineering in Medicine and Biology Society 8th Annual International Conference, Fort Worth, TX, USA, 7 November 1986; pp. 1289–1291.
106. Thomas, C.; Laurita, K.; Kavuru, M.; Vesselle, H.; Lee, D.; Sun, G.; Huebner, W. Biopotential Mapping System Description. Available online: https://bibbase.org/ (accessed on 29 October 2021).
107. Martel, S.; Lafontaine, S.; Bullivant, D.; Hunter, I.; Hunter, P. A Hardware Object-Oriented Cardiac Mapping System. In Proceedings of the IEEE Engineering in Medicine and Biology Society 17th Annual International Conference, Montreal, QC, Canada, 20–23 September 1995; p. 1647.
108. Einthoven, W. Le telecardiogramme. *Arch. Int. Physiol.* **1906**, *4*, 132–164.
109. Wagner, G. *Marriott's Practical Electrocardiography*, 11th ed.; Lippincott Williams & Wilkins: Philadelphia, PA, USA, 2008.
110. Milanic, M.; Jazbinsek, V.; Macleod, R.; Brooks, D.; Hren, R. Assessment of regularization techniques for electrocardiographic imaging. *J. Electrocardiol.* **2014**, *47*, 20–28. [CrossRef] [PubMed]
111. Bergquist, J.A.; Good, W.W.; Zenger, B.; Tate, J.D.; Rupp, L.C.; MacLeod, R.S. The electrocardiographic forward problem: A benchmark study. *Comput. Biol. Med.* **2021**, *134*, 104476. [CrossRef] [PubMed]
112. Rodrigo, M.; Climent, A.M.; Liberos, A.; Hernandez-Romero, I.; Arenal, A.; Bermejo, J.; Fernandez-Aviles, F.; Atienza, F.; Guillem, M.S. Solving Inaccuracies in Anatomical Models for Electrocardiographic Inverse Problem Resolution by Maximizing Reconstruction Quality. *IEEE Trans. Med. Imaging* **2018**, *37*, 733–740. [CrossRef]
113. Schuler, S.; Schaufelberger, M.; Bear, L.R.; Bergquist, J.A.; Cluitmans, M.J.M.; Coll-Font, J.; Önder, N.O.; Zenger, B.; Loewe, A.; MacLeod, R.S.; et al. Reducing Line-of-block Artifacts in Cardiac Activation Maps Estimated Using ECG Imaging: A Comparison of Source Models and Estimation Methods. *arXiv* **2021**, arXiv:2108.06602.

114. Burton, B.; Tate, J.; Erem, B.; Swenson, D.; Wang, D.; Brooks, D.; van Dam, P.; MacLeod, R. A Toolkit for Forward/Inverse Problems in Electrocardiography within the SCIRun Problem Solving Environment. In Proceedings of the 2011 Annual International Conference of the IEEE Engineering in Medicine and Biology Society, Boston, MA, USA, 30 August–3 September 2011; pp. 1–4.
115. Cuculich, P.; Robinson, C. Noninvasive Ablation of Ventricular Tachycardia. *N. Engl. J. Med.* **2018**, *378*, 1651–1652.
116. Xiu, D.; Karniadakis, G. The Wiener-Askey Polynomial Chaos for stochastic differential equations. *SIAM J. Sci. Comput.* **2002**, *24*, 619–644. [CrossRef]
117. Xiu, D. Efficient collocational approach for parametric uncertainty analysis. *Comm. Comput. Phys.* **2007**, *2*, 293–309.
118. Burk, K.M.; Narayan, A.; Orr, J.A. Efficient sampling for polynomial chaos-based uncertainty quantification and sensitivity analysis using weighted approximate Fekete points. *Int. J. Numer. Methods Biomed. Eng.* **2020**, *36*, e3395. [CrossRef] [PubMed]
119. Rasmussen, C.E. Gaussian processes in machine learning. In *Summer School on Machine Learning*; Springer: Berlin/Heidelberg, Germany, 2003; pp. 63–71.
120. Stein, M.L. *Interpolation of Spatial Data: Some Theory for Kriging*; Springer Science & Business Media: Berlin/Heidelberg, Germany, 2012.
121. Xiu, D.; Karniadakis, G. Modeling uncertainty in steady state diffusion problems via generalized polynomial chaos. *Comput. Methods Appl. Mech. Eng.* **2002**, *191*, 4927–4948. [CrossRef]
122. Geneser, S.; Xiu, D.; Kirby, R.; Sachse, F. Stochastic Markovian modeling of electrophysiology of ion channels: Reconstruction of standard deviations in macroscopic currents. *J. Theor. Biol.* **2007**, *245*, 627–637. [CrossRef] [PubMed]
123. Geneser, S.; MacLeod, R.; Kirby, R. Application of Stochastic Finite Element Methods to Study the Sensitivity of ECG Forward Modeling to Organ Conductivity. *IEEE Trans. Biomed. Eng.* **2008**, *55*, 31–40. [CrossRef]
124. Swenson, D.; Geneser, S.; Stinstra, J.; Kirby, R.; MacLeod, R. Cardiac Position Sensitivity Study in the Electrocardiographic Forward Problem Using Stochastic Collocation and BEM. *Annal. Biomed. Eng.* **2011**, *30*, 2900–2910. [CrossRef]
125. Rupp, L.C.; Bergquist, J.A.; Zenger, B.; Gillette, K.; Narayan, A.; Plank, G.; MacLeod, R.S. The Role of Myocardial Fiber Direction in Epicardial Activation Patterns via Uncertainty Quantification. *Comput. Cardiol.* **2021**, 1–4. in press.
126. Bergquist, J.A.; Zenger, B.; Rupp, L.C.; Narayan, A.; MacLeod, R.S. Uncertainty Quantification in Simulations of Myocardial Ischemia. *Comput. Cardiol.* **2021**, 1–4. in press.
127. Tate, J.D.; Good, W.W.; Zemzemi, N.; Boonstra, M.; van Dam, P.; Brooks, D.H.; Narayan, A.; MacLeod, R.S. Uncertainty Quantification of the Effects of Segmentation Variability in ECGI. In *Functional Imaging and Modeling of the Heart*; Ennis, D.B., Perotti, L.E., Wang, V.Y., Eds.; Springer International Publishing: Cham, Switzerland, 2021; pp. 515–522.
128. Kornreich, F.; Montague, T.; Kavadias, M.; Segers, J.; Rautaharju, P.; Horacek, B.; Taccardi, B. Qualitative and Quantitative Analysis of Characteristic Body Surface Potential Map Features in Anterior and Inferior Myocardial Infarction. *Am. J. Cardiol.* **1987**, *60*, 1230–1238. [CrossRef]
129. Uijen, G.; Heringa, A.; von Oosterom, A. Data reduction of body surface potential maps by means of orthogonal expansions. *IEEE Trans. Biomed. Eng.* **1984**, *31*, 706–714. [CrossRef]
130. Lux, R.; Green, L.; Abildskov, J. Statistical Representation and classification of electrocardiographic body surface potential maps. In *Computers in Cardiology*; IEEE Computer Society: Long Beach, CA, USA, 1984; pp. 251–254.
131. Farr, B.; Vondenbusch, B.; Silny, J.; Rau, G.; Effert, S. Localization of Significant Coronary Arterial Narrowings Using Body Surface Potential Mapping During Exercise Stress Testing. *Am. J. Cardiol.* **1987**, *5*, 528–530. [CrossRef]
132. Lux, R. Karhunen-Loeve Representation of ECG data. *J. Electrocardiol.* **1992**, *25*, 195. [CrossRef]
133. Trayanova, N.A.; Popescu, D.M.; Shade, J.K. Machine Learning in Arrhythmia and Electrophysiology. *Circ. Res.* **2021**, *128*, 544–566. [CrossRef] [PubMed]
134. Yao, X.; McCoy, R.G.; Friedman, P.A.; Shah, N.D.; Barry, B.A.; Behnken, E.M.; Inselman, J.W.; Attia, Z.I.; Noseworthy, P.A. ECG AI-Guided Screening for Low Ejection Fraction (EAGLE): Rationale and design of a pragmatic cluster randomized trial. *Am. Heart J.* **2020**, *219*, 31–36. [CrossRef]
135. Feeny, A.K.; Chung, M.K.; Madabhushi, A.; Attia, Z.I.; Cikes, M.; Firouznia, M.; Friedman, P.A.; Kalscheur, M.M.; Kapa, S.; Narayan, S.M.; et al. Artificial Intelligence and Machine Learning in Arrhythmias and Cardiac Electrophysiology. *Circ. Arrhythmia Electrophysiol.* **2020**, *13*, 873–890. [CrossRef]
136. Attia, Z.I.; Kapa, S.; Noseworthy, P.A.; Lopez-Jimenez, F.; Friedman, P.A. Artificial Intelligence ECG to Detect Left Ventricular Dysfunction in COVID-19: A Case Series. *Mayo Clin. Proc.* **2020**, *95*, 2464–2466. [CrossRef] [PubMed]
137. LeCun, Y.; Bengio, Y.; Hinton, G. Deep learning. *Nature* **2015**, *521*, 436–444. [CrossRef] [PubMed]
138. Brundage, J.N.; Suliafu, V.; Bergquist, J.A.; Zenger, B.; Rupp, L.C.; MacLeod, R.; Wang, B. Myocardial Ischemia Detection Using Body Surface ECG Recordings and Machine Shallow Learning. *IEEE Comput. Cardiol.* **2021**, in press.
139. Rim, B.; Sung, N.J.; Min, S.; Hong, M. Deep Learning in Physiological Signal Data: A Survey. *Sensors* **2020**, *20*, 969. [CrossRef]
140. Dhamala, J.; Ghimire, S.; Sapp, J.L.; Horáček, B.M.; Wang, L. Bayesian Optimization on Large Graphs via a Graph Convolutional Generative Model: Application in Cardiac Model Personalization. In *Medical Image Computing and Computer Assisted Intervention—MICCAI 2019*; Shen, D., Liu, T., Peters, T.M., Staib, L.H., Essert, C., Zhou, S., Yap, P.T., Khan, A., Eds.; Springer International Publishing: Cham, Switzerland, 2019; pp. 458–467.
141. Dosovitskiy, A.; Beyer, L.; Kolesnikov, A.; Weissenborn, D.; Zhai, X.; Unterthiner, T.; Dehghani, M.; Minderer, M.; Heigold, G.; Gelly, S.; et al. An Image is Worth 16x16 Words: Transformers for Image Recognition at Scale. *arXiv* **2020**, arXiv:2010.11929.

142. Vaswani, A.; Shazeer, N.; Parmar, N.; Uszkoreit, J.; Jones, L.; Gomez, A.N.; Kaiser, L.; Polosukhin, I. Attention Is All You Need. *arXiv* **2017**, arXiv:1706.03762.
143. Natarajan, A.; Chang, Y.; Mariani, S.; Rahman, A.; Boverman, G.; Vij, S.; Rubin, J. A Wide and Deep Transformer Neural Network for 12-Lead ECG Classification. In Proceedings of the IEEE Computers in Cardiology, Rimini, Italy, 13–16 September 2020.
144. Merino, M.; Gómez, I.; Molina, A. Envelopment filter and K-means for the detection of QRS waveforms in electrocardiogram. *Med. Eng. Phys.* **2015**, *37*, 605–609. [CrossRef] [PubMed]
145. Zhang, Y.; Yu, S. Single-lead noninvasive fetal ECG extraction by means of combining clustering and principal components analysis. *Med. Biol. Eng. Comput.* **2020**, *58*, 419–432. [CrossRef]
146. Xia, Y.; Han, J.; Wang, K. Quick detection of QRS complexes and R-waves using a wavelet transform and K-means clustering. *Biomed. Mater. Eng.* **2015**, *26*, S1059–S1065. [CrossRef]
147. Good, W.; Erem, B.; Zenger, B.; Coll-Font, J.; Bergquist, J.; Brooks, D.; MacLeod, R. Characterizing the transient electrocardiographic signature of ischemic stress using Laplacian Eigenmaps for dimensionality reduction. *Comput. Biol. Med.* **2020**, *127*, 104059. [CrossRef] [PubMed]
148. Zenger, B.; Good, W.W.; Bergquist, J.A.; Rupp, L.C.; Perez, M.D.; Stoddard, G.J.; Sharma, V.; Macleod, R.S. Transient Recovery of Epicardial and Torso ST-Segment Ischemic Signals During Cardiac Stress Tests: A Possible Physiological Mechanism. *J. Electrocardiol.* **2021**. [CrossRef]
149. Zenger, B.; Good, W.W.; Bergquist, J.A.; Rupp, L.C.; Perez, M.D.; Stoddard, G.J.; Sharma, V.; Macleod, R.S. Pharmacological and Simulated Exercise Cardiac Stress Tests Produce Different Ischemic Signatures in High-Resolution Experimental Mapping Studies. *J. Electrocardiol.* **2021**, *68*, 56–64. [CrossRef] [PubMed]
150. Kania, M.; Maniewski, R.; Kobylecka, M.; Zaczek, R.; Królicki, L.; Opolski, G.; Janusek, D. Prognostic value of the total cosine R to T measured in high resolution body surface potential mapping during exercise test. *Biomed. Signal Process. Control* **2015**, *20*, 135–141. [CrossRef]
151. Bauernfeind, T.; Préda, I.; Szakolczai, K.; Szucs, E.; Kiss, R.G.; Simonyi, G.; Kerecsen, G.; Duray, G.; Medvegy, M. Diagnostic value of the left atrial electrical potentials detected by body surface potential mapping in the prediction of coronary artery disease. *Int. J. Cardiol.* **2011**, *150*, 315–318. [CrossRef] [PubMed]
152. Kania, M.; Maniewski, R.; Zaczek, R.; Kobylecka, M.; Zbieć, A.; Królicki, L.; Opolski, G. High-Resolution Body Surface Potential Mapping in Exercise Assessment of Ischemic Heart Disease. *Ann. Biomed. Eng.* **2019**, *47*, 1300–1313. [CrossRef]
153. Daly, M.J.; Scott, P.; Owens, C.G.; Tomlin, A.; Smith, B.; Adgey, J. Body Surface Potential Mapping improves diagnosis of acute myocardial infarction in those with significant left main coronary artery stenosis. *Comput. Cardiol.* **2010**, *37*, 269–272.
154. Daly, M.J.; Scott, P.J.; Harbinson, M.T.; Adgey, J.A. Improving the Diagnosis of Culprit Left Circumflex Occlusion With Acute Myocardial Infarction in Patients With a Nondiagnostic 12-Lead ECG at Presentation: A Retrospective Cohort Study. *J. Am. Heart Assoc.* **2019**, *8*, 1–9. [CrossRef]
155. Hoekstra, J.W.; O'Neill, B.J.; Pride, Y.B.; Lefebvre, C.; Diercks, D.B.; Peacock, W.F.; Fermann, G.J.; Gibson, C.M.; Pinto, D.; Giglio, J.; et al. Acute Detection of ST-Elevation Myocardial Infarction Missed on Standard 12-Lead ECG With a Novel 80-Lead Real-Time Digital Body Surface Map: Primary Results From the Multicenter OCCULT MI Trial. *Ann. Emerg. Med.* **2009**, *54*, 779–788.e1. [CrossRef] [PubMed]
156. Wang, B.; Korhonen, P.; Tierala, I.; Hänninen, H.; Väänänen, H.; Toivonen, L. U wave features in body surface potential mapping in post-myocardial infarction patients. *Ann. Noninvasive Electrocardiol.* **2013**, *18*, 538–546. [CrossRef]
157. Kylmälä, M.M.; Konttila, T.; Vesterinen, P.; Kivistö, S.M.; Lauerma, K.; Lindholm, M.; Väänänen, H.; Stenroos, M.; Nieminen, M.S.; Hänninen, H.; et al. Assessment of myocardial infarct size with body surface potential mapping: Validation against contrast-enhanced cardiac magnetic resonance imaging. *Ann. Noninvasive Electrocardiol.* **2015**, *20*, 240–252. [CrossRef] [PubMed]
158. Guillem, M.S.; Climent, A.M.; Millet, J.; Arenal, Á.; Fernández-Avilés, F.; Jalife, J.; Atienza, F.; Berenfeld, O. Noninvasive localization of maximal frequency sites of atrial fibrillation by body surface potential mapping. *Circ. Arrhythmia Electrophysiol.* **2013**, *6*, 294–301. [CrossRef] [PubMed]
159. Marques, V.G.; Rodrigo, M.; Guillem, M.S.; Salinet, J. A robust wavelet-based approach for dominant frequency analysis of atrial fibrillation in body surface signals. *Physiol. Meas.* **2020**, *41*, 075004. [CrossRef] [PubMed]
160. Zeemering, S.; Lankveld, T.A.R.; Bonizzi, P.; Crijns, H.; Schotten, U. Principal component analysis of body surface potential mapping in atrial fibrillation patients suggests additional ECG lead locations. *Comput. Cardiol. CCAL* **2014**, *41*, 893–896.
161. Jurak, P.; Halamek, J.; Leinveber, P.; Vondra, V.; Soukup, L.; Vesely, P.; Sumbera, J.; Zeman, K.; Martinakova, L.; Jurakova, T.; et al. Ultra-high-frequency ECG measurement. In Proceedings of the Computing in Cardiology 2013, Zaragoza, Spain, 22–25 September 2013; pp. 783–786.
162. Jurak, P.; Matejkova, M.; Halamek, J.; Plesinger, F.; Viscor, I.; Vondra, V.; Lipoldova, J.; Novak, M.; Smisek, R.; Leinveber, P. Cardiac Resynchronization Guided by Ultra-High-Frequency ECG Maps. In Proceedings of the 2019 Computing in Cardiology (CinC), Singapore, 8–11 September 2019; pp. 1–4. [CrossRef]
163. Curila, K.; Jurak, P.; Halamek, J.; Prinzen, F.; Waldauf, P.; Kach, J.; Stros, P.; Plesinger, F.; Mizner, J.; Susankova, M.; et al. Ventricular activation pattern assessment during right ventricular pacing: Ultrahigh-frequency ECG study. *J. Cardiovasc. Electrophysiol.* **2021**, *32*. [CrossRef] [PubMed]

164. Postema, P.G.; van Dessel, P.F.H.M.; Kors, J.A.; Linnenbank, A.C.; van Herpen, G.; Ritsema van Eck, H.J.; van Geloven, N.; de Bakker, J.M.T.; Wilde, A.A.M.; Tan, H.L. Local Depolarization Abnormalities Are the Dominant Pathophysiologic Mechanism for Type 1 Electrocardiogram in Brugada Syndrome. *J. Am. Coll. Cardiol.* **2010**, *55*, 789–797. [CrossRef] [PubMed]
165. Fonseca-Guzmán, A.; Climent, A.M.; Millet, J.; Berné, P.; Brugada, J.; Ramos, R.; Brugada, R.; Guillem, M.S. Fragmentation in body surface potential mapping recordings from patients with Brugada syndrome. *Comput. Cardiol.* **2011**, *38*, 637–640.
166. Guillem, M.S.; Climent, A.M.; Millet, J.; Berné, P.; Ramos, R.; Brugada, J.; Brugada, R. Conduction abnormalities in the right ventricular outflow tract in Brugada syndrome detected body surface potential mapping. In Proceedings of the 2010 Annual International Conference of the IEEE Engineering in Medicine and Biology, Buenos Aires, Argentina, 31 August–4 September 2010; pp. 2537–2540. [CrossRef]
167. Daly, M.J.; Finlay, D.D.; Scott, P.J.; Nugent, C.D.; Adgey, A.A.J.; Harbinson, M.T. Pre-hospital body surface potential mapping improves early diagnosis of acute coronary artery occlusion in patients with ventricular fibrillation and cardiac arrest. *Resuscitation* **2013**, *84*, 37–41. [CrossRef]
168. Meo, M.; Denis, A.; Sacher, F.; Duchâteau, J.; Cheniti, G.; Puyo, S.; Bear, L.; Jaïs, P.; Hocini, M.; Haïssaguerre, M.; et al. Insights Into the Spatiotemporal Patterns of Complexity of Ventricular Fibrillation by Multilead Analysis of Body Surface Potential Maps. *Front. Physiol.* **2020**, *11*, 1–15. [CrossRef]
169. Douglas, R.A.G.; Samesima, N.; Filho, M.M.; Pedrosa, A.A.; Nishioka, S.A.D.; Pastore, C.A. Global and regional ventricular repolarization study by body surface potential mapping in patients with left bundle-branch block and heart failure undergoing cardiac resynchronization therapy. *Ann. Noninvasive Electrocardiol.* **2012**, *17*, 123–129. [CrossRef] [PubMed]
170. Meo, M.; Bonizzi, P.; Bear, L.R.; Cluitmans, M.; Abell, E.; Haïssaguerre, M.; Bernus, O.; Dubois, R. Body Surface Mapping of Ventricular Repolarization Heterogeneity: An Ex-vivo Multiparameter Study. *Front. Physiol.* **2020**, *11*, 933. [CrossRef]
171. Hren, R.; Horacek, B. Value of simulated body surface potential maps as templates in localizing sites of ectopic activation for radiofrequency ablation. *Physiol. Meas.* **1997**, *18*, 373. [CrossRef]
172. Barr, R.; Spach, M.; Herman-Giddens, G. Selection of the number and position of measuring locations for electrocardiography. *IEEE Trans. Biomed. Eng.* **1971**, *18*, 125–138. [CrossRef] [PubMed]
173. Bergquist, J.A.; Coll-Font, J.; Zenger, B.; Rupp, L.C.; Good, W.W.; Brooks, D.H.; MacLeod, R.S. Improving Localization of Cardiac Geometry Using ECGI. In Proceedings of the 2020 Computing in Cardiology, Rimini, Italy, 13–16 September 2020; pp. 1–4. [CrossRef]
174. Wissner, E.; Revishvili, A.; Metzner, A.; Tsyganov, A.; Kalinin, V.; Lemes, C.; Saguner, A.M.; Maurer, T.; Deiss, S.; Sopov, O.; et al. Noninvasive epicardial and endocardial mapping of premature ventricular contractions. *Europace* **2017**, *19*, 843–849. [CrossRef]
175. Erkapic, D.; Neumann, T. Ablation of premature ventricular complexes exclusively guided by three-dimensional noninvasive mapping. *Card Electrophysiol. Clin.* **2015**, *7*, 109–115. [CrossRef]
176. Potyagaylo, D.; Segel, M.; Schulze, W.H.W.; Dössel, O. Noninvasive Localization of Ectopic Foci: A New Optimization Approach for Simultaneous Reconstruction of Transmembrane Voltages and Epicardial Potentials BT. In *Functional Imaging and Modeling of the Heart*; Ourselin, S., Rueckert, D., Smith, N., Eds.; Springer: Berlin/Heidelberg, Germany, 2013; pp. 166–173.
177. Berger, T.; Pfeifer, B.; Hanser, F.F.; Hintringer, F.; Fischer, G.; Netzer, M.; Trieb, T.; Stuehlinger, M.; Dichtl, W.; Baumgartner, C.; et al. Single-Beat Noninvasive Imaging of Ventricular Endocardial and Epicardial Activation in Patients Undergoing CRT. *PLoS ONE* **2011**, *6*, e16255. [CrossRef] [PubMed]
178. Ghosh, S.; Silva, J.N.A.; Canham, R.M.; Bowman, T.M.; Zhang, J.; Rhee, E.K.; Woodard, P.K.; Rudy, Y. Electrophysiologic substrate and intraventricular left ventricular dyssynchrony in nonischemic heart failure patients undergoing cardiac resynchronization therapy. *Heart Rhythm* **2011**, *8*, 692–699. [CrossRef]
179. Ploux, S.; Lumens, J.; Whinnett, Z.; Montaudon, M.; Strom, M.; Ramanathan, C.; Derval, N.; Zemmoura, A.; Denis, A.; De Guillebon, M.; et al. Noninvasive Electrocardiographic Mapping to Improve Patient Selection for Cardiac Resynchronization Therapy. *J. Am. Coll. Cardiol.* **2013**, *61*, 2435–2443. [CrossRef]
180. Wan, Y.; Lewis, A.; Colasanto, M.; van Langeveld, M.; Kardon, G.; Hansen, C. A practical workflow for making anatomical atlases for biological research. *IEEE Comput. Graph. Appl.* **2012**, *32*, 70–80. [CrossRef]
181. Horáček, B.M.; Wang, L.; Dawoud, F.; Xu, J.; Sapp, J.L. Noninvasive electrocardiographic imaging of chronic myocardial infarct scar. *J. Electrocardiol.* **2015**, *48*, 952–958. [CrossRef]
182. Sapp, J.L.; Dawoud, F.; Clements, J.C.; Horáček, B.M. Inverse solution mapping of epicardial potentials: Quantitative comparison with epicardial contact mapping. *Circ. Arrythmia Electrophysiol.* **2012**, *5*, 1001–1009. [CrossRef] [PubMed]
183. Wang, L.; Gharbia, O.A.; Horáček, B.M.; Sapp, J.L. Noninvasive epicardial and endocardial electrocardiographic imaging of scar-related ventricular tachycardia. *J. Electrocardiol.* **2016**, *49*, 887–893. [CrossRef]
184. Parreira, L.; Carmo, P.; Adragao, P.; Nunes, S.; Soares, A.; Marinheiro, R.; Budanova, M.; Zubarev, S.; Chmelevsky, M.; Pinho, J.; et al. Electrocardiographic imaging (ECGI): What is the minimal number of leads needed to obtain a good spatial resolution? *J. Electrocardiol.* **2020**, *62*, 86–93. [CrossRef] [PubMed]
185. Cámara-Vázquez, M.Á.; Hernández-Romero, I.; Rodrigo, M.; Alonso-Atienza, F.; Figuera, C.; Morgado-Reyes, E.; Atienza, F.; Guillem, M.S.; Climent, A.M.; Barquero-Pérez, Ó. Electrocardiographic imaging including intracardiac information to achieve accurate global mapping during atrial fibrillation. *Biomed. Signal Process. Control* **2021**, *64*, 102354. [CrossRef]
186. Marques, V.G.; Rodrigo, M.; de la Salud Guillem, M.; Salinet, J. Characterization of atrial arrhythmias in body surface potential mapping: A computational study. *Comput. Biol. Med.* **2020**, *127*. [CrossRef] [PubMed]

187. Bokeriia, L.; Revishvili, A.; Kalinin, A.; Kalinin, V.; Liadzhina, O.; Fetisova, E. Hardware-software system for noninvasive electrocardiographic examination of heart based on inverse problem of electrocardiography. *Biomed. Eng.* **2008**, *42*, 273–279. [CrossRef]
188. Good, W.; Erem, B.; Coll-Font, J.; Zenger, B.; Horacek, B.; Brooks, D.; MacLeod, R. Novel Metric Using Laplacian Eigenmaps to Evaluate Ischemic Stress on the Torso Surface. *IEEE Comput. Cardiol.* **2018**, *45*, 1–4.
189. Alday, E.A.P.; Gu, A.; Shah, A.J.; Robichaux, C.; Wong, A.K.I.; Liu, C.; Liu, F.; Rad, A.B.; Elola, A.; Seyedi, S.; et al. Classification of 12-lead ECGs: The PhysioNet/Computing in Cardiology Challenge 2020. *Physiol. Meas.* **2020**, *41*. [CrossRef]
190. Giffard-Roisin, S.; Jackson, T.; Fovargue, L.; Lee, J.; Delingette, H.; Razavi, R.; Ayache, N.; Sermesant, M. Noninvasive Personalization of a Cardiac Electrophysiology Model from Body Surface Potential Mapping. *IEEE Trans. Biomed. Eng.* **2017**, *64*, 2206–2218. [CrossRef] [PubMed]
191. Giffard-Roisin, S.; Delingette, H.; Jackson, T.; Webb, J.; Fovargue, L.; Lee, J.; Rinaldi, C.A.; Razavi, R.; Ayache, N.; Sermesant, M. Transfer learning from simulations on a reference anatomy for ECGI in personalized cardiac resynchronization therapy. *IEEE Trans. Biomed. Eng.* **2018**, *66*, 343–353. [CrossRef]
192. Rababah, A.S.; Bear, L.R.; Dogrusoz, Y.S.; Good, W.; Bergquist, J.; Stoks, J.; MacLeod, R.; Rjoob, K.; Jennings, M.; Mclaughlin, J.; et al. The effect of interpolating low amplitude leads on the inverse reconstruction of cardiac electrical activity. *Comput. Biol. Med.* **2021**, *136*, 104666. [CrossRef] [PubMed]

Article

Excitation and Contraction of the Failing Human Heart In Situ and Effects of Cardiac Resynchronization Therapy: Application of Electrocardiographic Imaging and Speckle Tracking Echo-Cardiography

Christopher M. Andrews [1,*]**, Gautam K. Singh** [2] **and Yoram Rudy** [1]

[1] Cardiac Bioelectricity and Arrhythmia Center, Department of Biomedical Engineering, Washington University in St. Louis, St. Louis, MO 63130, USA; rudy@wustl.edu

[2] Division of Cardiology, Department of Pediatrics, Washington University in St. Louis, St. Louis, MO 63110, USA; singh_g@email.wustl.edu

* Correspondence: christopher.m.andrews@gmail.com

Citation: Andrews, C.M.; Singh, G.K.; Rudy, Y. Excitation and Contraction of the Failing Human Heart In Situ and Effects of Cardiac Resynchronization Therapy: Application of Electrocardiographic Imaging and Speckle Tracking Echo-Cardiography. *Hearts* **2021**, *2*, 331–349. https://doi.org/10.3390/hearts2030027

Academic Editors: Gaetano Santulli and Peter Macfarlane

Received: 1 June 2021
Accepted: 23 July 2021
Published: 23 July 2021

Publisher's Note: MDPI stays neutral with regard to jurisdictional claims in published maps and institutional affiliations.

Abstract: Despite the success of cardiac resynchronization therapy (CRT) for treating heart failure (HF), the rate of nonresponders remains 30%. Improvements to CRT require understanding of reverse remodeling and the relationship between electrical and mechanical measures of synchrony. The objective was to utilize electrocardiographic imaging (ECGI, a method for noninvasive cardiac electrophysiology mapping) and speckle tracking echocardiography (STE) to study the physiology of HF and reverse remodeling induced by CRT. We imaged 30 patients (63% male, mean age 63.7 years) longitudinally using ECGI and STE. We quantified CRT-induced remodeling of electromechanical parameters and evaluated a novel index, the electromechanical delay (EMD, the delay from activation to peak contraction). We also measured dyssynchrony using ECGI and STE and compared their effectiveness for predicting response to CRT. EMD values were elevated in HF patients compared to controls. However, the EMD values were dependent on the activation sequence (CRT-paced vs. un-paced), indicating that the EMD is not intrinsic to the local tissue, but is influenced by factors such as opposing wall contractions. After 6 months of CRT, patients had increased contraction in native rhythm compared to baseline pre-CRT (baseline: -8.55%, 6 months: -10.14%, $p = 0.008$). They also had prolonged repolarization at the location of the LV pacing lead. The pre-CRT delay between mean lateral LV and RV electrical activation time was the best predictor of beneficial reduction in LV end systolic volume by CRT (Spearman's Rho: -0.722, $p < 0.001$); it outperformed mechanical indices and 12-lead ECG criteria. HF patients have abnormal EMD. The EMD depends upon the activation sequence and is not predictive of response to CRT. ECGI-measured LV activation delay is an effective index for CRT patient selection. CRT causes persistent improvements in contractile function.

Keywords: electrocardiographic imaging (ECGI); heart failure (HF); cardiac resynchronization therapy (CRT); ultrasound; strain; speckle tracking echocardiography

1. Introduction

Electrocardiographic Imaging (ECGI) is a noninvasive imaging modality for studying cardiac electrophysiology [1]. ECGI noninvasively reconstructs epicardial potentials, which can be used to map epicardial activation and repolarization sequences and generate maps of electrophysiological scar substrate. Electrical information is valuable for understanding cardiac arrhythmias. However, cardiac structure and mechanics play an important role in many forms of heart disease. ECGI offers additional utility when paired with complementary imaging techniques. For example, we combined ECGI with late gadolinium enhancement MRI to correlate electrophysiological substrate with anatomical scars in arrhythmogenic right ventricular cardiomyopathy (ARVC) patients [2]. More recently, we combined ECGI with tagged MRI to study the electromechanics of the normal human heart in situ [3]. In the present manuscript, we describe our efforts to expand on this work by combining ECGI with speckle tracking echocardiography (STE) to study the electromechanics of heart failure (HF) and cardiac resynchronization therapy (CRT). In the present study, STE replaced the previously utilized MRI for the following reasons.

STE is an imaging technique which tracks the characteristic speckled patterns in echocardiographic images and computes deformation of the patterns to determine regional strain values [4,5]. STE offers several advantages over tagged MRI for studying electromechanics in the CRT patient population. First, many CRT devices are incompatible with MRI, which means that MRI cannot be used to evaluate strain in these patients post-implantation. Tagged MRI also requires a lengthy scan with many long breath holds, which may be too difficult for many HF patients. STE, in contrast, can be obtained from echocardiographic data, which are routinely collected from this patient population. Finally, STE also offers much higher framerates than tagged MRI, which can improve the temporal resolution for identifying features of strain curves.

HF is a progressive disease with high prevalence and high mortality that is a major contributor to healthcare costs [6]. HF is associated with adverse remodeling of the ventricular myocardium that includes chamber dilatation, slow conduction of electrical activation, decreased myocyte contraction amplitudes, altered intracellular Ca^{2+} handling, and prolonged action potential durations [7–12]. Approximately 30% of HF patients have dyssynchronous ventricular contraction, apparent by prolonged QRS duration on the ECG. Prolonged QRS duration and left ventricular ejection fraction (LVEF) are inversely correlated, and HF patients with QRS prolongation have higher all-cause mortality and may have higher incidence of sudden death than those with narrow QRS complexes [13]. CRT improves the synchrony of ventricular contraction by pacing both ventricles. Clinical trials found that CRT improves quality of life, reduces HF-related hospitalizations, prolongs patient survival, and may reduce the risk of sudden cardiac death [14–17]. CRT also corrects some of the pathological abnormalities in HF, a process known as reverse remodeling. Animal studies found that CRT restored cellular action potential duration (APD) and Ca^{2+} transients to normal levels [18,19]. Despite the widely reported benefits of CRT, the rate of patients who do not respond to the therapy has been remarkably stable at around 30% [20].

The QRS duration (an electrical index of ventricular dyssynchrony) is the most important clinical indication for CRT patient selection. A longer QRS duration, reflecting greater ventricular dyssynchrony, is generally a predictor of greater response to CRT. While various mechanical dyssynchrony assessments of dyssynchrony, such as STE, have been evaluated for their utility in CRT patient selection, they have generally failed to improve the rate of nonresponders. The EchoCRT trial, which evaluated CRT in patients with mechanical dyssynchrony and a narrow QRS complex, was stopped for futility with the conclusion that CRT was possibly detrimental in that patient cohort [21,22].

The disappointing performance of mechanical dyssynchrony indices in patient selection is an indication of gaps in our understanding of the electromechanics of HF and CRT. A better understanding of the physiology underlying response and non-response to CRT may advance future attempts to improve the therapy. The present study combines STE with ECGI to study the electromechanics of CRT in HF patients in situ. We imaged HF patients undergoing CRT (HF-CRT) longitudinally to determine baseline dyssynchrony prior to CRT, acute resynchronization at onset of CRT, and reverse remodeling after 6 months of CRT pacing. Previously published ECGI and STE data from a group of 20 healthy adults [3] served as control data for comparison.

2. Methods

2.1. Patient Cohort

We enrolled a representative group of 30 HF-CRT patients at Washington University in St. Louis—63% were male with a mean age 64.1 years, and 37% were female with a mean age 62.9 years. Any HF patients undergoing new CRT device implantation were eligible for inclusion in the study. Patient demographics, HF etiologies, and baseline EF and QRS durations are provided in Table 1. The study was approved by the Human Research Protection Office at Washington University in St. Louis (IRB ID: 201111090). All participants provided written informed consent.

Table 1. Study population baseline ejection fraction (EF), QRS duration (QRSd), heart failure etiology, and demographics. Baseline EF and QRSd values were measured during the study, prior to the onset of CRT pacing. ICM: Ischemic Cardiomyopathy. NICM: Nonischemic cardiomyopathy.

ID	EF (%)	QRSd (msec)	Etiology	Age	Sex	Race	Ethnicity
EM1	27	154	NICM	53.0	F	Caucasian	Not Hispanic
EM2	25	167	NICM	64.0	F	Caucasian	Not Hispanic
EM3	23	180	NICM	71.9	M	Caucasian	Not Hispanic
EM4	-	123	NICM	63.3	M	Caucasian	Not Hispanic
EM5	29	199	NICM	54.4	M	Caucasian	Not Hispanic
EM6	27	169	NICM	73.0	F	Caucasian	Not Hispanic
EM7	20	154	ICM	66.3	M	Caucasian	Not Hispanic
EM8	25	108	NICM	59.8	M	Caucasian	Not Hispanic
EM9	25	173	NICM	69.5	F	Caucasian	Not Hispanic
EM10	24	149	ICM	62.4	M	Caucasian	Not Hispanic
EM11	24	186	NICM	69.3	F	Caucasian	Not Hispanic
EM12	30	136	ICM	67.5	M	Caucasian	Not Hispanic
EM13	21	156	ICM	60.1	M	African American	Not Hispanic
EM14	29	184	NICM	54.5	F	Caucasian	Not Hispanic
EM15	20	162	ICM	77.6	M	Caucasian	Not Hispanic
EM16	23	134	NICM	64.0	F	Caucasian	Not Hispanic
EM17	24	200	NICM	52.5	M	Caucasian	Not Hispanic
EM18	24	184	ICM	78.7	F	Caucasian	Not Hispanic
EM19	24	204	NICM *	74.7	F	Caucasian	Not Hispanic
EM20	26	154	NICM	34.7	M	Caucasian	Not Hispanic
EM21	21	153	ICM	77.5	M	Caucasian	Not Hispanic
EM22	20	147	ICM	81.1	M	Caucasian	Not Hispanic
EM23	22	154	ICM	55.3	M	Caucasian	Not Hispanic
EM24	17	149	NICM	74.1	M	Caucasian	Not Hispanic
EM25	31	145	NICM	51.3	F	Caucasian	Not Hispanic
EM26	24	149	ICM	72.4	M	Caucasian	Not Hispanic
EM27	17	143	NICM	40.5	F	African American	Not Hispanic
EM28	23	142	NICM	62.4	M	Caucasian	Not Hispanic
EM29	25	162	NICM	61.4	M	African American	Not Hispanic
EM30	17	178	NICM	64.1	M	African American	Not Hispanic

- EM4 baseline EF data lost due to computer failure (patient excluded from response analysis). * Mild nonobstructive coronary artery disease.

2.2. Electrocardiographic Imaging (ECGI)

The ECGI method was described previously [1]. A schematic of the procedure is presented in Figure 1. Briefly, 256 uniformly distributed ECGs were simultaneously recorded from the torso using a portable recording system (ActiveTwo, BioSemi; the Netherlands). Patient-specific heart-torso geometries were obtained using a thoracic CT scan gated at 70% of the R-R interval while wearing the recording electrodes. The ECG recordings were combined with the heart-torso geometries to reconstruct unipolar epicardial electrograms noninvasively, using previously described methods [1]. Typically, 1000 electrograms were computed over the entire ventricular epicardium. Electrograms over the valve plane were excluded from further analysis.

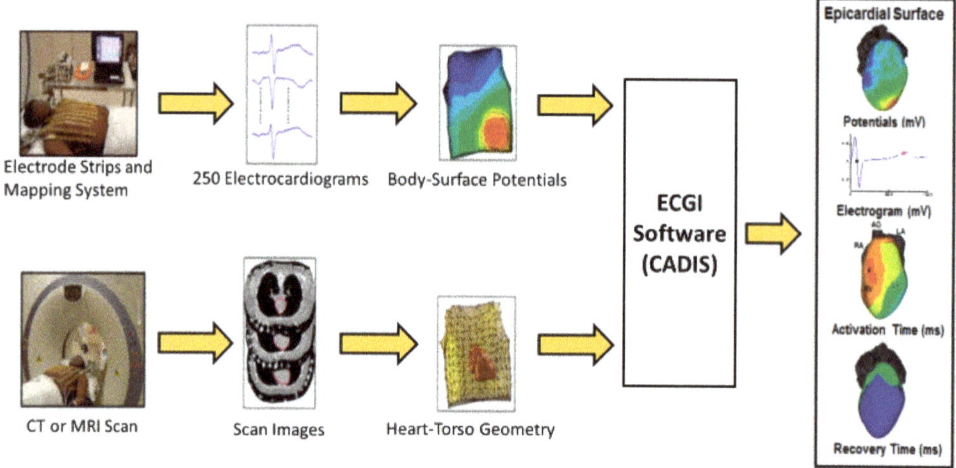

Figure 1. Schematic of the ECGI procedure. Body-surface potentials are recorded from the torso surface using a portable recording system (top). The heart-torso geometry is obtained using a computed tomography (CT) or magnetic resonance imaging (MRI) scan (bottom). The heart-torso geometry and torso potentials are combined and the inverse problem is solved to reconstruct unipolar epicardial electrograms. Electrograms are processed to determine local electrical parameters of interest (right frame).

2.3. Echocardiography

Echocardiography was performed using a commercially available ultrasound imaging system (Vivid 7, GE Healthcare, Milwaukee, WI, USA). Longitudinal strain was assessed in standard 4-chamber, 3-chamber (apical long axis), and 2-chamber apical views. Images were obtained using the maximum framerate that allowed for viewing the entire left ventricle (LV) chamber, with a typical frame rate of 70 frames per second. Strain curves in each apical view were computed using vendor speckle tracking software (EchoPAC, GE Healthcare, Milwaukee, WI, USA). To minimize the effect of noise on the timing of the strain curves, 3 beats were processed for each view and averaged. LV volumes and ejection fractions (also averaged over 3 beats) were determined using Simpson's biplane method. In cases where the 2-chamber view could not be obtained ($n = 5$), the 4-chamber volume parameters were used. We defined response as a decrease in LV end systolic volume (LVESV) $\geq$ 15% and/or an increase in ejection fraction (EF) $\geq$ 5% after 6 months of CRT. Responders and nonresponders were classified based on these echocardiographic criteria.

2.4. Longitudinal CRT Studies

The first ECGI and STE images were obtained after device implantation but prior to the onset of CRT pacing. This allowed us to determine the location of the pacing leads from the CT scan and image the patients prior to any CRT-induced reverse-remodeling. Patients were imaged in their native (un-paced) rhythm and at the onset of CRT pacing. Imaging was repeated in both paced and un-paced rhythms after 3 and 6 months of CRT pacing. To avoid the radiation exposure of additional CT scans during follow-up visits, the ECGI electrodes were placed in the same locations as the initial visit using images from the initial scan to guide electrode placement.

2.5. Analysis

2.5.1. Segmentation

Analysis of each imaging modality was performed blinded to the results of the other modalities. The LV was segmented using a modified version of the American Heart Association 17-Segment Model (Figure 2). The standard apical segments were replaced with anterolateral, inferolateral, anteroseptal, and inferoseptal apical segments. For the ECGI analysis, the lateral right ventricle (RV) was segmented using the same convention as the lateral LV. Electrogram parameters for each region were computed using the mean value from all electrograms within the region.

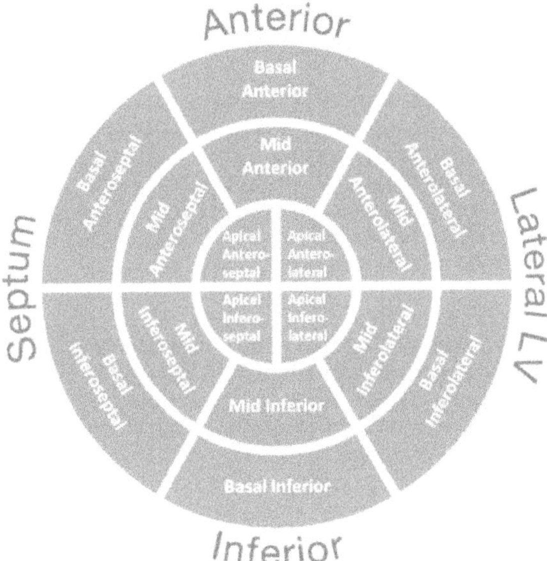

Figure 2. The LV was segmented using a modified version of the American Heart Association 17-Segment Model. Apical segments were modified from the standard model because ECGI images the epicardium which does not include any septal segments. The apical LV segments from the ECGI maps were divided into Apical Anterolateral and Apical Inferolateral segments.

2.5.2. Electrogram Analysis

Activation times were computed from epicardial electrograms as the time of steepest negative time-derivative of voltage ($-dV/dt_{max}$). Recovery times were computed as the steepest positive time-derivative during the T wave (dV/dt_{max}) [23,24]. Activation-recovery intervals (ARIs, a surrogate for local APD) were computed as the difference between recovery time and activation time. Activation and recovery maps were edited based on overall sequence and neighboring electrograms. Electrogram fractionation was quantified using the number of downward deflections between the QRS onset and T wave onset. Electrograms with more than 2 deflections were considered "fractionated" (see online supplement of [3] for additional details). ARIs and fractionation were compared across visits in native rhythm to avoid the impact of altered activation sequence on the indices. Parameters were computed at pacing sites by averaging values of the 10 electrograms closest to the pacing lead locations. The term "activation" refers to electrical excitation throughout the text.

2.5.3. Dyssynchrony Indices

We quantified electrical dyssynchrony from the 12-lead ECG using the QRS duration, computed as the latest QRS end in any lead minus the earliest QRS onset in any lead. We note that this definition differs from measurements that report the QRS duration as the maximum duration in any single lead. The total ventricular activation time was defined as the latest ECGI-determined activation time minus the earliest. We evaluated electrical dyssynchrony in ECGI using two previously reported indices. The LV activation delay was defined as the mean activation time in mid and basal lateral LV segments minus the mean activation time in mid and basal lateral RV segments [25]. LV activation dispersion was computed as the standard deviation of activation times within the LV [26].

We evaluated mechanical dyssynchrony using analogous indices applied to echocardiographic strain data. The mechanical delay was defined as the mean time of peak strain in mid and basal lateral LV segments minus the mean time of peak strain in mid and basal septal segments. Mechanical dispersion was defined as the standard deviation of contraction times within the LV.

2.5.4. Electromechanical Delay (EMD)

ECGI and echocardiographic parameters were aligned temporally using the body surface ECG from corresponding lead locations. Maxima or minima from the QRS voltage or voltage derivatives were used to align the ECG traces. The electromechanical delay (EMD) was defined as the time of peak strain within an LV segment minus the mean electrical activation time within the segment. Because ECGI does not image the septum, the EMD was not computed for septal segments.

2.5.5. Statistical Analysis

Comparisons between controls and the HF-CRT population were performed using Wilcoxon rank-sum tests. ECGI and STE data are presented for a representative control in Figure 3. Changes due to acute and chronic CRT pacing were assessed using Wilcoxon signed-rank tests. Correlations between parameters and volumetric reverse remodeling were performed using Spearman correlation coefficients.

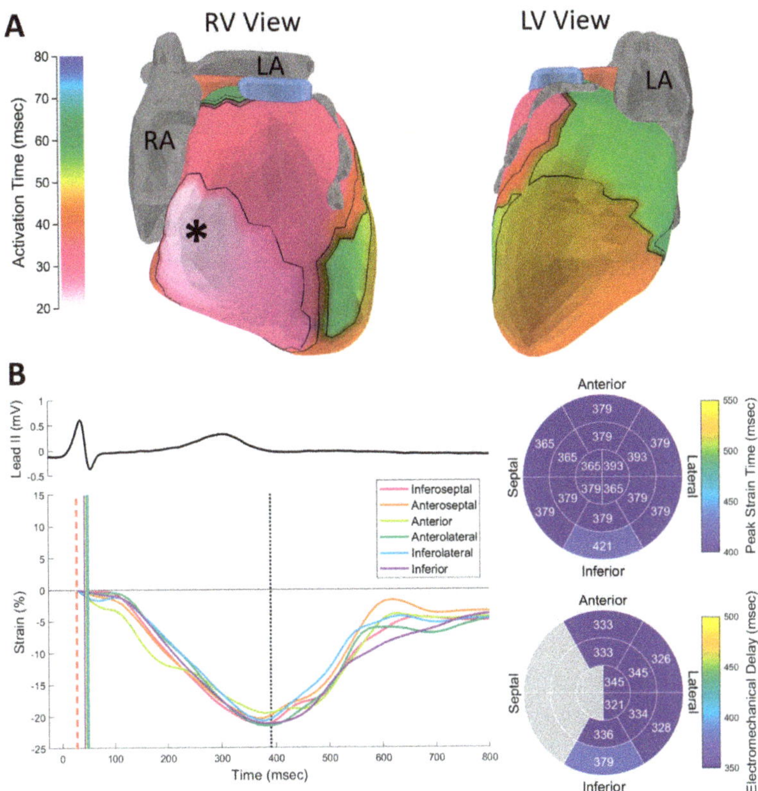

Figure 3. Healthy adult activation and contraction. (**A**) Activation isochrones. Atria and left anterior descending coronary artery are shown in gray. Right ventricular outflow tract is shown in blue. Left ventricular outflow tract is shown in pink. Asterisk indicates epicardial breakthrough site. (**B**) Speckle tracking echocardiography (STE) strain curves plotted below the ECG. Electrical activation times are indicated in the plot with vertical lines (dashed lines indicate right ventricular activation as an approximation of septal activation time). Dotted line indicates aortic valve closure. The timing of peak strain within anatomical segments (top bullseye plot) was homogeneous within the LV. Regional electromechanical delay (EMD) values (bottom bullseye plot) were computed by subtracting the electrical activation time from the time of peak strain within regions. EMD values were not computed for septal regions (shown in gray) because ECGI does not image the septum. RV: right ventricle; LV: left ventricle; RA: right atrium; LA: left atrium.

3. Results

3.1. Electrical Activation

The native (un-paced) pre-CRT epicardial activation sequence in HF-CRT patients was prolonged compared to controls (Table A1, Appendix A) and was acutely altered by the onset of CRT pacing (Table A2, Appendix A). The most common pattern of native epicardial activation in HF patients was normal RV activation with varying degrees of delay in the lateral LV. Figure 4 presents representative responders and nonresponders. RV activation patterns were generally consistent with control patients, with RV epicardial breakthrough and a radial activation spread from the breakthrough site to the rest of the RV. LV activation patterns were consistent with prior studies [25–27] and generally featured a "U-shaped" left bundle branch block activation pattern with a line of conduction block located between the epicardial aspect of the septum and the LV lateral wall. The best predictors of acute resynchronization efficacy were the baseline level of LV delay (long delay was predictive of effective resynchronization) and the location of the LV lead. The responders (patients

25 and 27) had high baseline electrical dyssynchrony with LV leads located in the region of late activation. The acute onset of CRT pacing resulted in significant improvement in the LV electrical delay parameter. In contrast, the first nonresponder (patient 12) had less baseline dyssynchrony and the onset of pacing resulted in less improvement in synchrony compared to the responders. The second nonresponder (patient 13) had high baseline dyssynchrony. However, the anterior location of the pacing lead reduced the effectiveness of resynchronizing the ventricles.

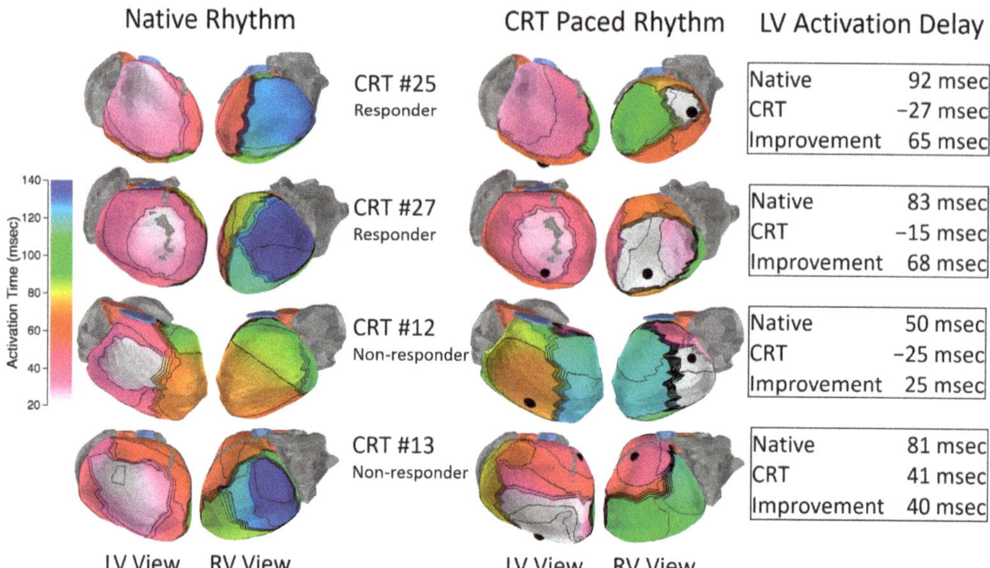

Figure 4. Activation isochrone maps in HF-CRT patients in native rhythm prior to CRT pacing (left) and at pacing onset (right). Pacing lead locations are indicated with black spheres. CRT pacing decreases LV activation delay absolute value ("Improvement"). Echocardiographic responders (top 2 rows) generally had high levels of dyssynchrony at baseline which was substantially improved by CRT pacing. Nonresponders often had less baseline dyssynchrony (row 3) or ineffective lead placement (row 4). RV: right ventricle; LV: left ventricle.

3.2. Contraction

Contraction in HF-CRT patients was dyssynchronous and impaired. HF-CRT patients had higher pre-systolic stretch and lower peak contraction amplitudes than controls (median -8.78% in HF-CRT vs. -20.65% in controls, $p < 0.001$, Table A1). Figure 5 (top) illustrates the native rhythm contraction sequence for a HF-CRT patient. Electrical activation of the lateral LV is delayed relative to the other segments (vertical lines). The lateral LV was stretched (early positive deflection) by the septal wall contraction prior to its own contraction and the lateral wall segments reached peak contraction after aortic valve closure (vertical dotted line). While the overall pattern of delayed LV contraction is apparent from the strain curves, it is difficult to extract markers of contraction timing from the plots because of the influence of opposing contraction and low amplitudes in some regions. In this patient, several of the septal wall segments reached peak strain after the lateral wall, despite the apparent overall pattern of delayed LV contraction. The mean electromechanical delays observed in HF patients closely matched control values. However, the dispersion of EMDs within the LV was larger (median 57 msec in HF-CRT vs. 25 msec in controls, $p < 0.001$, Table A1, Appendix A).

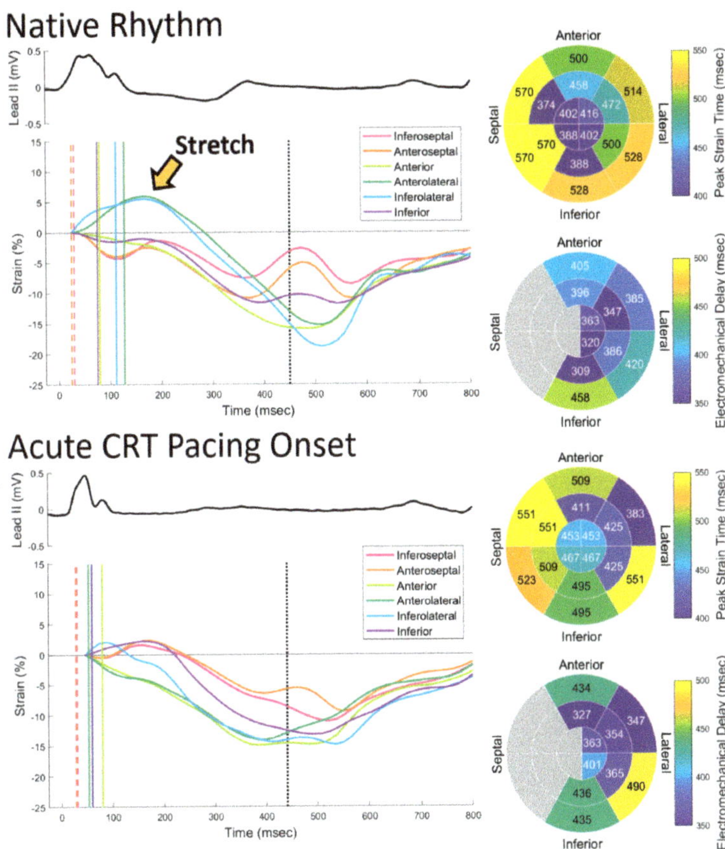

Figure 5. Native rhythm strains in HF-CRT patients (top) were dyssynchronous and lower in amplitude than controls. Lateral regions often stretched prior to contraction (arrow) and reached peak strain after aortic valve closure (dotted line). Many regions reached peak strain later than controls (top bullseye). The mean EMD in HF patients was the same as in controls, but values within the LV showed greater dispersion (bottom bullseye). The acute onset of CRT (bottom) decreased pre-systolic lateral wall stretch. Peak strain timing values did not capture synchrony improvements effectively. Regional EMDs were different for each activation sequence (native rhythm vs. CRT pacing), indicating that EMD is not a purely intrinsic property.

The onset of CRT pacing improved the overall appearance of synchrony in the STE strain curves (Figure 5, bottom). The improved mechanical synchrony was reflected in a decrease in the mechanical delay parameter (median 91 msec in native rhythm vs. −15 msec with CRT pacing, $p < 0.001$). However, the other mechanical dyssynchrony indices were not significantly changed (Table A2, Appendix A). The onset of CRT pacing reduced pre-systolic stretch of the lateral wall but mean peak strain values were unchanged (Table A2, Appendix A). EMD values within anatomical segments were altered by the onset of CRT pacing, indicating that the EMD is not an activation-sequence-independent property of the underlying tissue (Figure 5). There were no statistically significant changes in the mean EMD or the EMD dispersion caused by CRT pacing (Table A2).

3.3. Volumetric Reverse Remodeling (CRT Response)

Of the 30 HF-CRT patients, 22 patients completed the study and could be classified as responders (patients who had a decrease in LV end systolic volume (LVESV) $\geq 15\%$ and/or an increase in ejection fraction (EF) $\geq 5\%$ after 6 months of CRT) or nonresponders after 6 months of CRT. Out of these, seven (32%) were nonresponders, consistent with the rate of nonresponders observed in most CRT studies. We used Spearman's rank correlation coefficient to correlate baseline (prior to CRT) native rhythm electrical and mechanical parameters with change in LVESV (Table A3, Appendix A). In general, we found that electrical dyssynchrony parameters were more predictive of beneficial reverse remodeling than mechanical dyssynchrony and function parameters. Each ECGI index of electrical dyssynchrony (total activation time, LV electrical delay, and LV electrical dispersion) correlated with LVESV remodeling more strongly than body-surface QRS duration and all of the mechanical indices. The most predictive parameter was the LV electrical delay (Rho = -0.722, $p < 0.001$). In addition to baseline parameters, we correlated acute improvements in synchrony parameters at the onset of CRT pacing with change in LVESV (Table A4, Appendix A). In this evaluation, acute improvements to the LV activation delay and mechanical dispersion had the highest correlations with LVESV remodeling (Rho = 0.679 and Rho = 0.671, respectively). These correlations were both lower than the correlation between baseline LV delay and LVESV reverse remodeling.

While baseline dyssynchrony and LV lead placement were major factors in resynchronization, the presence of scars played a role in the efficacy of CRT pacing in several patients. Electrograms from scar regions are characteristically low-amplitude and fractionated. We identified the electrophysiological substrate of scars by computing the number of steep downward deflections in low voltage electrograms. HF-CRT patients had a higher percentage of fractionated electrograms than controls (median 1.77% in HF-CRT vs. 0.38% in controls, $p = 0.001$, Table A1, Appendix A). There were five patients with very high levels of fractionation (6 standard deviations above the control mean)—three were nonischemic, one was ischemic, and one had mild non-obstructive coronary artery disease. Patients 8 and 30 (Figure 6, top and bottom rows, respectively) demonstrate the importance of lead placement in relation to the scar. Patient 8 has an LV lead located in an inferolateral scar region, which delayed activation of the anterior LV. As a result, the resynchronization was less effective, and the patient was a nonresponder. Patient 30 had a scar region in the basal lateral LV, but the patient's LV lead was placed in the center of a region of healthy myocardium. This patient's electrical synchrony improved substantially, and the patient was a responder.

3.4. Electromechanical Remodeling

To assess changes in electrical and mechanical properties induced by chronic CRT pacing, we compared electrical and mechanical parameters in native rhythm prior to onset of CRT to native rhythm after 6 months of CRT. We found no significant changes in electrical or mechanical synchrony parameters when comparing native rhythm before and after chronic CRT pacing, indicating that the underlying level of dyssynchrony remains unchanged (Table A5, Appendix A). We also did not observe statistically significant changes in mean EMD or the dispersion of the EMD (Table A5, Appendix A). While patients' native (un-paced) dyssynchrony indices and EMDs were largely unchanged after 6 months of CRT pacing, we did observe a persistent improvement in native rhythm contraction magnitudes after 6 months of CRT (baseline median: -8.55%, remodeled median: -10.14%, $p = 0.008$) and sustained EF improvement (baseline median: 24%, remodeled median: 27%, $p < 0.001$). Figure 7 demonstrates the improved contraction in four HF-CRT patients.

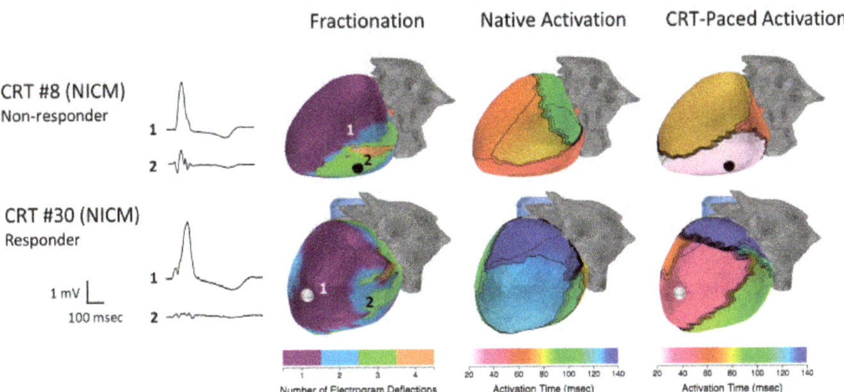

Figure 6. Left ventricular views of electrogram fractionation maps (first column), native rhythm activation (middle column), and CRT-paced activation (right column). Representative fractionated and un-fractionated electrograms are provided to the left of the maps. Numbers indicate electrogram locations. Pacing electrodes are indicated with black or white spheres. Pacing within regions of fractionation was less effective at activating nearby regions outside the scar (top row). Patients with large regions of fractionation could still be resynchronized effectively when paced outside of the fractionated region (bottom row). NICM: Nonischemic cardiomyopathy.

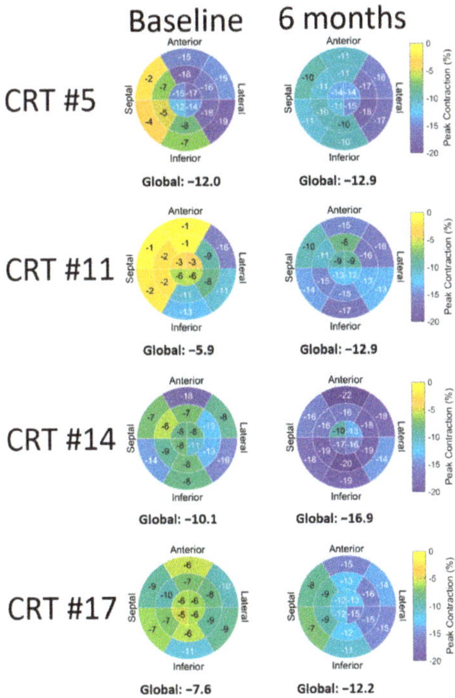

Figure 7. Peak contraction magnitudes improved during the course of CRT. Values at each visit were determined from un-paced native rhythm beats, indicating persistent improvements in contraction as a result of chronic CRT pacing. Global longitudinal strain values (in percent) are indicated below each bullseye plot.

Prior to the onset of CRT, HF patients' mean epicardial ARIs were prolonged by about 45 msec compared to controls (Table A1, Appendix A). After 6 months of CRT, mean epicardial ARIs were largely unchanged (Table A5, Appendix A). However, native rhythm ARIs near the LV pacing site were prolonged by a median value of 23 msec ($p = 0.002$). Figure 8 presents baseline and remodeled ARI maps for two patients.

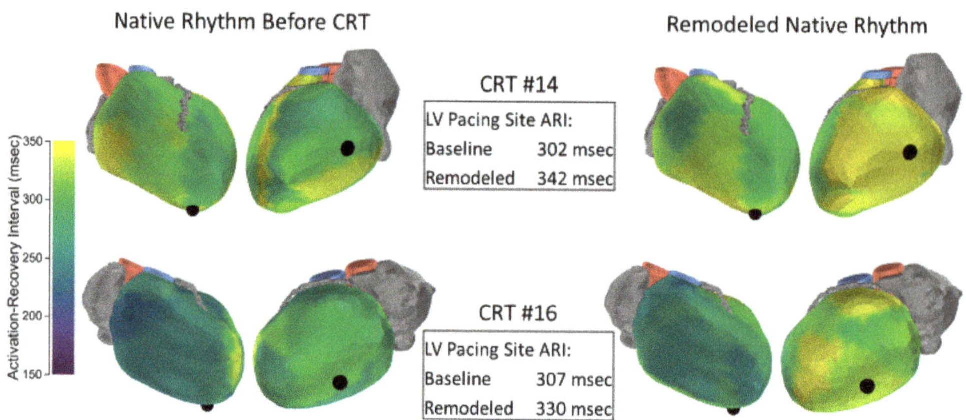

Figure 8. Native rhythm epicardial activation-recovery interval (ARI) maps in heart failure (**left**) and after 6 months of CRT pacing (**right**). After 6 months of CRT pacing, ARI values were prolonged at and around the location of the left ventricle (LV) pacing lead.

We correlated the changes in contraction and ARI prolongation at the LV pacing site with the change in LVESV. The change in contraction was correlated with LVESV change, but the LV ARI change was not (Table A6, Appendix A).

4. Discussion

This study combined ECGI and STE for the first time and applied this imaging longitudinally to a group of 30 HF-CRT patients to study the electromechanics of HF, the physiological effects of CRT-induced reverse remodeling, and the utility of ECGI and STE for predicting response to CRT. Using this combination of imaging modalities, we defined and measured the EMD, a novel measurement of excitation-contraction coupling. We found that HF-CRT patients had increased EMD dispersion compared to controls, but the EMD dispersion was not acutely corrected by the onset of pacing and did not significantly change after 6 months of CRT.

Because ECGI and STE are safe, noninvasive imaging techniques, we were able to image patients longitudinally. After 6 months of CRT, we found that patients' underlying dyssynchrony was unchanged, but increases in peak strain magnitudes revealed improved contraction. Additionally, the panoramic imaging of ECGI revealed prolonged repolarization at the LV pacing lead site. Because we performed echocardiographic evaluations at baseline and after 6 months of CRT, we were able to correlate LVESV reduction (volumetric reverse remodeling) with changes in electromechanical parameters. We found that increased strain magnitudes correlated with LVESV reduction, but the repolarization changes did not, suggesting that the repolarization changes are an effect of the LV epicardial pacing, rather than the reverse remodeling.

In our prospective evaluation of ECGI, STE, and conventional ECG dyssynchrony indices, we found that the ECGI LV activation delay parameter (the difference in mean RV and LV activation time) was the best predictor of response to CRT. Because we imaged patients at baseline and at the acute onset of CRT pacing, we were also able to correlate acute changes in dyssynchrony parameters with reverse remodeling. We found the strongest correlation between LV activation delay and LVESV reduction, and we hypothesize that this parameter could be used to optimize CRT device settings for maximal benefit to patients.

4.1. Reverse Remodeling

CRT is associated with a beneficial reverse-remodeling process which includes a reversal of LV chamber dilation. The volumetric reverse remodeling we observed in this study was consistent with previously reported studies. Notably, 32% of the patients in the present study were echocardiographic nonresponders to CRT, consistent with the nonresponder rate around 30% that is commonly reported [20].

We found that patients' levels of electrical and mechanical dyssynchrony after 6 months of CRT were largely unchanged in their native (un-paced) rhythms. Despite the persisting dyssynchrony, we observed increased contraction (strain) magnitudes and a small but statistically significant improvement in EF. The increase in contraction magnitudes that we observed could be an effect of the volumetric reverse remodeling. In simplified models, ventricular wall stress is proportional to chamber diameter as governed by the Law of Laplace [28]. Volumetric reverse remodeling can therefore decrease wall stress, which decreases the force the contractions are acting against. This may contribute to increased strain amplitudes. The correlation between LVESV decrease and contraction amplitude supports this as a possible mechanism, though it is possible that contractile improvements at the cellular level also contribute.

Multiple animal studies demonstrated ARI shortening as a result of CRT-induced reverse remodeling [18,19,29]. In our study, we found not only that mean ARIs do not decrease, but that ARIs near the LV pacing lead increased by a median value of 23 msec. This ARI prolongation was not correlated with LVESV remodeling, suggesting that this change is not driven by the reverse remodeling process, but rather is caused by the pacing pre-excitation. Prior ECGI studies showed that pre-excitation due to RV pacing and Wolff–Parkinson–White syndrome were both associated with prolonged ARI in the region of pre-excitation [30,31]. In the present study, we observed ARI prolongation at the LV pacing site, but not the RV pacing site. It is noteworthy that CRT devices are often programmed so that the RV is activated by the intrinsic conduction system (illustrated in Figure 4—note the unchanged RV activation pattern in patients 25 and 27). This may cause less pre-excitation at the RV lead location compared to the LV lead and consequently a minimal effect on ARI near the RV lead. The finding of ARI prolongation in the region of the LV pacing lead is consistent with a body-surface study of repolarization heterogeneity in HF-CRT patients. The population of that study showed increased repolarization heterogeneity following CRT implantation which declined following the acute resynchronization phase [32]. This raises the possibility that the ARI prolongation we observed near the LV lead could have decreased as CRT continued. The discrepancy between these human study results and animal data, which showed shortening of ARI prolongation after CRT [18,19,29], could be due to the species differences or the nature of the heart failure model, which was induced by RV tachypacing in the animal studies.

4.2. Response to CRT

Overall, this study found that electrical indices were more strongly correlated with LVESV reverse remodeling than STE indices. Despite promising initial evaluations, the multicenter PROSPECT trial found that no echocardiographic measures of dyssynchrony improved patient selection for CRT [21]. The EchoCRT trial, which evaluated CRT in patients with mechanical dyssynchrony and a narrow QRS complex, was stopped for futility with the conclusion that CRT was possibly detrimental in that patient cohort [22]. The activation sequence dependence of the EMD parameter that we observe is an indicator that the timing of peak strains within anatomical segments does not simply reflect the local contraction. Instead, the EMD is influenced by the entire contraction sequence including opposing wall contractions. This interaction likely decreases the effectiveness of STE indices of dyssynchrony for predicting CRT response. As expected, the baseline QRS duration was correlated with LVESV reverse remodeling. Both of the previously reported ECGI indices of dyssynchrony (the LV activation delay and the LV activation dispersion) were more strongly correlated with LVESV reverse remodeling than the QRS duration. Of all the dyssynchrony indices evaluated, the LV electrical activation delay was the best index for identifying potential responders to CRT.

The initial evaluations of ECGI in HF-CRT patients were retrospective studies of ischemic and nonischemic patients [25,26]. In the present study, we validate the results of those studies with prospective data. Our data are also consistent with another prospective ECGI study of HF-CRT patients that found the LV delay (referred to in that study as ventricular electrical uncoupling) was superior to QRS duration at identifying responders [33]. That study utilized a clinical composite endpoint (freedom from death, freedom from hospitalization, and NYHA functional class improvement after 6 months of CRT). By utilizing echocardiographic criteria for defining response in the present study, we were able to demonstrate that ECGI is predictive of CRT-induced reverse remodeling. In this study, we found that baseline (prior to CRT) electrical dyssynchrony was more predictive of improvement in LVESV remodeling than the improvement of synchrony at pacing onset. It would be valuable to explore whether using ECGI interactively to optimize lead placement and pacing timing could offer patients additional benefits.

An evaluation of ECGI in patients with pediatric congenital heart disease found that placing the LV lead in the site of latest activation resulted in favorable outcomes [34]. This study demonstrated the importance of the lead location in relation to electrophysiological scar. Prior studies using other imaging modalities demonstrated the importance of scar and its location relative to pacing leads for predicting CRT response [35]. An advantage of using ECGI for scar mapping is that regions of delayed activation, scar regions, and the baseline level of electrical dyssynchrony could all be imaged with a single modality.

4.3. Study Limitations

Because ECGI does not image the septum, electrical and electromechanical parameters for the septum could not be evaluated. While this study explored the relationship between acute change in ECGI-derived synchrony and reverse remodeling, there was no attempt to optimize CRT device settings using ECGI parameters. Future work could evaluate the impact of utilizing ECGI parameters to optimize device settings to maximize patient benefits. Clinical outcomes in HF-CRT patients are also impacted by medications [36] and comorbidities such as diabetes [37–39]. Future work should be sufficiently powered to account for these variables.

5. Conclusions

ECGI combined with STE is an effective tool for studying cardiac electromechanics. The delay between electrical activation and peak strain is abnormal in HF. CRT does not correct the EMD, but strain magnitudes improve as part of the reverse remodeling process. The LV activation delay measured by ECGI is an effective index for predicting response to CRT, and could have utility for patient selection or device optimization. CRT was less effective when the LV lead was located in a region of electrophysiological scar based on ECGI criteria. This suggests that ECGI could be used to guide LV lead placements. Because ECGI was not used to guide lead placement in the present study, this approach requires additional prospective validation.

Author Contributions: Conceptualization, C.M.A. and Y.R.; methodology, C.M.A., G.K.S., and Y.R.; software, C.M.A.; validation, C.M.A., G.K.S., and Y.R.; formal analysis, C.M.A.; investigation, C.M.A.; resources, G.K.S. and Y.R.; data curation, C.M.A.; writing—original draft preparation, C.M.A. and Y.R.; writing—review and editing, C.M.A. and Y.R.; visualization, C.M.A.; supervision, G.K.S. and Y.R.; project administration, C.M.A., G.K.S., and Y.R.; funding acquisition, Y.R. All authors have read and agreed to the published version of the manuscript.

Funding: This study was supported by NIH—National Heart, Lung and Blood Institute grants R01-HL-033343 and R01-HL-049054 (to YR) and by Washington University Institute of Clinical and Translational Sciences grant UL1-TR000448 from the National Center for Advancing Translational Sciences of the NIH. Dr. Rudy is the Fred Saigh Distinguished Professor at Washington University.

Institutional Review Board Statement: The study was conducted according to the guidelines of the Declaration of Helsinki and approved by Human Research Protection Office at Washington University in St. Louis.

Informed Consent Statement: Informed consent was obtained from all subjects involved in the study.

Data Availability Statement: The data presented in this study are available on request from the corresponding author. The data are not publicly available due to Protected Health Information (PHI) in the data.

Conflicts of Interest: Y.R. receives royalties from CardioInsight Technologies (CIT). CIT does not support any research conducted in Dr. Rudy's laboratory. The remaining authors have nothing to disclose.

Appendix A

Table A1. Control and HF-CRT baseline electrical and mechanical parameter comparison. All parameters were measured in native rhythm prior to CRT pacing. Groups were compared using Wilcoxon rank-sum tests. IQR: interquartile range; EM: electromechanical.

	Parameter	Control Median	Control IQR	Heart Failure Median	Heart Failure IQR	p
ECG	QRS duration (msec)	87	16	154	31	$p < 0.001$
Volume	Ejection fraction (%)	57	3	24	4	$p < 0.001$
	End systolic volume (mL)	46	20	122	56	$p < 0.001$
	End diastolic volume (mL)	111	36	160	72	0.002
	Stroke volume (mL)	63	18	39	19	$p < 0.001$
ECGI	Total activation time (msec)	42	12	123	33	$p < 0.001$
	LV activation delay (msec)	14	7	81	48	$p < 0.001$
	LV activation dispersion (msec)	9	3	24	9	$p < 0.001$
	Mean epicardial ARI (msec)	238	32	284	26	$p < 0.001$
	Mean LV ARI (msec)	240	23	292	25	$p < 0.001$
	Mean RV ARI (msec)	240	50	281	31	$p < 0.001$
	Percentage of fractionated electrograms (%)	0.38	1.03	1.77	2.11	0.001
	Fractionated deflections per electrogram	0.0038	0.0103	0.0179	0.0254	$p < 0.001$
Strain	Total strain time (msec)	98	42	196	67	$p < 0.001$
	Mechanical delay (msec)	12	53	91	112	$p < 0.001$
	Mechanical dispersion (msec)	28	10	70	19	$p < 0.001$
	Mean peak strain time (msec)	427	40	432	82	0.532
	Mean peak contraction (%)	−20.64	2.82	−8.78	2.42	$p < 0.001$
	Max pre-systolic stretch (%)	2.57	1.86	4.38	3.48	0.003
EM Delay	Mean electromechanical delay (msec)	375	40	373	50	0.940
	Electromechanical delay dispersion (msec)	25	15	57	19	$p < 0.001$

Table A2. The effect of acute CRT pacing onset on electrical and mechanical parameters in heart failure patients. Changes were evaluated using Wilcoxon signed-rank tests. IQR: interquartile range; EM: electromechanical.

	Parameter	Native Median	Paced Median	Paired Delta Median	Paired Delta IQR	p
ECG	QRS duration (msec)	154	128	−37	31	$p < 0.001$
Volume	Ejection fraction (%)	24	27	3	4	$p < 0.001$
	End systolic volume (mL)	122	114	1	20	0.914
	End diastolic volume (mL)	160	168	5	29	0.094
	Stroke volume (mL)	39	43	7	9	$p < 0.001$
ECGI	Total activation time (msec)	123	104	−19	29	0.004
	LV activation delay (msec)	81	10	−62	53	$p < 0.001$
	LV activation dispersion (msec)	24	29	11	2	0.057
Strain	Total strain time (msec)	196	182	−14	75	0.223
	Mechanical delay (msec)	91	−15	−96	91	$p < 0.001$
	Mechanical dispersion (msec)	70	63	−9	40	0.046
	Mean peak contraction (%)	−8.78	−8.78	0.22	1.97	0.549
	Max pre-systolic stretch (%)	4.38	3.6	−1.23	2.19	0.009
EM Delay	Mean electromechanical delay (msec)	373	391	10	50	0.097
	Electromechanical delay dispersion (msec)	57	61	8	33	0.039

Table A3. Correlations between the change in left ventricular end-systolic volume (LVESV) after 6 months of CRT pacing and baseline electrical and mechanical parameters in heart failure patients. All parameters were measured in native rhythm. Correlations were determined using Spearman's Rho tests.

	Parameter	Spearman's Rho	p
ECG	QRS duration (msec)	−0.486	0.023
ECGI	Total activation time (msec)	−0.583	0.005
	LV activation delay (msec)	−0.722	$p < 0.001$
	LV activation dispersion (msec)	−0.588	0.005
	Mean epicardial ARI (msec)	−0.110	0.625
	Mean LV ARI (msec)	0.168	0.454
	Mean RV ARI (msec)	−0.373	0.088
	Fractionated deflections per electrogram	0.241	0.280
	Fractionated deflections per electrogram at LV pacing site	0.387	0.075
Strain	Total peak strain time (msec)	−0.425	0.101
	Mechanical delay (msec)	0.132	0.639
	Mechanical dispersion (msec)	−0.585	0.019
	Mean peak contraction (%)	0.009	0.978
	Mean pre-systolic stretch (%)	−0.324	0.221
Electromechanical Delay	Mean electromechanical delay (msec)	−0.225	0.419
	Electromechanical delay dispersion (msec)	−0.204	0.466

Table A4. Correlations between the change in left ventricular end-systolic volume (LVESV) after 6 months of cardiac resynchronization therapy (CRT) pacing and the change in dyssynchrony parameters due to the acute onset of CRT pacing. Correlations were determined using Spearman's Rho tests.

	Parameter	Spearman's Rho	p
ECG	QRS duration (msec)	0.505	0.017
Volume	Ejection fraction (%)	0.025	0.934
ECGI	Total activation time (msec)	0.407	0.061
	LV activation delay absolute value (msec)	0.679	$p < 0.001$
	LV activation dispersion (msec)	0.545	0.010
Strain	Total peak strain time (msec)	0.583	0.023
	Mechanical delay absolute value (msec)	0.013	0.965
	Mechanical dispersion (msec)	0.671	0.008

Table A5. The effects of 6 months of cardiac resynchronization therapy pacing on heart failure patients' electrical and mechanical parameters. All parameters at each visit were computed from native rhythm beats. Changes were evaluated using Wilcoxon signed-rank tests.

	Parameter	Baseline Median	6 Months Median	Paired Delta Median	Delta IQR	p
ECG	QRS duration (msec)	154	157	5	11	0.164
Volume	Ejection fraction (%)	24	27	6	7	$p < 0.001$
	End systolic volume (mL)	125	121	−22	47	0.046
	End diastolic volume (mL)	167	171	−11	49	0.306
	Stroke volume (mL)	40	48	8	19	0.005
ECGI	Total activation time (msec)	123	115	−4	14	0.178
	LV activation delay (msec)	82	81	2	20	0.858
	LV activation dispersion (msec)	24	25	1	4	0.211
	Mean epicardial ARI (msec)	282	288	6	28	0.408
	Mean LV ARI (msec)	288	293	6	26	0.211
	Mean RV ARI (msec)	280	283	4	28	0.833
	LV pacing site ARI (msec)	280	306	23	36	0.002
	RV pacing site ARI (msec)	293	297	3	30	0.291
	Percentage of fractionated electrograms (%)	1.54	1.85	0.56	2.16	0.511
	Fractionated deflections per electrogram	0.0171	0.0235	0.0046	0.0492	0.565
Strain	Total peak strain time (msec)	196	210	14	77	0.891
	Mechanical delay (msec)	88	81	−6	65	1.000
	Mechanical dispersion (msec)	70	66	0	26	0.228
	Mean peak contraction (%)	−8.55	−10.14	−1.31	2.93	0.008
	Mean pre-systolic stretch (%)	4.59	3.48	−1.13	4.22	0.438
EM Delay	Mean electromechanical delay (msec)	367	381	14	45	0.169
	Electromechanical delay dispersion (msec)	57	50	−7	20	0.135

Table A6. Correlation between the change in left ventricular end-systolic volume after 6 months of cardiac resynchronization therapy (CRT) pacing and the changes in ARI and contraction remodeling. Correlations were determined using Spearman's Rho tests.

Parameter	Spearman's Rho	p
LV pacing site ARI (msec)	−0.206	0.356
Mean peak contraction (%)	0.618	0.013

References

1. Ramanathan, C.; Ghanem, R.N.; Jia, P.; Ryu, K.; Rudy, Y. Noninvasive electrocardiographic imaging for cardiac electrophysiology and arrhythmia. *Nat. Med.* **2004**, *10*, 422–428. [CrossRef]
2. Andrews, C.M.; Srinivasan, N.T.; Rosmini, S.; Bulluck, H.; Orini, M.; Jenkins, S.; Pantazis, A.; McKenna, W.J.; Moon, J.C.; Lambiase, P.D.; et al. Electrical and Structural Substrate of Arrhythmogenic Right Ventricular Cardiomyopathy Determined Using Noninvasive Electrocardiographic Imaging and Late Gadolinium Magnetic Resonance Imaging. *Circ. Arrhythmia Electrophysiol.* **2017**, *10*. [CrossRef] [PubMed]
3. Andrews, C.; Cupps, B.P.; Pasque, M.K.; Rudy, Y. Electromechanics of the Normal Human Heart In Situ. *Circ. Arrhythmia Electrophysiol.* **2019**, *12*. [CrossRef]
4. Blessberger, H.; Binder, T. Non-invasive imaging: Two dimensional speckle tracking echocardiography: Basic principles. *Heart* **2010**, *96*, 716–722. [CrossRef] [PubMed]
5. Collier, P.; Phelan, D.; Klein, A. A Test in Context: Myocardial Strain Measured by Speckle-Tracking Echocardiography. *J. Am. Coll. Cardiol.* **2017**, *69*, 1043–1056. [CrossRef] [PubMed]
6. Benjamin, E.J.; Virani, S.S.; Callaway, C.W.; Chamberlain, A.M.; Chang, A.R.; Cheng, S.; Chiuve, S.E.; Cushman, M.; Delling, F.N.; Deo, R.; et al. Heart Disease and Stroke Statistics—2018 Update: A Report From the American Heart Association. *Circulation* **2018**, *137*. [CrossRef]
7. Davies, C.; Davia, K.; Bennett, J.; Pepper, J.; Poole-Wilson, P.; Harding, S. Reduced contraction and altered frequency response of isolated ventricular myocytes from patients with heart failure. *Circulation* **1995**, *92*, 2540–2549. [CrossRef]
8. Tomaselli, G.F.; Zipes, D.P. What Causes Sudden Death in Heart Failure? *Circ. Res.* **2004**, *95*, 754–763. [CrossRef]
9. Nattel, S.; Maguy, A.; Le Bouter, S.; Yeh, Y.-H. Arrhythmogenic Ion-Channel Remodeling in the Heart: Heart Failure, Myocardial Infarction, and Atrial Fibrillation. *Physiol. Rev.* **2007**, *87*, 425–456. [CrossRef]
10. Eapen, Z.; Rogers, J.G. Strategies to attenuate pathological remodeling in heart failure. *Curr. Opin. Cardiol.* **2009**, *24*, 223–229. [CrossRef]
11. Kirk, J.A.; Kass, D.A. Cellular and Molecular Aspects of Dyssynchrony and Resynchronization. *Card. Electrophysiol. Clin.* **2015**, *7*, 585–597. [CrossRef]
12. Gambardella, J.; Trimarco, B.; Iaccarino, G. New Insights in Cardiac Calcium Handling and Excitation-Contraction Coupling. *Adv. Exp. Med. Biol.* **2018**, *1067*, 373–385. [CrossRef]
13. Kashani, A.; Barold, S.S. Significance of QRS Complex Duration in Patients With Heart Failure. *J. Am. Coll. Cardiol.* **2005**, *46*, 2183–2192. [CrossRef]
14. Abraham, W.T.; Fisher, W.G.; Smith, A.L.; Delurgio, D.B.; Leon, A.R.; Loh, E.; Kocovic, D.Z.; Packer, M.; Clavell, A.L.; Hayes, D.L.; et al. Cardiac Resynchronization in Chronic Heart Failure. *N. Engl. J. Med.* **2002**, *346*, 1845–1853. [CrossRef] [PubMed]
15. Cleland, J.G.; Daubert, J.-C.; Erdmann, E.; Freemantle, N.; Gras, D.; Kappenberger, L.; Tavazzi, L.; on behalf of The CARE-HF Study Investigators. Longer-term effects of cardiac resynchronization therapy on mortality in heart failure [the CArdiac REsynchronization-Heart Failure (CARE-HF) trial extension phase]. *Eur. Heart J.* **2006**, *27*, 1928–1932. [CrossRef] [PubMed]
16. Cleland, J.G.; Daubert, J.-C.; Erdmann, E.; Freemantle, N.; Gras, D.; Kappenberger, L.; Tavazzi, L. The effect of cardiac resynchronization on morbidity and mortality in heart failure. *N. Engl. J. Med.* **2005**, *352*, 1539–1549. [CrossRef] [PubMed]
17. Bristow, M.R.; Saxon, L.A.; Boehmer, J.; Krueger, S.; Kass, D.A.; De Marco, T.; Carson, P.; Dicarlo, L.; DeMets, D.; White, B.G.; et al. Cardiac-resynchronization therapy with or without an implantable defibrillator in advanced chronic heart failure. *N. Engl. J. Med.* **2004**, *350*, 2140–2150. [CrossRef] [PubMed]
18. Nishijima, Y.; Sridhar, A.; Viatchenko-Karpinski, S.; Shaw, C.; Bonagura, J.D.; Abraham, W.T.; Joshi, M.S.; Bauer, J.A.; Hamlin, R.L.; Györke, S.; et al. Chronic cardiac resynchronization therapy and reverse ventricular remodeling in a model of nonischemic cardiomyopathy. *Life Sci.* **2007**, *81*, 1152–1159. [CrossRef]
19. Aiba, T.; Hesketh, G.G.; Barth, A.S.; Liu, T.; Daya, S.; Chakir, K.; Dimaano, V.L.; Abraham, T.P.; O'Rourke, B.; Akar, F.G.; et al. Electrophysiological Consequences of Dyssynchronous Heart Failure and Its Restoration by Resynchronization Therapy. *Circulation* **2009**, *119*, 1220–1230. [CrossRef]
20. Daubert, C.; Behar, N.; Martins, R.P.; Mabo, P.; Leclercq, C. Avoiding non-responders to cardiac resynchronization therapy: A practical guide. *Eur. Heart J.* **2016**, ehw270. [CrossRef]
21. Chung, E.S.; Leon, A.R.; Tavazzi, L.; Sun, J.-P.; Nihoyannopoulos, P.; Merlino, J.; Abraham, W.T.; Ghio, S.; Leclercq, C.; Bax, J.J.; et al. Results of the Predictors of Response to CRT (PROSPECT) Trial. *Circulation* **2008**, *117*, 2608–2616. [CrossRef]
22. Beshai, J.F.; Grimm, R.A.; Nagueh, S.F.; Baker, J.H.; Beau, S.L.; Greenberg, S.M.; Pires, L.A.; Tchou, P.J. Cardiac-Resynchronization Therapy in Heart Failure with Narrow QRS Complexes. *N. Engl. J. Med.* **2007**, *357*, 2461–2471. [CrossRef]
23. Coronel, R.; De Bakker, J.M.T.; Wilms-Schopman, F.J.G.; Opthof, T.; Linnenbank, A.C.; Belterman, C.N.; Janse, M.J. Monophasic action potentials and activation recovery intervals as measures of ventricular action potential duration: Experimental evidence to resolve some controversies. *Heart Rhythm* **2006**, *3*, 1043–1050. [CrossRef]
24. Orini, M.; Srinivasan, N.; Graham, A.J.; Taggart, P.; Lambiase, P.D. Further Evidence on How to Measure Local Repolarization Time Using Intracardiac Unipolar Electrograms in the Intact Human Heart. *Circ. Arrhythmia Electrophysiol.* **2019**, *12*, e007733. [CrossRef] [PubMed]

25. Jia, P.; Ramanathan, C.; Ghanem, R.N.; Ryu, K.; Varma, N.; Rudy, Y. Electrocardiographic imaging of cardiac resynchronization therapy in heart failure: Observation of variable electrophysiologic responses. *Heart Rhythm* **2006**, *3*, 296–310. [CrossRef] [PubMed]
26. Ghosh, S.; Silva, J.N.A.; Canham, R.M.; Bowman, T.M.; Zhang, J.; Rhee, E.K.; Woodard, P.K.; Rudy, Y. Electrophysiological substrate and intraventricular LV dyssynchrony in non-ischemic heart failure patients undergoing cardiac resynchronization therapy. *Heart Rhythm* **2011**, *8*, 699. [CrossRef] [PubMed]
27. Auricchio, A.; Fantoni, C.; Regoli, F.; Carbucicchio, C.; Goette, A.; Geller, C.; Kloss, M.; Klein, H. Characterization of left ventricular activation in patients with heart failure and left bundle-branch block. *Circulation* **2004**, *109*, 1133–1139. [CrossRef]
28. Zhong, L.; Ghista, D.N.; Tan, R.S. Left ventricular wall stress compendium. *Comput. Methods Biomech. Biomed. Eng.* **2012**, *15*, 1015–1041. [CrossRef] [PubMed]
29. Aiba, T.; Tomaselli, G. Electrical remodeling in dyssynchrony and resynchronization. *J. Cardiovasc. Transl. Res.* **2012**, *5*, 170–179. [CrossRef]
30. Ghosh, S.; Rhee, E.K.; Avari, J.N.; Woodard, P.K.; Rudy, Y. Cardiac Memory in Patients With Wolff-Parkinson-White Syndrome: Noninvasive Imaging of Activation and Repolarization Before and After Catheter Ablation. *Circulation* **2008**, *118*, 907–915. [CrossRef]
31. Marrus, S.B.; Andrews, C.M.; Cooper, D.H.; Faddis, M.N.; Rudy, Y. Repolarization Changes Underlying Long-Term Cardiac Memory Due to Right Ventricular Pacing: Noninvasive Mapping With Electrocardiographic Imaging. *Circ. Arrhythmia Electrophysiol.* **2012**, *5*, 773–781. [CrossRef]
32. Cvijić, M.; Antolič, B.; Klemen, L.; Zupan, I. Repolarization heterogeneity in patients with cardiac resynchronization therapy and its relation to ventricular tachyarrhythmias. *Heart Rhythm* **2018**, *15*, 1784–1790. [CrossRef]
33. Ploux, S.; Lumens, J.; Whinnett, Z.; Montaudon, M.; Strom, M.; Ramanathan, C.; Derval, N.; Zemmoura, A.; Denis, A.; De Guillebon, M. Noninvasive electrocardiographic mapping to improve patient selection for cardiac resynchronization therapy: Beyond QRS duration and left bundle-branch block morphology. *J. Am. Coll. Cardiol.* **2013**, *61*, 2435–2443. [CrossRef]
34. Silva, J.N.; Ghosh, S.; Bowman, T.M.; Rhee, E.K.; Woodard, P.K.; Rudy, Y. Cardiac resynchronization therapy in pediatric congenital heart disease: Insights from noninvasive electrocardiographic imaging. *Heart Rhythm* **2009**, *6*, 1178–1185. [CrossRef]
35. Adelstein, E.C.; Saba, S. Scar burden by myocardial perfusion imaging predicts echocardiographic response to cardiac resynchronization therapy in ischemic cardiomyopathy. *Am. Heart J.* **2007**, *153*, 105–112. [CrossRef]
36. Hasin, T.; Davarashvili, I.; Michowitz, Y.; Farkash, R.; Presman, H.; Glikson, M.; Rav-Acha, M. Association of Guideline-Based Medical Therapy with Malignant Arrhythmias and Mortality among Heart Failure Patients Implanted with Cardioverter Defibrillator (ICD) or Cardiac Resynchronization-Defibrillator Device (CRTD). *J. Clin. Med.* **2021**, *10*, 1753. [CrossRef] [PubMed]
37. Kahr, P.C.; Trenson, S.; Schindler, M.; Kuster, J.; Kaufmann, P.; Tonko, J.; Hofer, D.; Inderbitzin, D.T.; Breitenstein, A.; Saguner, A.M.; et al. Differential effect of cardiac resynchronization therapy in patients with diabetes mellitus: A long-term retrospective cohort study. *ESC Heart Fail.* **2020**, *7*, 2773–2783. [CrossRef]
38. Kaya, E.; Senges, J.; Hochadel, M.; Eckardt, L.; Andresen, D.; Ince, H.; Spitzer, S.G.; Kleemann, T.; Maier, S.S.K.; Jung, W.; et al. Impact of diabetes on clinical outcome of patients with heart failure undergoing ICD and CRT procedures: Results from the German Device Registry. *ESC Heart Fail.* **2020**, *7*, 984–995. [CrossRef]
39. Sardu, C.; Marfella, R.; Santulli, G. Impact of diabetes mellitus on the clinical response to cardiac resynchronization therapy in elderly people. *J. Cardiovasc. Transl. Res.* **2014**, *7*, 362–368. [CrossRef]

Review

Current ECG Aspects of Interatrial Block

Antoni Bayés-de-Luna [1],*, Miquel Fiol-Sala [2], Manuel Martínez-Sellés [3,4] and Adrian Baranchuk [5]

[1]   Cardiovascular Research Foundation, Cardiovascular ICCC- Program, Research Institute Hospital de la Santa Creu i Sant Pau, IIB-Sant Pau, 08025 Barcelona, Spain

[2]   Illes Balears Health Research Institut (IdISBa), Hospital Son Espases, 07120 Palma, Spain; miquelfiol@yahoo.es

[3]   Cardiology Department, Hospital General Universitario Gregorio Marañón, Calle Doctor Esquerdo, 46, 28007 Madrid, Spain; mmselles@secardiologia.es

[4]   Medical School of the Complutense, University of Madrid, 28040 Madrid, Spain

[5]   Department of Cardiology, Queen's University, Kingston, ON K7L 3N6, Canada; adrian.Baranchuk@kingstonhsc.ca

*   Correspondence: abayes@santpau.cat; Tel.: +34-93-556-56-12

Abstract: Interatrial blocks like other types of block may be of first degree or partial second degree, also named transient atrial block or atrial aberrancy, and third degree or advanced. In first degree, partial interatrial block (P-IAB), the electrical impulse is conducted to the left atrium, through the Bachmann's region, but with delay. The ECG shows a P-wave $\geq$ 120 ms. In third-degree, advanced interatrial block (A-IAB), the electrical impulse is blocked in the upper part of the interatrial septum (Bachmann region); the breakthrough to LA has to be performed retrogradely from the AV junction zone. This explains the p $\pm$ in leads II, III and aVF. In typical cases of A-IAB, the P-wave morphology is biphasic ($\pm$) in leads II, III and aVF, because the left atrium is activated retrogradely and, therefore, the last part of the atrial activation falls in the negative hemifield of leads II, III and aVF. Recently, some atypical cases of A-IAB have been described. The presence of A-IAB is a risk factor for atrial fibrillation, stroke, dementia, and premature death.

Keywords: interatrial block; partial interatrial block; advanced interatrial block; atypical patterns

Citation: Bayés-de-Luna, A.; Fiol-Sala, M.; Martínez-Sellés, M.; Baranchuk, A. Current ECG Aspects of Interatrial Block. *Hearts* **2021**, *2*, 419–432. https://doi.org/10.3390/hearts2030033

Academic Editor: Peter Macfarlane

Received: 19 July 2021
Accepted: 6 September 2021
Published: 8 September 2021

Publisher's Note: MDPI stays neutral with regard to jurisdictional claims in published maps and institutional affiliations.

1. Introduction

The diagnosis of different types of atrioventricular (AV) and ventricular blocks is well known and is explained in most cardiology and clinical electrocardiography books [1–4]. However, the ECG diagnosis of blocks at the atrial level based in changes of morphology and duration of the P-wave, although established many years ago [5], is sometimes even not mentioned in some books of clinical ECG [6]. The P-wave still remains the *ECG Cinderella*. We will review in this chapter the most important aspects related with blocks at the atrial level.

Usually, atrial blocks are located at the interatrial level and, like other types of blocks, could be classified as first degree or partial (P-IAB), second degree or transient, and third degree or advanced (A-IAB) (Figure 1).

Bachmann [7] published the first ECG of partial-IAB in 1944, and Puech [8] published the first case of advanced-IAB in 1956. In the following three decades, only small series or isolated cases were published [9–12]. Finally, in 1985 [5], our group published the ECG-VCG criteria of A-IAB in a series of 88 cases, and three years later in 1988 [13] we published for the first time that patients with A-IAB, compared with a control group of patients with P-IAB, were associated with a much higher incidence of atrial fibrillation (AF) during the follow-up.

In the following 15 years not many papers were published. Only our group [14,15] and the groups of Spodick [16], Garcia-Cosio [17] and Platonov [18] published some papers on this topic.

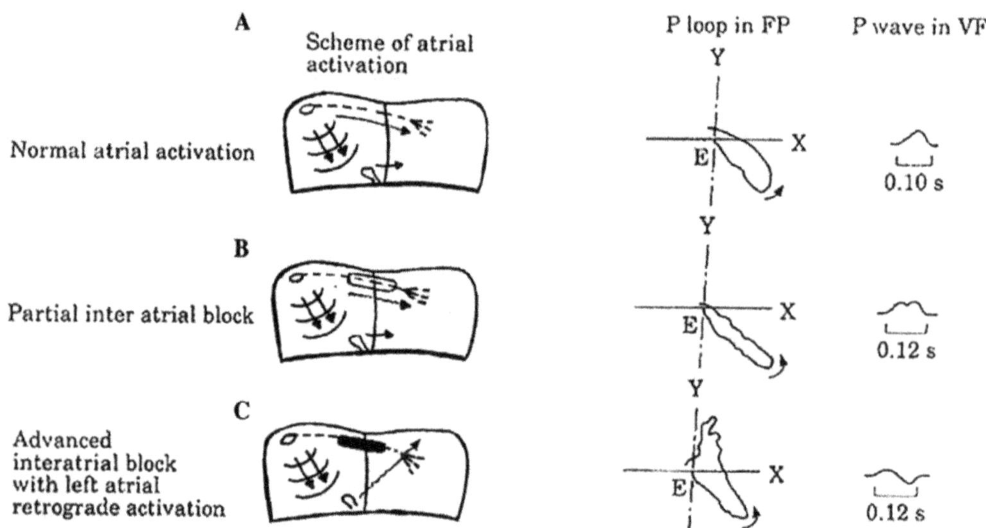

Figure 1. (**A**) Diagram of atrial conduction under normal circumstances; (**B**) partial interatrial block, and (**C**) typical advanced interatrial block with left atrial retrograde activation (IAB with LARA).

The poor knowledge about IAB is probably because it was considered that IAB was equivalent to left atrial enlargement. In 2012 we published a consensus document [19] that stated that $p \geq 120$ ms was explained by a delay of conduction between the right and left atrium, accomplishing the three conduction principles that are required for an ECG pattern to be considered a block: (i) the ECG pattern may be transient [20]; (ii) the ECG pattern may be experimentally reproduced [21,22]; and (iii) the ECG pattern may exist in the absence of cardiac enlargement and ischemic heart disease [5].

Furthermore, in the last few years, we have demonstrated some other important features associated with IAB knowledge: (i) ECG criteria for typical and atypical patterns of IAB [23]; (ii) atrial fibrosis is the most frequent anatomic substrate [24,25]; (iii) clinical associations with not only AF, but also stroke, dementia and mortality [26].

2. ECG Diagnosis of Interatrial Blocks (IAB)

2.1. How to Measure P-Wave Duration?

The measurement of the P-wave [27] and the diagnosis of IAB may be conducted at a first glance if the ECG recording and the morphology of the P-wave in leads II, III, and aVF are clear and free of artifact or pacing spikes. Sometimes, measurements can be difficult due to low voltages and consistency on the measurements is key. Analyzing digital ECG images using amplification is paramount. The use of ECG recording systems that allow at least three simultaneous channels (six or twelve channels) is ideal.

Particular attention should be given to the six leads of the frontal plane, if possible, at the same time, because this is where we will better identify the changes of the P-wave duration and of morphology produced by IAB (biphasic P-wave).

It is also recommended to measure the p terminal force in V1 (PtfV1) [28] as it is considered an extremely specific criterion for the diagnosis of associated left atrial enlargement (LAE). However, the morphology of PtfV1 presents significant variations depending on the location of the V1 precordial electrode [29]. This may explain, at least in part, the recent negative results obtained using the value of PtfV1 as a risk factor for ischemic stroke [30].

To perform a good P-wave duration measurement, it is important to check and define the interval between the earliest detection of the P-wave (onset) in any lead of the frontal plane, and the latest one (offset). Once these two points are defined with lines, the P-wave duration can be measured using calipers or semi-automatic calipers such as the ones provided by Geogebra program, although in clinical practice the duration and morphology of the P-wave may be obtained at first glance.

2.2. First Degree (Partial) Interatrial Block

The electrical impulse is conducted normally from the right atrium to the left atrium, through the Bachmann's region, but with delayed conduction. Therefore, the ECG shows a P-wave $\geq$ 120 ms (Figure 1) (Table 1).

Table 1. ECG classification of interatrial blocks (IAB).

1. Partial Interatrial Block (First Degree) (P-IAB) P-wave $\geq$ 120 ms without negative terminal component in the inferior leads
2. Advanced interatrial block (third degree) (A-IAB) Typical pattern P-wave $\geq$ 120 ms with biphasic morphology in leads II, III aVF ($\pm$) Atypical A-IAB may be atypical by morphology or by duration: (i): Morphological criteria Type 1: P-wave $\geq$ 120 ms with biphasic morphology in leads III and aVF and the final component of the P-wave in lead II is isodiphasic. Type 2: P-wave $\geq$ 120 ms. The second part of The P-wave is biphasic ($\mp$). This means that the global P-wave is triphasic ($+ - +$). Type 3: P-wave $\geq$ 120 ms. The first part of P-wave in leads III and aVF is isoelectric, but the last part is negative. Therefore, it is necessary to perform differential diagnoses with junctional rhythm. (ii) Duration criteria P-wave < 120 ms with typical morphology (Biphasic ($\pm$) P-wave in leads II, III, and aVF)
3. Second degree: The presence of ECG pattern of IAB is intermittent

2.3. Third-Degree (Advanced) Interatrial Block

The electrical impulse is blocked in the upper part of the interatrial septum (Bachmann's region), and as the rest of the septum is predominantly connective tissue, the breakthrough to LA is achieved retrogradely from the AV junction zone or other near zones [5,18]. This explains the $p \pm$ in leads II, III and aVF (Figure 1C) (Table 1).

There is an experimental validation of this phenomenon. As we have already commented, Waldo et al. [21], first in 1971, demonstrated that by cutting the Bachmann's region at both sides of the septum in dogs the biphasic pattern can be reproduced. More recently, Guerra et al. [22] provoked transient A-IAB by applying ice in an open chest (pig model), reproducing transient morphology from normal P-wave to P-IAB and A-IAB and resolution after normalization of temperature (Figure 2).

2.3.1. Typical ECG Pattern

The ECG of typical IAB shows (Figures 3C and 4) a P-wave $\geq$ 120 ms with biphasic ($\pm$) morphology in leads II, III, and aVF, due to caudocranial activation of the left atrium. Our group defined the ECG–VCG criteria in a large series of patients and demonstrated that this type of pattern is associated with LAE in 90% of the cases [5].

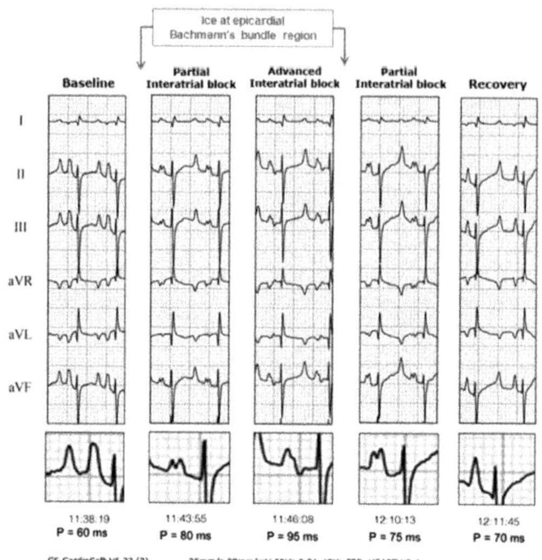

Figure 2. Transient changes of P-wave recorded in lead II after applying ice to the Bachmann's region. Surface electrocardiogram (ECG) of an open-chest anesthetized healthy adult swine, before, during, and after direct application of ice at the transversus sinus of the pericardium (Bachmann's region). A change in P-wave duration and morphology, constituting a transient interatrial block (IAB), is observed as rapidly evolving from partial to advanced IAB (A-IAB). Subsequently, as the ice melts, the ECG pattern normalizes. Please note that P-wave duration in pigs is different (shorter) than in humans. (Taken from reference 22).

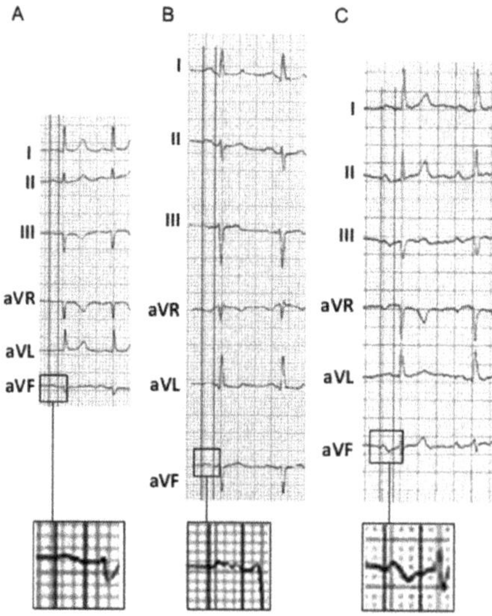

Figure 3. Examples of the 3 types of atrial activity: (**A**) normal P-wave (P-wave duration <120 ms), (**B**) partial interatrial block (P-wave duration ≥120 ms), (**C**) advanced interatrial block (P-wave duration ≥120 ms with biphasic morphology (+/− in leads II, III, and aVF).

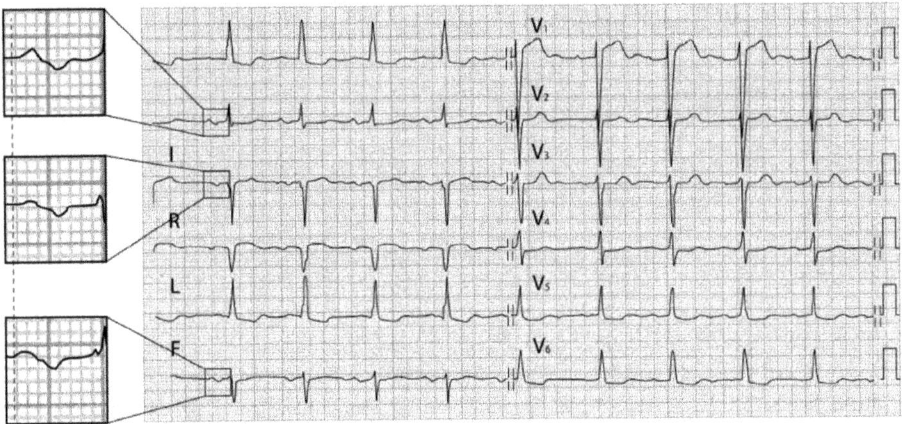

Figure 4. Typical ECG of advanced interatrial block (P-wave ± in leads II, III, and aVF and duration ≥120 ms) in a patient with ischemic heart disease. When amplified (left) we can see the beginning and the end of the P-wave in the three leads.

Figure 5 shows (A,B) the typical ECG and VCG pattern, with ± P-wave morphology in leads II and III, and an open VCG loops in frontal plane (FP) and right sagittal plane (RSP) with the final part of the loops upwards. (B). This figure also shows (C) that the atrial stimulus moves first downwards (HRA-LRA) and after upwards (LRA-high esophageal).

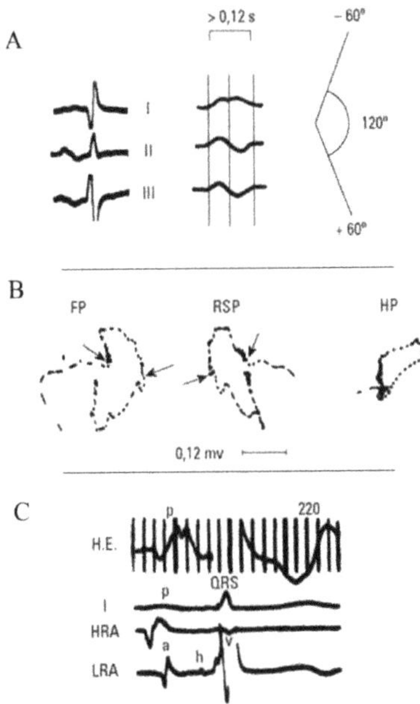

Figure 5. (**A**). P-wave ± morphology in leads II and III typical of advanced IAB with retrograde conduction to the left atrium. Note how the ÂP and the angle between the direction of the activation in the first and second parts of the P-wave are measured. (**B**). Note also the open *P* loops with the last part upwards (FP and RSP). (**C**). Intra-esophageal ECG (HE) and endocavitary registrations (HRA: high right atrium; LRA: low right atrium) demonstrate that the electrical stimulus moves first downwards (HRA–LRA) and then upwards (LRA–HE). (Taken from reference 5).

Figure 6 shows the endocardial mapping in a case of A-IAB, with the caudocranial activation of the LA.

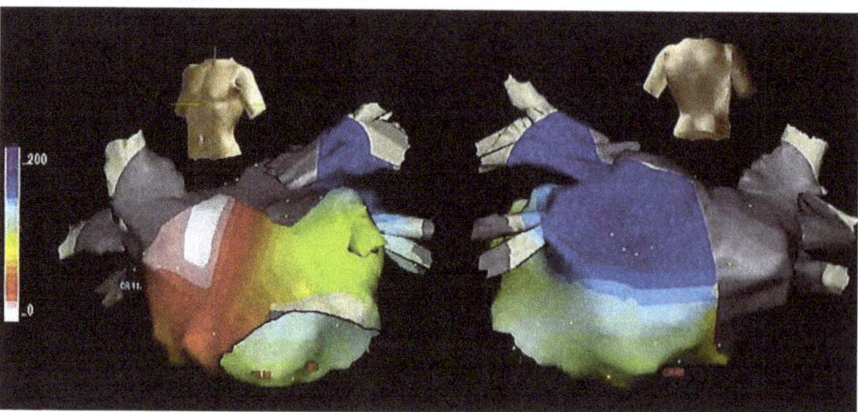

Figure 6. Virtual anatomic rendering of the LA in a patient with typical biphasic ($\pm$) P-wave in leads II, III, and aVF suggestive of A-IAB (Bachmann's region block). Note that early left atrial activation (white) occurs at the high septal wall, as expected for Bachmann region conduction. Activation does not progress through the left atrial roof because of the presence of a large zone of low voltage (gray) that diverts activation toward the low septal (orange-yellow) then the low posterior (green) and finally the high posterior (violet) left atrial wall. (Taken from reference 19).

We demonstrated in 1988 (Figure 7) [13] that this type of A-IAB is frequently associated with atrial fibrillation and/or atrial flutter during follow-up, especially in patients with prior heart disease. This association has been named *Bayes' syndrome* [27,31].

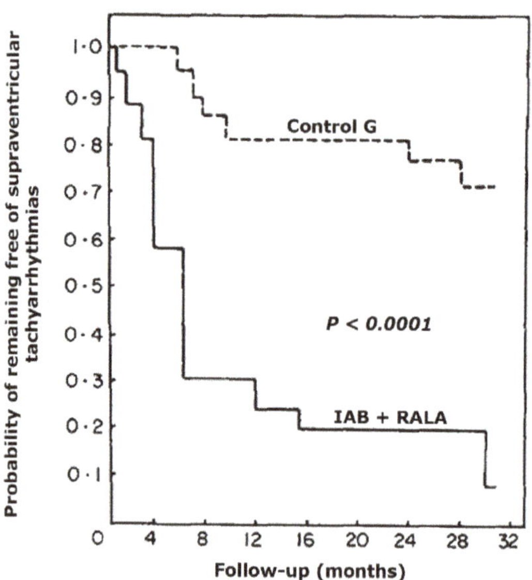

Figure 7. Probability of remaining free of supraventricular tachyarrhythmias (atrial flutter and atrial fibrillation) in patients with advanced interatrial block (IAB) and controls (partial IAB). (Take from reference 13).

2.3.2. Atypical Patterns

During the review of ECGs belonging to different cohorts (Heart Failure [32], BAYES Registry [33,34], REGICOR [35], Centenarians [36]), we realize that some P-wave ECG patterns are very similar to typical A-IAB patterns, but present some minor differences of morphology and duration of P-wave. We consider that these are of atypical patterns of A-IAB (Figures 8–11) (Table 1).

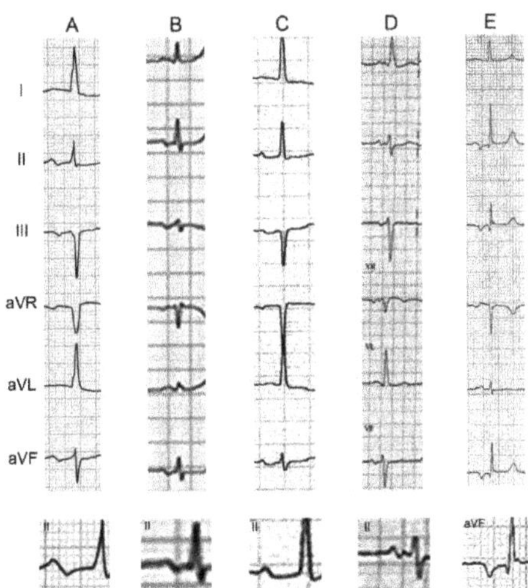

Figure 8. (**A**). Typical A-IAB. (**B**). Atypical A-IAB by duration. (**C**). Type 1 atypical A-IAB due to morphology. The P-wave is biphasic in leads III, and aVF, but the terminal component of the P wave in lead II is isodiphasic. (**D**). Type II atypical A-IAB. The P-wave is biphasic in leads III and aVF, but triphasic in lead II (+ − +). (**E**). Type III atypical A-IAB. The P-wave morphology is negative in leads III and aVF, and biphasic in lead II with the initial component of the P-wave in leads III and aVF isodiphasic. (Taken from reference 23).

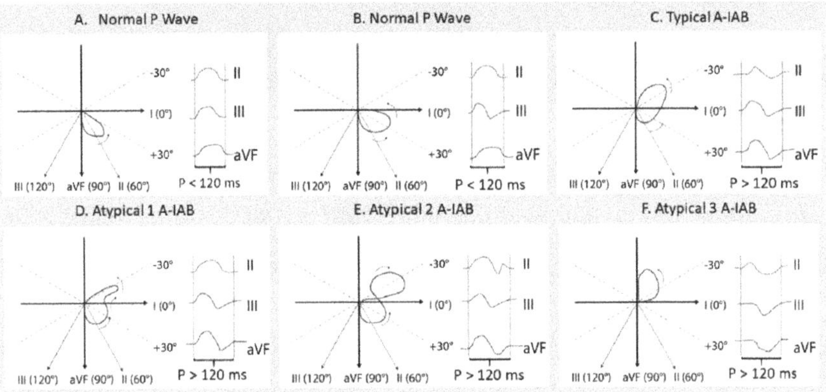

Figure 9. (**A,B**): Normal P-waves. In panel B, there is a biphasic (±) pattern in lead III. This is considered normal because the last part of the P loop falls in the negative hemifield of lead III, that start at +30° but it is positive in leads II and aVF, because the P loop falls in positive hemifield of these leads that starts −30° and 0° (**C**) typical A-IAB. The second part of the P loop falls in the negative hemifield of leads II, III and aVF. (**D–F**): the 3 atypical A-IAB patterns by morphology. (Taken from reference 23).

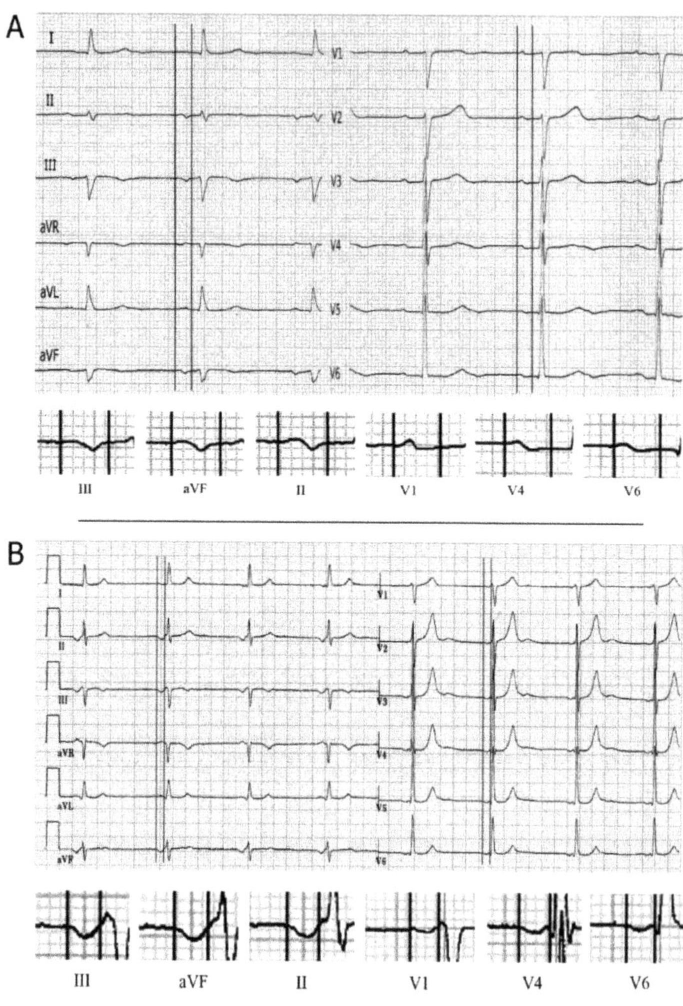

Figure 10. (**A**) P-wave in a case of A-IAB atypical type 3 by morphology. Note that the P-wave is negative in III and aVF, but with a first part isodiphasic as may be seen with vertical lines. It may be confused with junctional rhythm, but in this case (see **B**) all the P-wave is negative and, furthermore, the P-wave is also negative in V4–V6. (Taken from reference 37).

A-IAB exists when there is evidence that at least part of the left atrium is activated retrogradely. This may be ensured if the last part of aVF is negative because this represents that the last part of the loop fall in the negative hemifield of aVF (beyond 0°) (Figure 9C–F). This is the key point to assure that there is retrograde activation of the LA.

In typical cases of A-IAB, the P-wave morphology is ± in leads II, III and aVF because the last part of the atrial activation falls in the negative hemifield of leads II, III and aVF, and the duration of the P-wave is ≥120 ms.

However, there are atypical patterns of A-IAB with some morphological changes in leads II, III and aVF, but always with final negative component in aVF and *p* duration ≥120 ms, or a P-wave pattern ± in leads II, III and aVF but with duration of P-wave less than 120 ms.

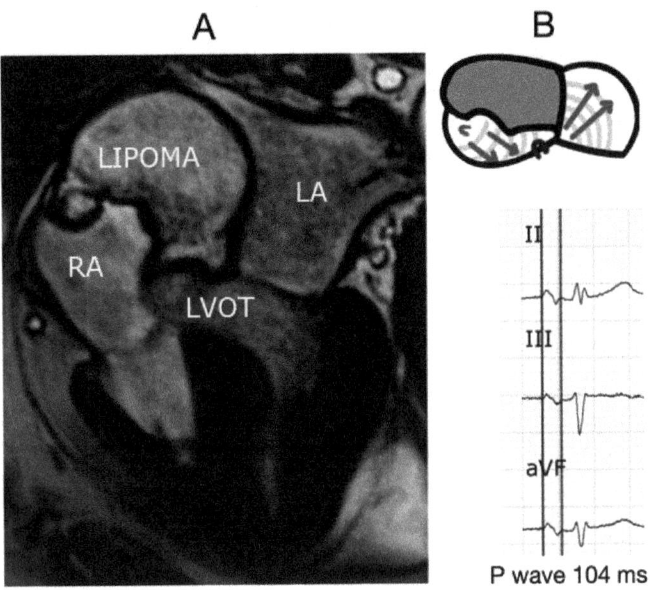

Figure 11. (**A**,**B**) A case of a 73-year old man found to have a large lipoma (4 × 5 cm) located on the interatrial septum. Cardiac magnetic resonance imaging allowed for complete characterization. As the atrial fibrosis is not extended in all atria, the duration of the P-wave in spite of A-IAB is <120 ms. (Taken from reference 38).

2.3.3. Atypical A-IAB Due to Changes in P-Wave Morphology

(i)   **Type I:** The P-wave in leads III and aVF remain biphasic (+/−. However, the terminal component of the P-wave in lead II is isoelectric (Figures 8C and 9D). Therefore, P-wave in lead II presents positive pattern.

(ii)   **Type II:** The second part of the P-wave in lead II is biphasic (∓), therefore, the P-wave is triphasic (+ − +) (Figures 8D and 9E).

(iii)   **Type III** (Figures 8E and 9F) The P-wave morphology in leads III and aVF is completely negative, but started being isodiphasic, and the P-wave in lead II is biphasic (±). This pattern requires a differential diagnosis with junctional rhythm (Figure 10) [37]. If the polarity of the P-waves in leads V5-V6 is positive, then atypical A-IAB is diagnosed, while if it is negative, then junctional rhythm is the final diagnosis.

2.3.4. Atypical A-IAB Due to Changes in P-Wave Duration

In some A-IAB the sinus impulse has to follow a longer path to reach the LA in the absence of fibrosis, as happens in the rare circumstances of an atrial tumor; this blocks the Bachmann bundle. In these cases, the morphology ± of A-IAB may be explained by the presence of an interatrial block in the Bachmann region due to tumor, but the duration of the P-wave may still last less than 120 ms, because there is not fibrosis in the rest of the atria, including the supplementary routers [18,38] (Table 1) (Figure 11).

2.4. Second-Degree Atrial Block

Second-degree interatrial block appears transiently and is usually advanced and, more rarely, partial. It is also known as atrial aberrancy [39,40] when it manifests as P-waves that vary in shape from one beat to another (Figure 12).

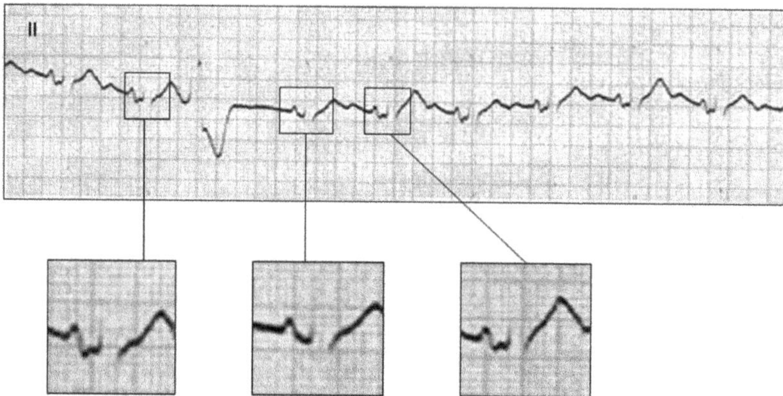

Figure 12. Lead II ECG strip from an 82-year-old man with frequent PVCs. The first two beats show typical advanced IAB (P-wave in lead II is biphasic, with P-wave duration >120 ms). After the premature ventricular contraction (PVC), there is a pause followed by a P-wave of normal duration and morphology. The next P wave again depicts advanced IAB. This case serves as an example of second degree IAB induced by a pause after a PVC.

This type of interatrial block is more frequently seen in the following circumstances: (a) induced by atrial or ventricular premature complexes (Figure 12); (b) may transiently appear and disappear if frequent ECGs are registered [27,28] (Figure 13).

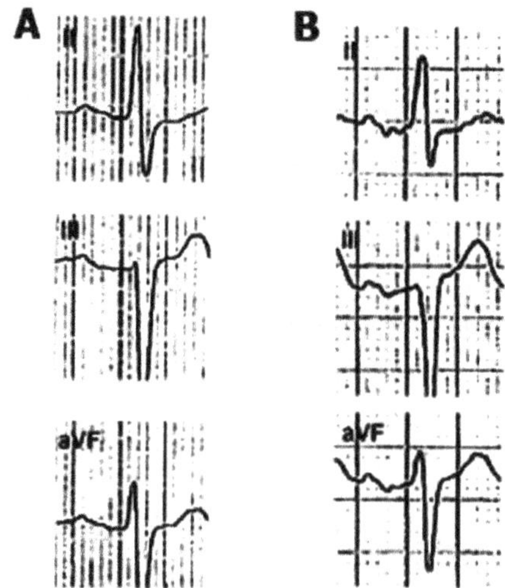

Figure 13. (**A**) Leads II, III and aVF of a 77-year-old man with hypertrophic cardiomyopathy. Heart rate 70 bpm: P-wave 160 ms (partial IAB). (**B**) Same patient was hospitalized due to a febrile episode (39°). The heart rate increased to 100 bpm, and the P-wave depicts a typical pattern of advanced IAB (biphasic morphology in leads II, III, and aVF) and duration of 175 ms. The advanced IAB pattern is associated with a tachycardia-dependent (Phase 3) block. This ECG pattern normalized after fever was controlled and heart rate decreased.

Sometimes the interatrial block is progressive (Figure 14). The P-wave changes from normal to P-IAB and finally A-IAB over time, indicating a progressive fibrosis of the atrium [41].

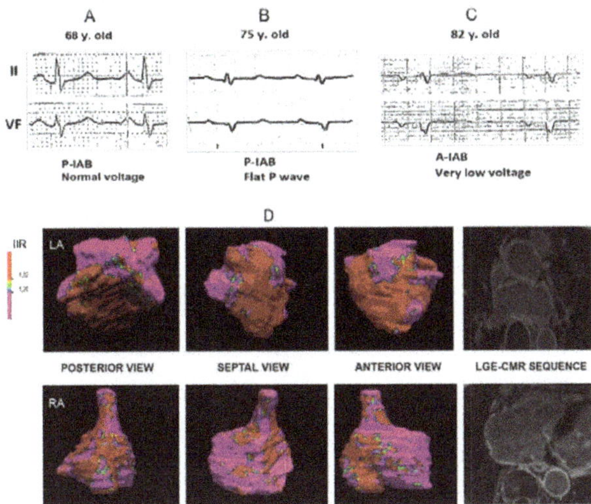

Figure 14. A case that describes the temporal evolution (**A**–**C**) of IAB. (**A**) P-wave of partial IAB to a definitive A-IAB pattern (**C**) with extremely low voltages. See in the lower part (**D**), magnetic resonance image with advanced fibrosis.

3. Clinical Implications

The first demonstration that A-IAB was a risk factor for supraventricular arrhythmias was the publication by Bayés de Luna et al. [5] who demonstrated that patients with A-IAB presented in a short-term follow-up, a high incidence of atrial fibrillation/flutter compared to the control group who presented with P-IAB. In the following 15 years, it some groups published articles that confirmed these results [42,43].

Baranchuk et al. coined the term *Bayes' Syndrome* [27] to describe the association of A-IAB and atrial fibrillation/flutter; and many groups [44–50] became also interested in this relationship. In recent years, has been demonstrated that A-IAB (and AF) are risk factors for atrial tachyarrhythmias [13], ischemic stroke [33,37,51], cognitive impairment and dementia [43,52,53] and mortality [35,54].

P-IAB may also be a risk factor for complications, especially if the duration of the P-wave $\geq$ 150 ms conveys a three-fold all-cause mortality risk [52].

From a hemodynamic point of view, the IAB produces only small consequences, especially if P-wave duration is not very much altered. On the contrary, A-IAB produces clear changes of activation of the atria which result in hemodynamic consequences.

The challenge is to decide if anticoagulation may be useful to decrease all these complications. A randomized trial is mandatory (oral anticoagulants vs. placebo or aspirin) to determine if this treatment alternative may be beneficial.

Author Contributions: Writing—original draft preparation, A.B.-d.-L., M.F.-S., M.M.-S. and A.B.; writing—review and editing, A.B.-d.-L., M.F.-S., M.M.-S. and A.B.; visualization, A.B.-d.-L., M.F.-S., M.M.-S. and A.B.; supervision, A.B.-d.-L., M.F.-S., M.M.-S. and A.B. All authors have read and agreed to the published version of the manuscript.

Funding: Not applicable.

Institutional Review Board Statement: Not applicable.

Informed Consent Statement: Not applicable.

Data Availability Statement: Not applicable.

Conflicts of Interest: The authors declare no conflict of interest.

References

1. Eugene, B. Foreword. In *Clinical Electrocardiography. A Textbook*, 2nd ed.; Futura Publishing Company, Inc.: Armonk, NY, USA, 1998.
2. Fuster, V.; Narula, J.; Harrington, R.A.; Eapen, Z.J. (Eds.) *Hurst's The Heart*, 14th ed.; McGraw-Hill Education: New York, NY, USA, 2017.
3. Camm, J.; Lüscher, T.F.; Maurer, G.; Serruys, P.W. (Eds.) *The ESC Textbook of Cardiovascular Medicine*, 3rd ed.; Oxford University Press: Oxford, UK, 2019.
4. Bayés de Luna, A. *Clinical Electrocardiography*; Wiley-Blackwell: Hoboken, NJ, USA, 2012.
5. Bayés de Luna, A.; de Ribot, R.F.; Trilla, E.; Julia, J.; Garcia, J.; Sadurni, J.; Riba, J.; Sagues, F. Electrocardiographic and vectorcardiographic study of interatrial conduction disturbances with left atrial retrograde activation. *J. Electrocardiol.* **1985**, *18*, 1–13. [CrossRef]
6. Strauss, D.G.; Schocken, D.D. *Marriott's Practical Electrocardiography*, 13th ed.; Wolters Kluwer: Philadelphia, PA, USA, 2021.
7. Bachmann, G. The significance of splitting of the P-wave in the ECG. *Ann. Intern. Med.* **1941**, *14*, 1702–1709. [CrossRef]
8. Puech, P. *L'activité Électrique Auriculaire. Normale et Pathologique*; Masson & Editeurs: Paris, France, 1956.
9. Castillo Fenoy, A.; Vernant, P. Les troubles de la conduction interauriculaire pour bloc du faisceau du Bachmann. *Arch. Mal. Coeur. Vaiss.* **1971**, *64*, 1490–1503.
10. Di Biase, M.; Rizzon, P. Blocco interatriale con attivazione caudocraneale dell'atrio sinistro. *G. Ital. Cardiol.* **1975**, *5*, 323.
11. García Civera, R.; Llácer Escorihuela, A.; Benages Martínez, A.; López Merino, V. Estudio de la activación auricular y de la conducción AV en el bloqueo del haz de Bachmann en el corazón humano. [Study of auricular activation and A-V conduction in Bachmann bundle block in the human heart]. *Rev. Esp. Cardiol.* **1972**, *25*, 341. [PubMed]
12. Bayés de Luna, A.; Bonnin, O.; Ferriz, J.; Fort De Ribot, R.; Julia, J.; Oter, R.; Trilla, E.; Roman, M.; Vernis, J.; Vilaplana, J.; et al. Trastorno de conducción intraauricular con conducción retrógrada auricular izquierda. Estudio electrocardiológico y clínico a propósito de 24 casos. *Rev. Esp. Cardiol.* **1978**, *31*, 173–178. [PubMed]
13. Bayés de Luna, A.; Cladellas, M.; Oter, R.; Torner, P.; Guindo, J.; Martí, V.; Rivera, I.; Iturralde, P. Interatrial conduction block and retrograde activation of the left atrium and paroxysmal supraventricular tachycarrhythmias. *Eur. Heart J.* **1988**, *9*, 1112. [CrossRef]
14. Bayés de Luna, A.; Oter, M.C.; Guindo, J. Interatrial conduction block with retrograde activation of the left atrium and paroxysmal supraventricular tachyarrhythmic treatment. *Int. J. Cardiol.* **1989**, *22*, 147–150. [CrossRef]
15. Bayés de Luna, A.; Guindo, J.; Viñolas, X.; Martinez-Rubio, A.; Oter, R.; Bayés-Genís, A. Third-degree interatrial block and supraventricular tachyarrhythmias. *EP Eur.* **1999**, *1*, 43–46. [CrossRef]
16. Ariyarajah, V.; Puri, P.; Apiyasawat, S.; Spodick, D.H. Interatrial block: A novel risk factor for embolic stroke? *Ann. Noninvasive Electrocardiol.* **2007**, *12*, 15–20. [CrossRef]
17. Garcia Cosio, F.; Martín-Peñato, A.; Pastor, A.; Núñez, A.; Montero, M.A.; Cantale, C.P.; Schames, S. Atrial activation mapping in sinus rhythm in the clinical electrophysiology laboratory. Observations in Bachmann's bundle block. *J. Cardiovasc. Electrophysiol.* **2004**, *15*, 524–531. [CrossRef]
18. Platonov, P.G.; Mitrofanova, L.B.; Chirreikin, L.V.; Olsson, S.B. Morphology of inter-atrial conduction routes in patients with atrial fibrillation. *Europace* **2002**, *4*, 183–192. [CrossRef]
19. Bayés de Luna, A.; Platonov, P.; Cosio, F.G.; Cygankiewicz, I.; Pastore, C.; Baranowski, R.; Bayés-Genis, A.; Guindo, J.; Viñolas, X.; Garcia-Niebla, J.; et al. Interatrial blocks. A separate entity from left atrial enlargement: A consensus report. *J. Electrocardiol.* **2012**, *45*, 445–451. [CrossRef] [PubMed]
20. Bayés de Luna, A.; Baranchuk, A.; Niño Pulido, C.; Martínez-Sellés, M.; Bayés-Genís, A.; Elosua, R.; Elizari, M.V. Second-Degree Interatrial block: Brief review and concept. *Ann. Noninvasive Electrocardiol.* **2018**, *23*, e12583. [CrossRef] [PubMed]
21. Waldo, A.; Bush, H.L., Jr.; Gelband, H.; Zorn, G.L., Jr.; Vitikainen, K.J.; Hoffman, B.F. Effects on the canine P waves of discrete lesions in the specialized atrial tracts. *Circ. Res.* **1971**, *29*, 452–467. [CrossRef] [PubMed]
22. Guerra, J.; Vilahur, G.; Bayés de Luna, A.; Cabrera, J.A.; Martínez-Sellés, M.; Mendieta, G.; Baranchuk, A.; Sánchez-Quintana, D. Interatrial block can occur in the absence of left atrial enlargement: New experimental model. *Pacing Clin. Electrophysiol.* **2020**, *43*, 427. [CrossRef] [PubMed]
23. Bayés de Luna, A.; Escobar-Robledo, L.A.; Aristizabal, D.; Weir Restrepo, D.; Mendieta Badimon, G.; Màssó-van Roessel, A.; Elosua, R.; Bayés-Genís, A.; Martínez-Sellés, M.; Baranchuk, A. Atypical advanced interatrial block: Definition electrocardiographic recognition. *J. Electrocardiol.* **2018**, *51*, 1091–1093. [CrossRef] [PubMed]
24. Kottkamp, H. Human atrial fibrillation substrate: Towards a specific fibrotic atrial cardiomyopathy. *Eur. Heart J.* **2013**, *34*, 2731–2738. [CrossRef]
25. Bayés de Luna, A.; Martínez-Sellés, M.; Elosua, R.; Bayés-Genís, A.; Mendieta, G.; Baranchuk, A.; Breithardt, G. Relation of Advanced Interatrial block to risk of atrial fibrillation and stroke. *Am. J. Cardiol.* **2020**, *125*, 1745–1748. [CrossRef] [PubMed]
26. Bayés de Luna, A.; Martínez-Sellés, M.; Bayés-Genís, A.; Elosua, R.; Baranchuk, A. Lo que todo clínico debe conocer. What every clinician should know about Bayés Syndrome. *Rev. Esp. Cardiol. (Engl. Ed.)* **2020**, *73*, 758–762. [CrossRef]
27. Baranchuk, A. *Interatrial Block and Supraventricular Arrhythmias. Clinical Implications of Bayés' Syndrome*; Cardiotext Publishing: Minneapolis, MN, USA, 2017.
28. Morris, J.J., Jr.; Estes, E.H., Jr.; Whalen, R.E.; Thompson, H.K., Jr.; Mcintosh, H.D. P wave analysis in valvular heart disease. *Circulation* **1964**, *29*, 242. [CrossRef] [PubMed]

29. Rasmussen, M.U.; Fabricius-Bjerre, A.; Kumarathurai, P.; Larsen, B.S.; Dominguez, H.; Kanters, J.K.; Sajadieh, A. Common source of miscalculation and misclassification of P-wave negativity and P-wave terminal force in lead V1. *J. Electrocardiol.* **2019**, *53*, 85–88. [CrossRef] [PubMed]

30. Sajeev, J.K.; Koshy, A.N.; Dewey, H.; Kalman, J.M.; Bhatia, M.; Roberts, L.; Cooke, J.C.; Frost, T.; Denver, R.; Teh, A.W. Poor reliability of P-wave terminal force V1 in ischemic stroke. *J. Electrocardiol.* **2019**, *52*, 47–52. [CrossRef]

31. Bacharova, L.; Wagner, G.S. The time for naming the interatrial block syndrome: Bayes syndrome. *J. Electrocardiol.* **2015**, *48*, 133–134. [CrossRef] [PubMed]

32. Escobar-Robledo, L.A.; Bayés de Luna, A.; Lupón, J.; Baranchuk, A.; Moliner, P.; Martínez-Sellés, M.; Zamora, E.; de Antonio, M.; Domingo, M.; Cediel, G.; et al. Advanced Interatrial Block Predicts New-onset Atrial Fibrillation and Ischemic Stroke in Patients with Heart Failure: The "Bayes Syndrome-HF" Study. *Int. J. Cardiol.* **2018**, *271*, 174–180. [CrossRef] [PubMed]

33. Martínez-Sellés, M.; Elosua, R.; Ibarrola, M.; de Andrés, M.; Díez-Villanueva, P.; Bayés-Genis, A.; Baranchuk, A.; Bayés-de-Luna, A.; BAYES Registry Investigators. Advanced interatrial block and P-wave duration are associated with atrial fibrillation and stroke in older adults with heart disease: The BAYES registry. *Europace* **2020**, *22*, 1001–1008. [CrossRef] [PubMed]

34. Martínez-Sellés, M.; Martínez-Larrú, E.; Ibarrola, M.; Santos, A.; Díez-Villanueva, P.; Bayés-Genis, A.; Baranchuk, A.; Bayés-de-Luna, A.; Elosua, R. Interatrial block and cognitive impairment in the BAYES prospective registry. *Int. J. Cardiol.* **2020**, *321*, 95–98. [CrossRef] [PubMed]

35. Massó-van Roessel, A.; Escobar-Robledo, L.A.; Dégano, I.R.; Grau, M.; Sala, J.; Ramos, R.; Marrugat, J.; Bayés de Luna, A.; Elosua, R. Analysis of the association between electrocardiographic P-wave characteristics and atrial fibrillation in the REGICOR Study. *Rev. Esp. Cardiol. (Engl. Ed.)* **2017**, *70*, 841–847. [CrossRef]

36. Martínez-Sellés, M.; Massó-van Roessel, A.; Álvarez-Garcia, J.; Garcia de la Villa, B.; Cruz-Jentoft, A.; Vidán, M.T.; López, J.; Felix-Redondo, F.J.; Durán, J.M.; Bayés-Genís, A.; et al. (The Investigators of the Cardiac and Clinical Characterization of Centenarians (4C) registry). Interatrial block and atrial arrhythmias in centenarians: Prevalence, associations, and clinical implications. *Heart Rhythm.* **2016**, *13*, 645–651. [CrossRef]

37. de Luna, A.B.; Platonov, P.G.; García-Niebla, J.; Baranchuk, A. Atypical Advanced Interatrial block or junctional rhythm? *J. Electrocardiol.* **2019**, *5*, 85–86. [CrossRef]

38. Gentille-Lorente, D.I.; Scott, L.; Escobar-Robledo, L.A.; Mesa-Maya, M.A.; Carreras-Costa, F.; Baranchuk, A.; Martínez-Sellés, M.; Elosua, R.; Bayés-Genís, A.; Bayés-de-Luna, A. Atypical advanced interatrial block due to giant atrial lipoma. *Pacing Clin. Electrophysiol.* **2021**, *44*, 737–739. [CrossRef]

39. Chung, E.K. Aberrant atrial conduction: Unrecognized electrocardiographic entity. *Br. Heart J.* **1972**, *34*, 341–346. [CrossRef] [PubMed]

40. Julià, J.; Bayés de Luna, A.; Candell, J.; Fiol, M.; Pons, G.; Obrador, D.; Oca, F.; Trilla, E.; Vilaplana, J.; Wilke, M. Aberrancia auricular: A propósito de 21 casos. *Rev. Esp. Cardiol.* **1978**, *31*, 207. [PubMed]

41. Benito, E.M.; Bayés de Luna, A.; Baranchuk, A.; Mont, L. Extensive atrial fibrosis assessed by late gadolinium enhancement cardiovascular magnetic resonance associated with advanced interatrial block electrocardiogram pattern. *Europace* **2017**, *19*, 377. [CrossRef] [PubMed]

42. Agarwal, Y.K.; Aronow, W.S.; Levy, J.A.; Spodick, D.H. Association of interatrial block with development of atrial fibrillation. *Am. J. Cardiol.* **2003**, *91*, 882. [CrossRef]

43. Holmqvist, F.; Platonov, P.; Carlson, J.; Zareba, W.; Moss, A.J.; MADIT II Investigators. Abnormal P wave morphology is a predictor of atrial fibrillation in MADIT II patients. *Ann. Noninvasive Electrocardiol.* **2010**, *15*, 63–72. [CrossRef]

44. Enriquez, A.; Conde, D.; Hopman, W.; Mondragon, I.; Chiale, P.A.; Bayés de Luna, A.; Baranchuk, A. Advanced interatrial block is associated with recurrence of atrial fibrillation post pharmacological cardioversion. *Cardiovasc. Ther.* **2014**, *32*, 52–56. [CrossRef]

45. Enriquez, A.; Sarrias, A.; Villuendas, R.; Ali, F.S.; Conde, D.; Hopman, W.M.; Redfearn, D.P.; Michael, K.; Simpson, C.; Bayés de Luna, A.; et al. New-onset atrial fibrillation after cavotricuspid isthmus ablation: Identification of advanced interatrial block is key. *Europace* **2015**, *17*, 1289–1293. [CrossRef]

46. Sadiq Ali, F.; Enriquez, A.; Conde, D.; Redfearn, D.; Michael, K.; Simpson, C.; Abdollah, H.; Bayes de Luna, A.; Hopman, W.; Baranchuk, A. Advanced Interatrial Block Predicts New Onset Atrial Fibrillation in Patients with Severe Heart Failure and Cardiac Resynchronization Therapy. *Ann. Noninvasive Electrocardiol.* **2015**, *20*, 586–591. [CrossRef]

47. Alexander, B.; MacHaalany, J.; Lam, B.; van Rooy, H.; Haseeb, S.; Kuchtaruk, A.; Glover, B.; Bayés de Luna, A.; Baranchuk, A. Comparison of the extent of coronary artery disease in patients with versus without interatrial block and implications for new-onset atrial fibrillation. *Am. J. Cardiol.* **2017**, *119*, 1162–1165. [CrossRef]

48. Wu, J.T.; Wang, S.L.; Chu, Y.J.; Long, D.Y.; Dong, J.Z.; Fan, X.W.; Yang, H.T.; Duan, H.Y.; Yan, L.J.; Qian, P. CHADS2 and CHA2DS2-VASc scores predict the risk of ischemic stroke outcome in patients with interatrial block without atrial fibrillation. *J. Atheroscler. Thromb.* **2017**, *24*, 176–184. [CrossRef]

49. O'Neal, W.T.; Kamel, H.; Zhang, Z.M.; Chen, L.Y.; Alonso, A.; Soliman, E.Z. Advanced interatrial block and ischemic stroke. The atherosclerosis risk in communities study. *Neurology* **2016**, *87*, 352–356. [CrossRef] [PubMed]

50. Skov, M.W.; Ghouse, J.; Kühl, J.T.; Platonov, P.G.; Graff, C.; Fuchs, A.; Rasmussen, P.V.; Pietersen, Ä.; Nordestgaard, B.G.; Torp-Pedersen, G.; et al. Risk Prediction of Atrial Fibrillation Based on Electrocardiographic Interatrial Block. *J. Am. Heart Assoc.* **2018**, *7*, e008247. [CrossRef]

51. O'Neal, W.T.; Zhang, Z.M.; Loehr, L.R.; Chen, L.Y.; Alonso, A.; Soliman, E.Z. Electrocardiographic advanced interatrial block and atrial fibrillation risk in the general population. *Am. J. Cardiol.* **2016**, *117*, 1755–1759. [CrossRef] [PubMed]

52. Magnani, J.W.; Gorodeski, E.Z.; Johnson, V.M.; Sullivan, L.M.; Hamburg, N.M.; Benjamin, E.J.; Ellinor, P.T. P wave duration is associated with cardiovascular and all-cause mortality outcomes: The National Health and Nutrition Examination Survey. *Heart Rhythm.* **2011**, *8*, 93–100. [CrossRef]

53. Maheshwari, A.; Norby, F.L.; Soliman, E.Z.; Alraies, M.C.; Adabag, S.; O'Neal, W.T.; Alonso, A.; Chen, L.Y. Relation of prolonged P-wave duration to risk of sudden cardiac death in the general population (from the Atherosclerosis risk in Communities Study). *Am. J. Cardiol.* **2017**, *119*, 1302–1306. [CrossRef] [PubMed]

54. Gutierrez, A.; Norby, F.L.; Maheshwari, A.; Rooney, M.R.; Gottesman, R.F.; Mosley, T.H.; Lutsey, P.L.; Oldenburg, N.; Soliman, E.Z.; Alonso, A.; et al. Association of abnormal P-wave indices with dementia and cognitive decline over 25 years: ARIC-NCS (The Atherosclerosis Risk in Communities Neurocognitive Study). *J. Am. Heart Assoc.* **2019**, *8*, e014553. [CrossRef] [PubMed]

MDPI

St. Alban-Anlage 66

4052 Basel

Switzerland

Tel. +41 61 683 77 34

Fax +41 61 302 89 18

www.mdpi.com

Hearts Editorial Office

E-mail: hearts@mdpi.com

www.mdpi.com/journal/hearts